Birkhäuser

Trends in the History of Science

Trends in the History of Science is a series devoted to the publication of volumes arising from workshops and conferences in all areas of current research in the history of science, primarily with a focus on the history of mathematics, physics, and their applications. Its aim is to make current developments available to the community as rapidly as possible without compromising quality, and to archive those developments for reference purposes. Proposals for volumes can be submitted using the online book project submission form at our website www.birkhauser-science.com.

More information about this series at http://www.springer.com/series/11668

Hubertus Fischer • Volker R. Remmert •
Joachim Wolschke-Bulmahn
Editors

Gardens, Knowledge and the Sciences in the Early Modern Period

Birkhäuser

Editors
Hubertus Fischer
Centre of Garden Art and Landscape Architecture
Leibniz University Hannover
Hannover, Germany

Volker R. Remmert
Interdisciplinary Centre for Science and Technology Studies (IZWT)
Wuppertal University
Wuppertal, Germany

Joachim Wolschke-Bulmahn
Centre of Garden Art and Landscape Architecture
Leibniz University Hannover
Hannover, Germany

ISSN 2297-2951 ISSN 2297-296X (electronic)
Trends in the History of Science
ISBN 978-3-319-26340-3 ISBN 978-3-319-26342-7 (eBook)
DOI 10.1007/978-3-319-26342-7

Library of Congress Control Number: 2016940500

Printed on acid-free paper

This book is published under the trade name Birkhäuser
The registered company is Springer International Publishing AG Switzerland (www.birkhauser-science.com)

Contents

Introduction 1
Hubertus Fischer, Volker R. Remmert, and Joachim Wolschke-Bulmahn

Part I Scientization and Knowledge About Nature

The Art of Garden and Landscape Design and the Mathematical Sciences in the Early Modern Period 9
Volker R. Remmert

"Without Design, or Fate, or Force": Why Couldn't John Evelyn Complete the *Elysium Britannicum*? 29
Michael Leslie

The Power of the Sun-King at the Potager du Roi 55
Chandra Mukerji

Part II Mathematical Sciences and the Art of Gardening

The Organ of the Villa d'Este in Tivoli and the Standards of Pneumatic Engineering in the Renaissance 77
Simone M. Kaiser and Matteo Valleriani

Sundials on the Quirinal: Astronomy and the Early Modern Garden . . . 103
Denis Ribouillault

Jacques Lemercier's *Scenografia* of Montjeu: Architectural Prints, Cartography, and Landscape in 1620 135
Anthony Gerbino

Utopia, Science and Garden Art in the Early Modern Era 153
Hubertus Fischer

Part III Botany Between Art and Science

Botanical Illustration and the Idea of the Garden in the Sixteenth Century Between Imitation and Imagination 183
Alessandro Tosi

Gardens on Canvas and Paper: Cataloguing Botanical Abundance in Late Medici Tuscany 211
Irina Schmiedel

Reconstructing Order: The Spatial Arrangements of Plants in the Hortus Botanicus of Leiden University in Its First Years 235
Gregory Grämiger

Garden Visits, Observations, Reading and Excerpts: Martin Fogel (1634–1675) and His Techniques of Acquiring Knowledge 253
Carola Piepenbring-Thomas

Part IV Botanical Knowledge and Horticulture

Watering the Renaissance Garden: Horticultural Theory and Irrigation Practice in Sixteenth-Century Tuscany 269
Anatole Tchikine

Gardening Nature, Gardening Knowledge: The Parallel Activities of Stabilizing Knowledge and Gardens in the Early Modern Period 289
Alette Fleischer

Gardening Knowledge Through the Circulation of Agricultural Treatises in Portugal From the Sixteenth to Eighteenth Centuries 305
Ana Duarte Rodrigues

Commerce and Erudition: Civic Self-Representation Through Botany and Horticulture in Germany, Sixteenth to Eighteenth Centuries 319
Iris Lauterbach

Part V Perspective

Landscape Design and the Natural Sciences in Germany and the United States in the Early Twentieth Century: "Reactionary Modernism"? 345
Joachim Wolschke-Bulmahn

Index 367

Introduction

Hubertus Fischer, Volker R. Remmert, and Joachim Wolschke-Bulmahn

Whether in the field of medicine, mathematics, botany or architecture: a systematic investigation into the interconnections between knowledge and gardens has yet to be undertaken. Despite numerous case studies, basic analyses are not even available for individual knowledge constellations, for generating and systematising, transferring and applying specific forms of knowledge to horticulture and garden art. The interdisciplinary symposium "Gardening and Knowledge. Landscape Design and the Sciences in the Early Modern Period", funded by the Volkswagen Foundation, was intended to fill a sensitive gap spanning the sixteenth to eighteenth centuries. It was jointly organised by the Centre of Garden Art and Landscape Architecture (CGL) of the Leibniz Universität Hannover and the Interdisciplinary Centre for Science and Technology Studies (IZTW) of the Bergische Universität Wuppertal. This volume is the result of the symposium held at the Leibniz Universität in September 2012.

The outstanding contributions made by the mathematical sciences and botany to the genesis and development of early modern garden art and garden culture constitute the main focus of the present volume. In comparison, medicine—as the third traditionally important system of knowledge within garden culture—and architecture figure much less importantly here. In accordance with early modern understanding, the mathematical sciences are regarded here in both their 'pure'

H. Fischer • J. Wolschke-Bulmahn
Centre of Garden Art and Landscape Architecture, Leibniz University Hannover, Herrenhäuser Straße 8, 30419 Hannover, Germany
e-mail: wolschke-bulmahn@ila.uni-hannover.de

V.R. Remmert (✉)
Interdisciplinary Centre for Science and Technology Studies (IZWT), Wuppertal University, Gaussstr. 20, 42119 Wuppertal, Germany
e-mail: remmert@uni-wuppertal.de

H. Fischer et al. (eds.), *Gardens, Knowledge and the Sciences in the Early Modern Period*, Trends in the History of Science, DOI 10.1007/978-3-319-26342-7_1

forms, such as geometry and arithmetic, and in their 'mixed' forms, such as optics, music, gnomonics, astronomy, cartography, architecture, etc. Almost all facets of the mathematical sciences, in this broad conception, are dealt with in the contributions to this volume.

The same is largely true of botany. As a general knowledge of plants, it still embraced those sections, such as plant morphology, plant physiology or plant systematics, which only later developed into disciplines in their own right. While botanical illustration, the rendering of a plant's form and structure as exactly as possible, represents a kind of early stage of plant morphology, the various attempts at cataloguing and arranging them spatially can be understood as the beginnings of plant systematics. However, both the illustration and the ordering and cataloguing of plants are still in an area located between science and art, between systematic, aesthetic and sociocultural criteria. Thus, one could at the same time speak of a 'mixed' botany.

With their many facets, the mathematical sciences and botany point to the increasingly 'scientific' approach that was being adopted in garden art and garden culture in the Early Modern period. This approach or process did not, however, happen in an isolated manner; it took place in interaction with philosophical, religious, political and social processes and positions. As this volume forcefully shows, such processes may make use of the knowledge of nature manifested in gardens for various purposes: for princely representation and bourgeois self-representation, for religious symbolisation with moral and propagandist intent, for a philosophically inspired encyclopaedist approach or a political demonstration of power.

The sources from which the contributions are derived include written and pictorial testimonies, but also those that unite text and picture—engraving or drawing. Amongst them are manuscripts and archival material that are presented here or have been specially evaluated for the first time. It should come as no surprise that specialist works and handbooks on horticulture and garden art, on mathematics, perspective, gnomonics and hydraulics are among these or that botanical books, hand-written and printed lists of plants and catalogues, herbaria and florilegia have been drawn upon. While agricultural and architectural treatises look further afield, a diary and personal library point to a special case. Among the written sources, the card index on the one hand and the literary utopia on the other will probably appear to be the most unusual. Among the pictorial sources, the example of scenography will perhaps stand out most obviously, although for aesthetic reasons, the painted pictures of flowers really threaten to steal the show.

The use of the sources is as varied as the sources themselves. Nonetheless, there are principle questions that are addressed in this volume. Among them is the question—or in fact a whole complex of questions—as to how knowledge of nature and technology relating to gardens is generated, handed down, systematised and transferred, as well as being adapted or functionalised. The engineering knowledge of the Renaissance gained from practical experience can certainly prove to be superior to the knowledge codified in ancient texts. On the other hand, the limitations of technological innovations reveal themselves, for example, when dealing with the irrigation of earlier gardens, since the manifold requirements and tasks in plant cultivation can only be managed by means of laborious manual work and the gardeners' expert knowledge of water, right down to the subtle details.

The degree of complexity in acquiring, collecting and ordering botanical and garden cultural knowledge in individual cases is suggested by the way in which diaries, experiments, book collections, and the knowledge read and noted on index cards are interconnected, as seen in the excerpts and notes from oral sources, letters and newspapers. In contrast, the question as to which garden knowledge was present and available when and in which country, generally points towards the knowledge mostly collected in books in particular areas and regions.

The geographical dimension addressed here includes Portugal, which has hitherto received little attention. Several contributions are devoted to Italy, especially to Tuscan and Roman garden culture, but Great Britain, France, the Netherlands and the large German-speaking areas are also examined. In a way, however, these are 'artificial' boundaries, for trade, travel and correspondence break down such boundaries in particular rhythms, not just within Europe but also further abroad. This is true of the bourgeois garden culture of the Imperial free cities of southern Germany, which not only profit from a lively commercial and scholarly exchange with Italy, but, as the case of the Fugger family in Augsburg shows, also gain access to rare exotic plants and flowers through overseas trade connections. Even more impressive is a Dutch example from the seventeenth century, in which a statesman assembles the flora of the Mediterranean, Turkey, Asia, the Cape of Good Hope, Peru and Suriname in the five acres of his garden.

From the more global aspects and enquiries, the examples mentioned above take us to the volume's actual content. It is divided into four sections and a final chapter leading into the twentieth century. The first section, "Scientization and Knowledge about Nature" embraces an innovative contribution and two examples of studies relating to the principle enquiry into the process of making garden art more scientific in the Early Modern. In the second section, "Mathematical Sciences and the Art of Garden", the 'mixed' mathematical sciences are investigated, whilst in the sections "Botany between Art and Science" and "Botanic Knowledge and Horticulture", botany is examined in its relationship to garden art and garden culture. The question, not easily answered, as to whether the reception of 'modern' scientific theories in landscape architecture in the first decades of the twentieth century actually represents a path to modernity or rather implies a "reactionary modernism", to refer to Jeffrey Herff's work, is addressed in the final chapter.

At the beginning of the process of scientization one finds what is referred to today as the 'Scientific Revolution'. This raises the mathematical sciences to the status of leading sciences ("Leitwissenschaften" according to sociologist Nobert Elias), which are also employed as instruments of power and sources of technological and socio-political knowledge. In the first contribution, Volker R. Remmert sketches out the opportunities this process opened up in early modern garden art for gaining mastery of and manipulating nature, right down to the transformation of gardens into model sites of political representation. At the same time, he summarises the status of research and, by stating numerous unanswered questions, he formulates areas for which future research is needed. Those consequences which can derive from the possibility of the sciences progressing are detailed by Michael Leslie using the example of John Evelyn's *Elysium Britannicum*. In this case, they lead to a dilemma that cannot be solved and is ultimately responsible for the fact that the *Elysium* remained unfinished. It is the dilemma between Epicureanism and the limitlessness of science on the one hand and the limits that religion, the state and tradition lay down on the other. For a religious man and someone of good birth, this means that science proves to be self-contradictory, for, with its own field of study being limitless, it strives to, or has to, escape the danger of crossing boundaries. On the basis of *potager du roi* in Versailles, Chandra Mukerij explores how the knowledge of nature can become a tool of politics. Her re-reading of Jean de la Quintinie's *Instructions pour les jardins fruitiers et potagers* (1692) discloses the semantics through which knowledge of nature and political power are interlinked and permeate one another. Knowledge of nature can be employed to gain direct control of the land; it can be used to train and develop human energies, skills and duties in the sense of a Christian caring for the soil. But as the study of the powers acting within nature, it can also serve to demonstrate the absolute and compelling power of the sun (and of the Sun King).

The second section steps into the wide field of the various applications of mathematical and technological knowledge in Early Modern garden art. This is first illustrated through Italy. Simone M. Kaiser and Matteo Valleriani turn their attention to the hydraulic organ in the Villa d'Este in Tivoli. With the help of an unpublished contemporary description and meticulous comparisons, they are able to show that the art of engineering in the sixteenth century did indeed turn to ancient texts for inspiration, but in fact represented the result of continuous practical technological improvements rather than the outcome of reinterpreting such texts. In another area of the sciences employed in gardens, that of astronomy and gnomonics, Denis Ribouillault likewise succeeds in correcting traditional views. By revealing the symbolic and moral function of sun dials, citing the examples of those in the papal gardens of Monte Cavallo and the Jesuit gardens of Sant'Andrea Quirinale, he simultaneously puts into perspective the significance of the scientific debates at the end of the sixteenth and beginning of the seventeenth centuries (heliocentrism, Galileo) for the understanding of time and the iconography of Roman garden art of this age. Anthony Gerbino turns to another field when examining Jacques Lemercier's remarkable "scenografia" of the Montjeu castle and garden of the early seventeenth century. This concerns one of the earliest project sketches of a French garden within a seignorial landscape. It is shown how, in this sketch, mapping and surveying in the field, architectural graphics and cartography are fused to produce a new form of architectural representation, a form that had found imitators by no later than the 1650s. This section is concluded with Hubertus Fischer's examination of humanist civic utopias against the backdrop of Renaissance town planning theory. From this theory, he develops the idea of architecture as a discipline that is formed in the urban space and its gardens. He then investigates the increasingly scientific approach towards utopias and the function of garden culture in utopian towns through the middle of the seventeenth century.

The botanical section of this volume is opened by Alessandro Tosi's analysis of botanical illustrations so important for the sixteenth century. It reveals, on the one hand, the imitation of nature in the sense of a new 'naturalism'—in as exact terms as possible, as demanded by the new scientific spirit—and on the other—linked to this and promoted by it—an imagination which lets botanical illustration in this age vacillate between science and art. A similar direction is followed by the contribution of Irina Schmiedel, albeit with an eye towards a later age and a different object. With the help of the visual and verbal cataloguing of Tuscan flora, for which Cosimo III. Medici commissioned artists and botanists, she explains how beauty and rarity converge with the demonstration of knowledge and wealth. This raises the question of whether "gardens on canvas and paper" have promoted scientific progress in the sense of a taxonomic debate or have, in fact, hindered it. Gregory Grämiger grapples with the problem of plant systematics in the next contribution. He likewise uses catalogues to reconstruct the original arrangement of the plants in the Leyden Hortus Botanicus. It is thereby revealed that very different systems exerted an influence on their arrangement, including those of smell and taste or such as corresponded to the sections of a book on flora. Finally, it could have also simply been their value, a very unbotanical aspect, that determined where they were to be placed. Using the example of the personal library, the card index and the travel diary of Martin Fogel, a Hamburg doctor and scholar of the seventeenth century, Carola Piepenbring-Thomas examines a special case of how and whereby botanical and garden cultural knowledge could be acquired and organised. The reader is offered informative insights into how knowledge is generated by methods and media that supplement and support one another mutually in a complex way.

The second botany section reverts to aspects of the first chapter, but above all extends the field of enquiry and its objects. The first contribution by Anatole Tchikine explores a question that is as central as it is neglected. How was the irrigation of Italian gardens actually carried out from the perspective of plant cultivation in theory and practice? This meticulous source-based analysis produces a 'manuale' of horticultural knowledge: namely, numerous individually adjusted and circumspect methods and techniques of manual horticultural work that cannot be replaced by technological innovations that at the same time have an influence on the shaping of the garden.

The contribution by Alette Fleischer is devoted to the question as to how and by means of which media (herbaria, books, illustrations, lists of plants) 'nature' was converted into mobile, unchanging and combinable (botanical) knowledge, so as to then circulate freely in space and time as accumulated knowledge. These theoretical reflections are underpinned by the examples of the Dutch botanist Jacob Breyne and the developmental process of his botanical books and of the collections of plants in the gardens of Jerome van Beverningh from the second half of the seventeenth century. Ana Duarte Rodrigues is concerned exclusively with the circulation of garden knowledge in Portugal from the sixteenth to eighteenth centuries, using the literature of agricultural treatises present in the country's most important libraries. She deals less with the knowledge itself than with the tradition of treatises and their translation and circulation in Portugal as witnessed by their external characteristics. Similar investigations in other countries would provide the opportunity to undertake a comparative analysis and a more in-depth study. Iris Lauterbach closes the botanical section with her study of "Trade and Scholarship". This deals with the same period considered in the examination of the treatise circulation in Portugal, but in contrast to some of the earlier contributions, this one looks into civic gardens in Germany rather than at courtly gardens. Often owned by doctors, apothecaries and merchants, these gardens played a significant role in the development of botany and plant cultivation on account of the owners' wide-ranging trade relations and international scholarly exchange. The fact that these gardens formed important elements of urban bourgeois self-representation links them to this era's general culture of representation.

Michael Leslie's reflections on John Evelyn's "Elysium Britannicum" make just as clear as Irina Schmiedel's questions about the visual and verbal plant catalogues commissioned by Cosimo III. Medici that the increasingly 'science-oriented' approach in garden culture is not a process that is free of contradictions. It would require examination of greater depth to establish how developments in 'science' played out in garden art and garden culture. Against this backdrop, it is worth looking at the beginning of the twentieth century, when garden and landscape design in Germany and the United States were becoming intensively aware of more recent developments in the natural sciences, ecology, plant geography and plant sociology. For Joachim Wolschke-Bulmahn, this poses the question of whether one should not also speak, to use Jeffrey Herff's term, of a "reactionary modernism" in garden design in Germany in the early twentieth century. From these provisional findings, one might perhaps draw the conclusion that, in the course of future research into the dynamic connection between "gardens, knowledge and science" one should also investigate more penetratingly the forces of resistance and contradiction.

Part I

Scientization and Knowledge About Nature

The Art of Garden and Landscape Design and the Mathematical Sciences in the Early Modern Period

Volker R. Remmert

Abstract

The mathematical sciences of the early modern period comprised many fields of knowledge, and those such as astronomy, geography, optics, music, practical geometry, acoustics, architecture and arithmetic were often deliberately oriented towards practical applications. Between the mid-sixteenth and mid-eighteenth centuries, practitioners of the mathematical sciences and garden and landscape designers shared the conviction that nature could be controlled and manipulated, and the methods used and the knowledge acquired in the mathematical sciences opened up new ways to do this. These potentialities affected the realm of landscape design and gardening in various formative ways that reached directly into the political sphere by offering new possibilities for political representation, of which there are numerous noteworthy examples, including the gardens of Versailles, perhaps the most magnificent representational gardens in seventeenth-century Europe.

The Mathematical Sciences in Early Modern Europe

In early modern Europe, the term *mathematical sciences* was used to describe those fields of knowledge concerned with measure, number, and weight, reflecting the much quoted passage from the *Wisdom of Solomon* 11, 20: "But thou hast ordered all things in measure and number and weight". The mathematical sciences consisted of *mathematicae purae*, which dealt with quantities, continuous and discrete as in geometry and arithmetic, and *mathematicae mixtae*, which dealt not

---

This essay draws on my publications (Remmert 2004, 2007, 2008). I am indebted to Ben Kern for his critical reading of the paper.

V.R. Remmert (✉)
Interdisciplinary Centre for Science and Technology Studies (IZWT), Wuppertal University, Gaussstr. 20, 42119 Wuppertal, Germany
e-mail: remmert@uni-wuppertal.de

H. Fischer et al. (eds.), *Gardens, Knowledge and the Sciences in the Early Modern Period*, Trends in the History of Science, DOI 10.1007/978-3-319-26342-7_2

only with quantities but also with qualities, as in geography, optics, music, cosmography, astronomy, gnomonics, architecture, etc. The mathematical sciences, which played such a central role in the Scientific Revolution, were often directly related to practical applications such as navigation, surveying and the production of scientific instruments and sundials. It was not until the eighteenth and early nineteenth centuries that the various branches of the *mathematical sciences* became independent from one another and developed into separate disciplines.

Prior to the Scientific Revolution, the mathematical sciences had been ranked low in the hierarchy of academic disciplines compared to theology and philosophy, and especially natural philosophy. This changed in the seventeenth century when mathematization and mathematical modes of explanation began to dominate many fields of inquiry and areas of society, and the "Cinderella" at the bottom of the list came to be perceived of as the "Belle of the Ball". The mathematical sciences, and in particular the closely related "new physics", started to assume the role of leading academic disciplines (Leitwissenschaften). Moreover, they began to be used, like modern sciences and mathematics, as potential instruments of power and a source of valuable technical and socio-political knowledge. This change in the social and epistemological status of the mathematical sciences was a key prerequisite of the Scientific Revolution, and was thus fundamental to developments in the history of science and early modern European history as a whole, not least because it led to the scientization of modern societies (Biagioli 1993; Dear 1995; Knobloch 1989: 13–20; Remmert 1998: 79–90 and 187–200). Landscape and garden designers were affected by this change, and were drawn to the new possibilities the mathematical sciences offered or seemed to offer.

What Happens in a Garden?

What was to happen in a prestigious early modern garden once it had been planted and made ready for its users? The question is obviously unanswerable, especially because the making of a garden is usually an unending process. However, the use of the mathematical sciences would have certainly provided a garden with much entertainment and delight. Music and acoustical effects, fountains and sundials, geometrical forms, automata and many other things derived from the mathematical sciences that could be found in early modern gardens would naturally be treated in a thorough or encyclopaedic course of the mathematical sciences (e.g., Schott 1661). The frontispiece (Fig. 1) of the *Aerarium philosophiae mathematicae* (*Treasure chest of the mathematical sciences*) by the Jesuit Mario Bettini (1582–1657) illustrates this nicely (Bettini 1648). The garden presented here shows the correspondence between garden design and the possibilities that the mathematical sciences opened up: forms, symmetries, fountains, architecture and perspectives become a setting for demonstrations of optical illusions, musical performances and the display of mathematical devices and diagrams.

Other contemporary frontispieces also testify to the close link between gardening and the early modern mathematical sciences, examples being those of the Jesuit

Fig. 1 Mario Bettini, *Aerarium philosophiae mathematicae*, Bologna 1648, frontispiece by Francesco Curti. © Herzog August Bibliothek Wolfenbüttel

Gaspar Schott's *Cursus mathematicus* (Schott 1661) and another work by Mario Bettini, the *Apiaria universae philosophiae mathematicae* (Bettini 1642). The preface of Bettini's *Apiaria* makes this even more explicit in its mention of the *hortus mathematicus*, the garden of the mathematical sciences. This metaphor might not simply be a piece of early modern rhetoric aimed at legitimizing the mathematical sciences. The *hortus mathematicus* is understood as a field in which the theory and practice of gardening and the mathematical sciences interact. Gaspar Schott, in the frontispiece of his *Cursus mathematicus*, and Bettini, in the frontispiece of his *Aerarium*, made expert use of the notion that everything needed to create a garden of amusement and edification—geometry, architecture, perspective, optics, music, etc.—could be derived from the mathematical sciences (Remmert 2004, 2011: 214–228). If designed accordingly, the fashionable gardens of the rich and powerful would demonstrate the sophistication and hence the prestige of their owners; but at the same time, they would be promoting the mathematical sciences among the most influential members of society, a prospect of which Schott and Bettini were undoubtedly aware. However, a question remains as to whether Bettini's metaphor of the *hortus mathematicus* was in fact anything more than a trick to legitimize the mathematical sciences. Before addressing this, a number of other issues relevant to the general theme need to be considered.

What Is a Gardener?

In the seventeenth and early eighteenth centuries, authors writing on gardening often stressed the importance of the mathematical sciences for gardening and landscape design. For example, Jacques Boyceau, superintendent of the royal gardens under Louis XIII, in his posthumously published *Traité du Jardinage* of 1638, was emphatic in demanding that young gardeners be thoroughly instructed in geometry, architecture, arithmetic and perspective. By training young gardeners in the arts and the relevant mathematical sciences, Boyceau's intent was to emancipate gardening from the crafts ("travail de la main") and raise it to the status of an art—a shift that architecture had achieved much earlier. But Boyceau's agenda went further: he wanted to turn gardening into what might be called a scientific enterprise—a science of nature (Boyceau 1638). Boyceau's programme was very successful and had a significant impact on young gardeners such as André Le Nôtre, the creator of the gardens of Versailles. Bringing the mathematical sciences into gardening was not only intended to offer new design possibilities, but also had another, equally important aim, namely to enhance the status of gardening, *Jardinage*, among the arts and the disciplines, and of course, at court.

The ideal qualifications of a gardener were widely discussed in gardening tracts of the seventeenth and early eighteenth centuries (Evelyn 2001: 33–35; Hesse 1710: 13–16; Mandirola 1660: 1f; Stromer von Reichenbach 1671: 6–12). Wolf Albrecht Stromer von Reichenbach (1636–1702) of Nuremberg gave a typical exposition in his 1671 book, *Der Italiänische Blumen- und Pomeranzen-Garten*. In the chapter

"Von dem Gärtner/und wie er beschaffen seyn soll" ("Of the Gardiner, and how he is to be qualified"), he summarized:

> [...] an honest gardener has to be familiar with a variety of disciplines. Instead of only being well acquainted with nature, he has to know a bit of geometry and surveying, understand architecture, comprehend the course of the heavens and the constellations, and he also has to be able to sketch a ground plan. Otherwise he should rather be called a stupid peasant instead of a gardener.[1]

Stromer von Reichenbach's point, that gardening should extend the range and types of knowledge upon which it was based, was also made by other authors.

Heinrich Hesse, gardener of the Mainz Elector, emphasized this in his "Teutscher Gärtner" ("German Gardener") of 1710, requiring that whoever "wants to become an honest gardener has to properly learn the art in a princely garden".[2] With respect to the mathematical sciences, he should have a good understanding of hydraulics and the uses of water, he should be familiar with surveying and be able to build good sundials. In view of these requirements, Hesse's stipulation that he should also be able to write and calculate comes as a surprise.

The inadequacy of a gardener's education was a similar concern in England. A published memorandum from the Worshipful Company of Gardeners in London deplored the insufficient training and competences of many (purported) gardeners, stressing their lack of knowledge in the mathematical sciences even on a very elementary level: "These Pretenders are most of them so unskillful in *Mathematicks*, that they know not common proportions, nor to take the profiles of any ground" (Worshipful Company of Gardeners 1700: 8).

Nevertheless, by the late seventeenth century, the mathematical sciences had become a central authority upon which gardening could claim to be an almost academic enterprise, which, while not yet a profession in its own right, was often practised as a combination of skills informed by knowledge of the mathematical sciences. The landscape designer Daniel Marot (1661–1752) was considered to be "a very ingenious Mathematician" (Harris 1699: 47), and in the eighteenth century, the German Johann Hermann Knoop (ca. 1700–1769) called himself a "hortulanus, mathematicus et scientiarum amator" (gardener, lover of mathematics and science) (de Jong 1988: 21). When in 1709, Dézallier d'Argenville (1680–1765), voicing the views of the late André Le Nôtre, expected a gardener to be familiar with geometry

[1] "Mit wenigen Worten/ein rechtschaffener Gärtner muß mit vielen Wissenschaften begabet seyn/ und nicht nur allein einen guten Naturkündiger geben/sondern auch etwas von der Geometria und Feldmeßkunst wissen/die Architectur verstehen/den Himmels-Lauff und die Constellationes erkennen/und einen guten Hand- und Grund-Riß aufziehen können; dann sonsten er vielmehr ein unverständiger Bauer/als ein Kunst-Gärtner/benamet werden mag." (Stromer von Reichenbach 1671: 9)

[2] "Was aber ein rechtschaffener Gärtner seyn wil/der muß seine Kunst vorhero in einem Chur=oder Fürstlichen Garten recht und wohl lernen [...]." (Hesse 1710: 13–15)

("il faut être un peu Geometre"), he was mentioning only one of the many fields within the mathematical sciences that could be and were being applied to garden making (Dézallier d'Argenville 1709: 16).

The topic of a gardener's qualifications is, of course, closely linked to the process of professionalization in the practice of gardening and landscape design. While this process has only recently been studied closely (Schweizer 2012), it is clear that in the seventeenth and early eighteenth centuries, knowledge of the mathematical sciences and their many branches played a key role in two respects. On the one hand, scientization and mathematization transformed the practice of garden and landscape design and emancipated it from, for example, its old rival, architecture. The reference to and use of the mathematical sciences was instrumental not only in opening up new options and creating new wonders and mirabilia, but also, and perhaps even more so, in fashioning gardening as a new profession with an autonomous standing in the hierarchy of the arts and at court (Remmert 2004), and even in redefining it as a new science, the "science of gardening" ("science du Jardinage"), as Jacques-François Blondel put it in the mid-eighteenth century (Blondel 1747: 71). On the other hand, knowledge gleaned from the mathematical sciences facilitated new ways for gardeners to control and dominate nature. Because gardens played an important role in the political culture of the times, using them to display domination over nature and promoting techniques of controlling it was very much in the interest of the ruling class. Thus, the new options for gardening reached right into the political sphere by offering new forms for the representation and exhibition of power, as evidenced, for example, at Versailles, the gardens of which were, perhaps, the most magnificent example of political representation in seventeenth-century Europe (Mukerji 1997; Vérin 1991).

To sum up: from the 1640s to the early eighteenth century, many authors writing on gardening and landscape design were convinced that the use of the mathematical sciences was essential if the art of gardening and landscape design was to become a reliable and calculable enterprise, and something that could be understood as a science of nature (de Jong 2000: 21; Garrigues 2001: 24f; Remmert 2004: 15; cf. Seller 2001: 251). But the most significant connection between gardening and the mathematical sciences lay in their joint belief that man had dominion over nature and in the inexorable progress that the arts and the sciences would make in bringing this about.

Control of Nature and Belief in Progress

Between the mid-sixteenth and mid-eighteenth centuries, practitioners of the mathematical sciences and of gardening and landscape design shared the conviction that they could control nature. With respect to gardening, the gentler understanding was that nature needed a helping hand to become perfect and reach the ultimate level of the "third nature" (Hunt 2000: 39f; de Jong 1988: 18–20; Lauterbach 2004: 222–231).

Jan van der Groen testified to this when he stated that "nature, at times being awkward, can be set straight through our art, it can be guided and put into order."[3] William Wotton, who attributed to John Evelyn the notion of gardening as a science of nature, celebrated the gardener's power over nature in his *Reflections upon Ancient and Modern Learning*: "Other Men can only *follow* Nature; the Gardener alone *leads* it, and hastens or slackens its pace according as suits best with his Designs or Inclinations." (Wotton 1697: 306f). In the mid-eighteenth century, Jacques-François Blondel put it quite simply: "Nature wants to be helped" (Blondel 1747: 71: "La Nature veut être aidée."). Heinrich Hesse, in his *Teutscher Gärtner*, did not write much about this topic in the book itself, but the title engraving makes clear that gardens are the epitome of control over nature (Fig. 2). The garden is represented as the highest form of nature ("third nature"), surrounded by well-ordered farmland ("second nature"), while nature ("first nature") seems to have disappeared, presumably because it was completely under control.

Such views express an understanding of gardening as not only an expanding field of knowledge, but also as a field of constant progress. This belief, traceable in Francis Bacon's famous remark that gardening is the "Greater Perfection" (Bacon 1625: 266), is nicely reflected in the printer's preface to the 1711 edition of Dézallier d'Argenville's *La Théorie et la Pratique du Jardinage*, which says that the topic of gardening could never be exhausted because nature would forever be "cultivated by New Arts".[4]

These "New Arts" that provided gardeners and landscape designers with ever better techniques to exert power over nature would, of course, include the mathematical sciences (de Jong 1988: 20f), and not without reason, since both practices saw themselves at the cutting edge of progress and the mastery of nature. In 1603, the Jesuit mathematician, Christoph Grienberger (1564–1636), a disciple and collaborator of the highly respected mathematician, Christoph Clavius, declared in a lecture at the *Collegio Romano* that there could be no doubt that the movements of the heavens could be translated into mathematicians' laws (Mathematicorum leges). He went on to say that the elements of nature were also ruled by the mathematical sciences and that often mathematicians liberated these elements from their dungeons (carceribus) and "introduced them into the gardens and palaces of kings" (in Regum hortos ac palatia introducuntur) (Gorman 2003: 17 and 77–79). Later, Grienberger's fellow Jesuits Mario Bettini and Gaspar Schott took up this association with gardens in the metaphor of the *hortus mathematicus* mentioned above.

---

[3] "Aber auff unser vornehmen zu kommen/sagen wir/daß die Natur/die sich manchmal ungeschicklich erzeigt/durch Kunst kann auffgerichtet/geleitet und in Ordnung gebracht werden." (Groen 1669, page one of introduction). On Groen see (de Jong 1988: 16ff).

[4] "[...]; mais quand on composéroit encore plusieurs autres Volumes sur cette Matiére-là, on ne pourroit jamais l'épuiser, ni même l'exprimer au juste dans quel qu'une de ses Parties, parce qu'elle varie tous les jours, sur mille diférens sujets que la Nature, cultivée par les Nouveaux Arts [...]." (Dézallier d'Argenville 1711: page 4 of the letter of dedication)

Des ersten Buchs
Erster Theil/
Des
Teutschen Gärtners/
Handelt von
Anlegung und Zubereitung eines Lust-
Gartens/ unter unserm Teutschen Climate,
Deßgleichen von
Pflantzung und Wartung der Bulbosischen/
Tuberosischen und andern Gewächsen.
Aus eigener Erfahrung mit Fleiß zusammen getragen/
und den Garten-Liebhabern zum Besten in folgende
Capitel auffgesetzet
Durch
Heinrich Hessen.

Fig. 2 Heinrich Hesse: *Teutscher Gärtner*, Leipzig 1710, engraved title page. © Herzog August Bibliothek Wolfenbüttel

Although it was quite unusual for advocates of the mathematical sciences to make allusions to gardens in such an explicit way, detecting "mechanics and geometry in the workings of nature", as Christiaan Huygens was prone to do, or achieving "dominion over nature through the mathematical sciences", as Christian Wolff put it, was a matter of course for early modern mathematicians and scientists.[5] Therefore, it is not unexpected that some of the most prestigious mathematicians of the early modern period were either involved in garden projects or followed their latest developments. Huygens regularly provided his brother-in-law Philips Doublet (1633–1707) with news about gardening culture in France, reported on the latest advances in fountain technology, in which he had a serious interest, and in 1661, wrote about his meeting with André Le Nôtre, this "grand enventeur de jardinages" (de Jong 2005: 41). Leibniz, in the summer of 1696, wrote to Andreas Du Mont (†1697) about his suggestion for the fountain at Herrenhausen, which he claimed was superior to others without being excessively expensive,[6] and in 1749, Leonard Euler (1707–1783) made a detailed study of the workings of the Great Fountain in Sanssouci, which Frederick the Great wanted to reach unprecedented heights (Eckert 2002).

The results of these involvements are difficult to gauge, given the differences between intentions and outcomes. However, such examples indicate that the close link between garden design and the mathematical sciences was not limited to the usefulness of, for example, geometry in garden layouts. The mechanistic worldview of the early modern mathematical sciences had infiltrated deeply into the culture of gardening and landscape design, and, along with the overlapping field of engineering, they had become a leading authority for the increasingly academic status of gardening, as mentioned earlier. With respect to politics, since gardens and the mathematical sciences were equally suited to represent the absolute power of princes and kings, both could be perfectly tuned to demonstrate a sovereign's control over nature. Chandra Mukerji made this point for French formal gardens, which, according to her, "were meant to represent nature as it was understood by science, full of order and diversity" (Mukerji 1990: 671).

The Mathematical Sciences and Garden Design

The geometric garden is often considered as a typical instance of the mathematization of garden design. Abraham Bosse stretched this to extremes in an illustration in his *Maniere universelle de M. Desargues pour praticquer la Perspective* (Bosse 1648) (Fig. 3). Here, the garden is subjugated by geometry to an extent that it is

[5] Huygens: "Il n'y a point de matière qui puisse m'agréer davantage que la méchanique et la géométrie qu'on voit dans les ouvrages de la nature." Quoted from (Baridon 2003: 83). Wolff: "so bringt uns die Mathematick zu der vollkommensten Erkäntnis aller möglichen Dinge in der Welt. Da nun ferner diese Erkäntnis uns geschickt machet die Kräffte der Natur nach unserem Gefallen zu unserem Nutzen in dem Grade anzuwenden, den wir verlangen; so erlangen wir durch die Mathematick die Herrschaft über die Natur." (Wolff 1716: 864)

[6] Leibniz to Du Mont, July 11, 1696.

Fig. 3 Abraham Bosse, *Maniere universelle de M. Desargues pour praticquer la Perspective*, Paris 1648, engraving, p. 149. © Herzog August Bibliothek Wolfenbüttel

impossible to decipher whether this is supposed to be a real garden or the imagined geometrical space of a *hortus mathematicus*. One can easily imagine how Saint-Simon, critic of the gardens of Versailles, would have been provoked by such a "violence against nature".[7]

A wider view would, however, show that many early modern gardens were structured by more than just geometry, indeed by a whole array of mathematical sciences. John Dixon Hunt, in his masterly treatment of the practice of garden theory, has observed that "the thrust of garden art [...] was its celebration of the superiority of human creation and art over the natural world. It should delight the eye and the mind with representations by which men and women demonstrated both their creative skills and their understanding of the natural and cultural worlds" (Hunt 2000: 85). These characteristics align closely with those of the early modern mathematical sciences, which were often likewise appreciated for their capacity to entertain and delight (Remmert 1998: 165–170; Stafford 1994), and had shown how nature could be understood, thus opening up new ways to manipulate it. These new options not only had a profound effect on landscape and garden design, but also led early modern mathematicians themselves to become involved in garden projects: for example, many of the scientific experiments at Kassel were closely related to the demands of garden design (Gaulke 2010), and, as mentioned earlier, Leibniz, at Herrenhausen, and Euler, at Sanssouci, were each engaged in the contest to make the highest reaching garden fountain in Europe (Bredekamp 2012: 65ff; Eckert 2002).

Stromer von Reichenbach had said that in order to design a garden, "an honest gardener has to be familiar with a variety of disciplines". Decisions have to be made concerning the gardens' organization of space, the planting, the artistic features and the technical equipment. Knowledge of geometry becomes essential to the layout, knowledge of architecture is required to design pavilions and green houses, hydraulics is needed for the making of fountains and water-works, and familiarity with perspective and acoustics (including the effects of echoes) is called for if the experience of a garden is to be purposeful. All these fields of knowledge—geometry, architecture, hydraulics, perspective and acoustics—are contained within the canon of the mathematical sciences.

The practical realities of gardening usually require that, before the planting and equipment can be installed, the grounds have to be surveyed and the land modified to suit the design, and this too requires knowledge of the mathematical sciences. As at Versailles, as Chandra Mukerji has described, these tasks were often entrusted to military engineers who had the necessary expertise in moving vast amounts of soil and changing landscapes (Mukerji 1997). Along with familiarity with mathematical instruments, the disciplines needed for military engineering—fortification,

[7] (Saint-Simon 1857: 467): "La violence qui y a été faite partout à la nature repousse et dégoûte malgré soi."

geodesy, practical geometry, etc.—are all also contained within the mathematical sciences.

Meanwhile, work was being carried out at a more theoretical level. One of the most determined propagators of the use of the mathematical sciences in gardening was the English virtuoso John Evelyn (1620–1706), well-known as a co-founder of the Royal Society and for his diary.[8] On his Grand Tour of Italy and France, Evelyn developed a keen interest in gardens and began to immerse himself in the literature on the subject. Beginning in the 1650s, he became a key figure in introducing continental writings on gardens to England. In 1659, Evelyn discussed with Robert Boyle the idea of founding an institution for "the promotion of experimental knowledge", which was inspired by Francis Bacon's fictional ideal college, "Solomon's House". Evelyn suggested: "[W]hy might not some gentlemen, whose geniuses are greatly suitable, & who desire nothing more than to give a good example, preserve science, & cultivate themselves, join together in a society, & resolve upon some orders & oeconomy, to be mutually observed, such as shall best become the end of their union [...]."[9] Without going into the details of Evelyn's proposal (cf. Darley 2006: 156f; Leslie 1998: 147; Remmert 2007: 26–28), it included the recommendation: "Every one [is] to cultivate his own garden." (Evelyn 1879: III, 266). The reference to cultivating the self and the injunction to cultivate one's own garden (as a garden of knowledge) make clear that, for Evelyn, the promotion of knowledge was analogous to gardening, a conception no doubt prompted again by Bacon (Bacon 1625: 266; cf. Hyde 2005: chapter 4).

Evelyn came to see the garden as a heterotopia, that is, as a space that exists as a physical reality and is part of the world. But at the same time, it was a place to retreat from that world where one could critically and systematically study the same, and was thus simultaneously positioned outside of it. The pleasure garden and the microcosm of learning, the kitchen garden and the laboratory, the real and the imaginary garden could no longer be separated (Foucault 1986; Remmert 2007: 33f).

Evelyn spent almost 50 years of his long life writing and constantly re-writing his compendium on gardens, *Elysium Britannicum, or The Royal Gardens* (Evelyn 2001). Only small parts of this magnificent opus were published in his lifetime, and it was not until 2001 that a printed edition of the full manuscript was made available.[10] Evelyn's somewhat ambitious aim was to transform gardening into a science of nature. In order to do this, the *Elysium Britannicum* draws on a whole range of mathematical sciences—geometry, optics, astronomy, astrology, perspective, architecture, acoustics, the science of waters and so on. In the chapter *Of a Gardiner, and how he is to be qualified*, he emphasizes the importance of the

[8] On Evelyn see (Darley 2006; Harris and Hunter 2003; O'Malley and Wolschke-Bulmahn 1998; Parry 1992).

[9] Evelyn to Robert Boyle, September 3, 1659 (Evelyn 1879: III, 262).

[10] On the edition and the shortcomings of the commentary see the review by Thomas (2001).

mathematical sciences to gardeners: "[W]hat *Plato* caused to be inscribed upon the *Architrave* of his Schoole dore, would be set with as much reason over that of our Garden, Αγεωμέτρητος *nemo* [Let no one ignorant of geometry enter]." (Evelyn 2001: 33f). Time and again, Evelyn stressed that "it were much to be wished that our Gardiner had more than a Superficiall Skill in the Mathematicks" (Evelyn 2001: 100). Throughout the manuscript, he repeatedly points to the theoretic frameworks necessary to create the objects and effects in a garden: for instance, about fountains and hydraulics, he says: "And because of all expedients for these Effects, the *Hydraulique* have bin ever {The most} esteemed, & do so naturally {accord} with fountaines, & the motions created by them, it will be very requisite that our Gardiner turn a little back, & {refresh} his memorie with some rules & praecognit's, set downe in cap: 9th: where we have discoursed at large, concerning Fountaines & Waterworks" (Evelyn 2001: 231). In requiring that the good gardener be knowledgeable in the mathematical sciences, Evelyn is clearly speaking to the educated, and in the fourth chapter of the *Elysium Britannicum*, he is explicit, saying that he did "not pretend here to write to *Cabbage-planters*; but to the best refined of our Nation who delight in Gardens, and aspire to the perfections of the Arte [...]" (Evelyn 2001: 42; cf. Hunter 1998: 103f).

The sources Evelyn drew on have not yet been carefully examined, but the manuscript abounds with references to volumes of work by contemporary practitioners of the mathematical sciences, such as Bernard Lamy, Salomon de Caus, Marin Mersenne and William Oughtred, and, above all, the heavy Latin folios of Jesuits authors such as Mario Bettini, Athanasius Kircher and Gaspar Schott (Remmert 2013). The Jesuit publications were essential to the teaching of the whole range of mathematical sciences, including their practical branches, which were as relevant to landscape and garden design as they were to fortification building and military engineering (Harris 1995; Romano 1999; Vanpaemel 2012).

Jesuits were not the only authors writing on the practical branches of the mathematical sciences. Many others were publishing books and booklets on topics such as practical geometry, surveying, fortifications and so on. Practical geometry was particularly useful because it provided the knowledge and methods essential to designing the organizational structure of a garden and deducing how that structure could be realized in a given landscape. This not only pertained to the making of French formal gardens, but also to the English landscape gardens of the eighteenth and nineteenth centuries, even when their use of geometry did not figure very prominently (Clark 2010; cf. Phibbs 2006). Measuring and surveying were, of course crucial to garden and landscape design. André Le Nôtre paid close attention to these aspects, and it is well known that he worked with very precise measurements (Baridon 2003: 113–119; Thompson 2006: 75–95). Accuracy demanded expertise in handling surveying instruments, not only to meet the requirements of exactitude, but also because small blunders in measurement could lead to considerable extra costs (Garrigues 2001: 24f; Vérin 1991: 139f).

During the eighteenth century, authors writing on gardening, for example, Dézallier d'Argenville in his *Théorie et Pratique du Jardinage* of 1709, Stephen Switzer in his *Ichnographia Rustica* of 1718 and Batty Langley in his *New*

Principles of Gardening of 1728, included the relevant parts of mathematical knowledge in their books, as Evelyn had done. But during the seventeenth century, in order to learn the necessary techniques, landscape designers and gardeners often had to rely on the then widely available primers written for young military officers, such as Sébastien Leclerc's *Pratique de la geometrie* (Paris 1669, 1682, 1691). This use of military textbooks has been described by Chandra Mukerji in her analysis of the relationship between garden and landscape design and military engineering during the seventeenth century in her study of Versailles (Mukerji 1997).

In the meantime, quite a few seventeenth and early eighteenth century books on mathematics were referring to the connections between mathematics and the practice of gardening. A case in point is the *Géométrie pratique* by Alain Manesson-Mallet (1630–1706), "maitre de mathématiques" at the court of Louis XIV in Versailles, which was published in Paris in 1702. The *Géométrie pratique* covers 1500 pages in four volumes. Volume 1 deals with elementary geometry (Euclid); volume 2 with trigonometry; 3 with planimetry; and 4 with stereometry. Many of the more than 500 engravings are related to gardening. Their mathematical diagrams are most often accompanied by depictions of palaces and gardens so that the connection between mathematics and architecture and garden design is repeatedly made clear visually throughout the *Géométrie pratique* (Fig. 4). The engravings, with their numerous views of Versailles in reverence to Manesson-Mallet's patron and king, Louis XIV, also illustrate many general aspects of garden design, showing that, for Manesson-Mallet, there was an important concurrence between the practice of gardening and the practical branches of the mathematical sciences (Remmert 2008: 55f).

While mathematicians such as Manesson-Mallet demonstrated their interest in gardens, authors writing on gardening, for example, Dézallier d'Argenville, Switzer and Langley as mentioned above, started to incorporate elements of the mathematical sciences into their works that might prove useful to garden design. Antoine-Joseph Dézallier d'Argenville (1680–1765) begins his *La Théorie et la Pratique du Jardinage* (1709) with a chapter on geometric methods (*Préliminaire de quelques Pratiques de Geometrie, décrites sur le Papier, avec la maniere de les raporter fidélement sur le Terrain*), which addresses the question of transferring a design into reality on the land. He does not expect a gardener to read books on practical geometry, but he gives a digest of the fundamentals because a gardener needs to have some knowledge of geometry ("il faut être un peu Geometre"). Stephen Switzer (1682–1745) in his *Ichnographia Rustica* (1718) follows this example and clearly states that the gardener "ought to be a tolerable good Mathematician, Historian, and Architect" (Switzer 1718: III, 9). Among the "mathematical" chapters that fill half of the second volume of his compendium, one is titled "Instrumental Instructions, and particularly the Use of Surveying Instruments" (Switzer 1718: II, chap. 3). Batty Langley (1696–1751) in his *New Principles of Gardening* (1728) pursued a similar line and "thought it necessary, [. . .] to lay down all the most useful *Elements of Geometry*, necessary to be well understood by every *good Gardener* that's employed in *making and laying out Gardens*" (Langley 1728: viif).

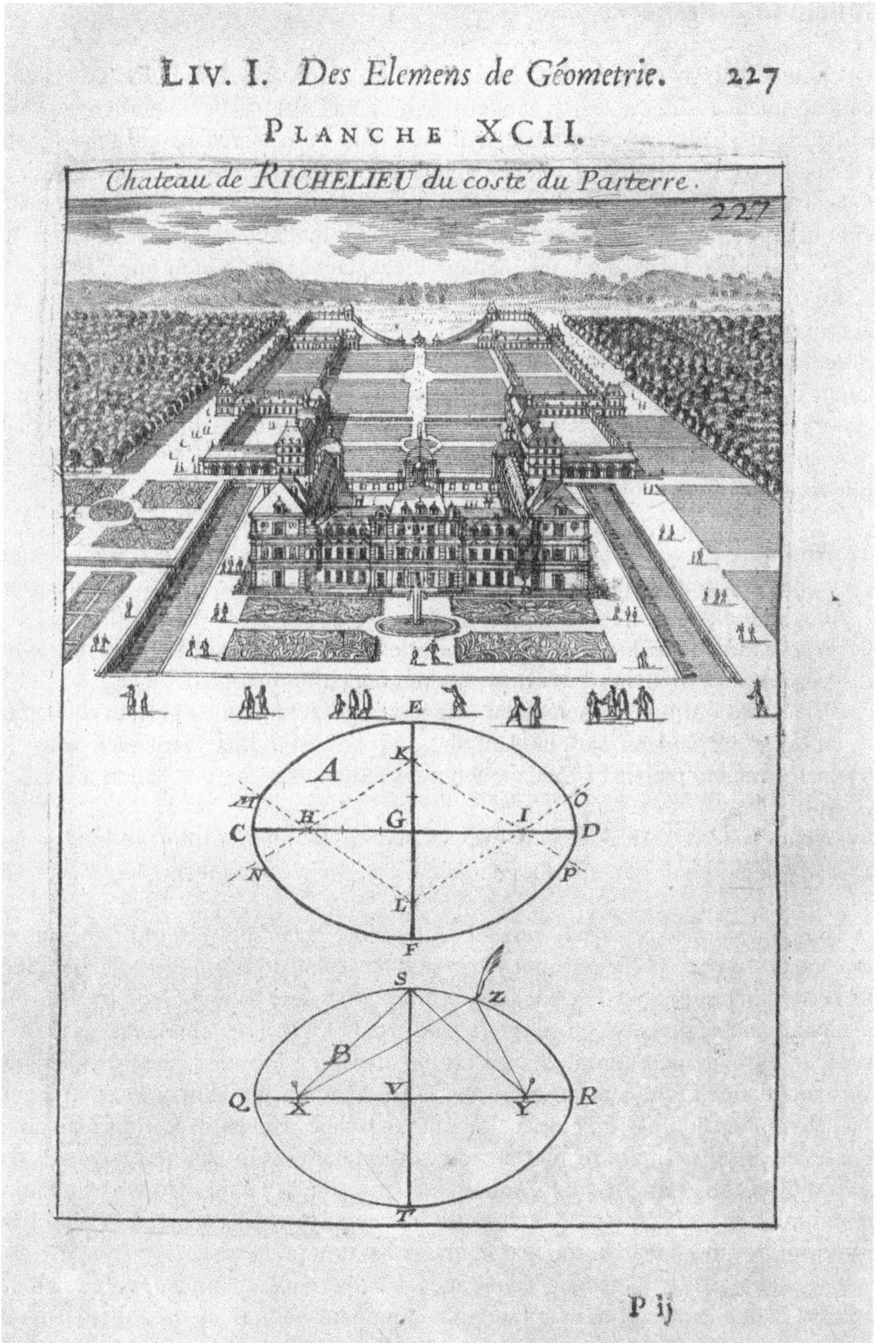

Fig. 4 Manesson-Mallet, Alain: *La géométrie pratique*, Paris 1702, engraving, I, 227. © Herzog August Bibliothek Wolfenbüttel

Concluding Remarks

A detailed analysis of the relationship between landscape design and the mathematical sciences in the early modern period has yet to be undertaken. More generally, it is obvious that there is a vast field of research to be done on the intersection of the history of science, mathematics and technology and the history of landscape design and gardening. Studying such intersections is, however, not without problems. Which aspects of early modern mathematics, science and technology should be focused on: the mathematical sciences with their many branches, as, for instance, acoustics, perspective, geometry; technology in its various guises (automata or water-works; instruments and techniques for measurement and surveying, etc.)—all related to the grand topic of mathematization; the crafts and practical arts; the ways in which the relevant knowledge was acquired, transferred, and organised; to give only a few examples from a long list?

A thorough analysis has to consider all of these issues. In particular, the following questions would need to be addressed:

(1) Who were the garden makers, which aspects of the mathematical sciences and technologies influenced their decisions, and how and for what reasons were they influenced?
(2) Who codified (and wrote down) the relevant forms of knowledge? For what reasons, and what are the forms of communication involved?
(3) What role did the mathematical sciences and technologies play in the actual making of gardens and landscapes, and how did this feed back into the mathematical sciences and technology? (Gaulke 2010; Valleriani 2011: 130–132)
(4) What epistemological aspects were at play in the relationship between garden and landscape design and the mathematical sciences and technology?

These questions are closely related to the more general topics of mathematization and scientization, the advances of which have shaped European societies since the early modern period. The sociologist Nico Stehr sees scientization as a process of mutual exchange between scientific and technological development on the one hand, and social and cultural change on the other, the former typically generating constant change in the latter (Stehr 1994). The relationship between early modern sciences and technology and the making of landscapes and gardens, thus explained, is a noteworthy example of this process. This relationship has not, however, yet been subject to systematic study, although, in my opinion, it might offer stimulating opportunities to historians of both fields, especially when considered within the framework of the scientization and mathematization processes.

The making of gardens and landscapes was, perhaps, the most visible demonstration of the early modern impulse to dominate nature. In this, there was an obvious alignment with the mathematical sciences. Practitioners of both sought the patronage of princes who saw and fashioned themselves as masters of nature. But further, they shared the conviction that by understanding and controlling nature, man would be able to recreate paradise on earth. These similarities of purpose and aspiration, and the resulting interaction of garden design and the mathematical sciences, did not, however, last longer than from the mid-sixteenth century to the mid-eighteenth century. The application of the mathematical sciences in garden and landscape design had run its course—their increasingly rigid mechanistic worldview had now proved inappropriate to the realities of gardening. In tune with this, Denis Diderot, in his essay *Thoughts on the Interpretation of Nature* of 1751, pointedly observed, "that the domain of mathematics is a world purely of the intellect, where what are taken to be absolute truths cease entirely to be so when applied to the world we live in" (Diderot 1999: 35). The mathematical sciences ceased to be the "leading discipline" (Leitwissenschaft) of gardening, which in 1765 was characterized in the *Encyclopédie* thus: "Gardening is the art of planting, of decorating and cultivating [in] every kind of garden; it is part of botany."[11]

Bibliography

Sources

Bacon, Francis. 1625. The Essayes or Counsels, Civill and Morall. London.

Bettini, Mario. 1642. Apiaria universae philosophiae mathematicae. 3 vols. Bologna.

Bettini, Mario. 1648. Aerarium philosophiae mathematicae. Bologna.

Blondel, Jacques-François. 1747. Discours sur la maniere d'étudier l'Architecture, & les arts qui sont relatifs à celui de bâtir, prononcé par M. Blondel Architecte à Paris, à l'ouverture de son deuxième cours public sur l'Architecture, le 16 Juin 1747. Mercure de France 53: 57-73.

Bosse, Abraham. 1648. Maniere universelle de M. Desargues pour praticquer la Perspective. Paris.

Boyceau, Jacques. 1638. Traité du Jardinage selon les raisons de la nature et de l'art. Paris.

Dézallier d'Argenville, Antoine-Joseph. 1709. La Théorie et la Pratique du Jardinage. Paris.

Dézallier d'Argenville, Antoine-Joseph. 1711. La Théorie et la Pratique du Jardinage. The Hague.

Diderot, Denis. 1999. Thoughts on the Interpretation of Nature and Other Philosophical Works. Introduced and annotated by David Adams. Manchester: Clinamen Press.

Evelyn, John. 1879. The Diary of John Evelyn to which are added a selection from his familiar letters. 4 vols., ed. William Bray. London.

Evelyn, John. 2001. Elysium Britannicum, or The Royal Gardens. Ed. John E. Ingram. Philadelphia.

Groen, Jan van der. 1669. Le Jardinier Hollandois/Der Niederländische Gärtner. Amsterdam. [bilingual in parallel columns]

Harris, Walter. 1699. A Description of the King's Royal Palace and Gardens at Loo, together with a Short Account of Holland. London.

Hesse, Heinrich. 1710. Teutscher Gärtner. Leipzig.

Langley, Batty. 1728. New Principles of Gardening. London.

Mandirola, Agostino. 1660. Manuale de' Giardinieri. Vicenza.

Manesson-Mallet, Alain. 1702. La géométrie pratique. 4 vols. Paris.

Saint-Simon, Louis de Rouvroy de. 1857. Mémoires complets. Vol. XII. Paris.

Schott, Gaspar. 1661. Cursus mathematicus. Würzburg.

[11] Article *Jardinage* in the *Encyclopédie*, vol. VIII, Paris 1765, 459f: "[...] le jardinage est l'art de planter, de décorer& de cultiver toutes sortes des jardins; il fait partie de la Botanique."

Stromer von Reichenbach, Wolf Albrecht. 1671. Der Italiänische Blumen- und Pomeranzen-Garten F. Augustini Mandirolae. Nuremberg.
Switzer, Stephen. 1718. Ichnographia Rustica. 3 vols. London.
Wolff, Christian. 1716 [reprinted 1965]. Mathematisches Lexicon. Leipzig [Hildesheim, Olms].
Wotton, William. 1697^2 [1694]. Reflections upon Ancient and Modern Learning. London.
Worshipful Company of Gardeners. 1700. Adam armed, or An Essay. Early English Books Online (EEBO).http://eebo.chadwyck.com/search/full_rec?SOURCE=pgthumbs.cfg&ACTION=ByID&ID=47682791&FILE=../session/1393007415_26152&SEARCHSCREEN=CITATIONS&SEARCHCONFIG=var_spell.cfg&DISPLAY=AUTHOR. Accessed February 19, 2007.

Secondary Literature

Baridon, Michel. 2003. Histoire des Jardins de Versailles. Versailles: Actes Sud.
Biagioli, Mario. 1993. Galileo, Courtier: The Practice of Science in the Culture of Absolutism. Chicago et al.: University of Chicago Press.
Bredekamp, Horst. 2012. Leibniz und die Revolution der Gartenkunst. Herrenhausen, Versailles und die Philosophie der Blätter. Berlin: Wagenbach.
Clark, Ronald. 2010. Die Regeln des "Natürlichen". Der konstruierte Landschaftsgarten bei Christian Schaumburg (1788-1868). Stadt+Grün 11: 39-44.
Darley, Gillian. 2006. John Evelyn: Living for Ingenuity. New Haven et al.: Yale University Press.
Dear, Peter R. 1995. Discipline & Experience: The Mathematical Way in the Scientific Revolution. Chicago et al.: University of Chicago Press.
Eckert, Michael. 2002. Euler and the Fountains of Sanssouci. Archive for History of Exact Science 56: 451-468.
Foucault, Michel. 1986. Of Other Spaces. Diacritics 16: 22-27.
Garrigues, Dominique. 2001. Jardins et jardiniers de Versailles au Grand Siècle. Seyssel: Champ Vallon.
Gaulke, Karsten. 2010. Experimentelle Naturlehre und Gartenkunst: Was haben Vakuumpumpen und Dampfmaschinen mit den Wasserkünsten zu tun?. In Hortus ex machina. Der Bergpark Wilhelmshöhe im Dreiklang von Kunst, Natur und Technik, eds. Sandra Kress and Jennifer Verhoeven, 156-166. Stuttgart: Theiss.
Gorman, Michael John. 2003. Mathematics and Modesty in the Society of Jesus: The Problems of Christoph Grienberger. In The New Sciences and Jesuit Science, ed. Mordechai Feingold, 1-120. Dordrecht/Boston/London: Kluwer Academic Publishers.
Harris, Frances, and Michael Hunter. 2003. John Evelyn and His Milieu. London: The British Library.
Harris, Steven J. 1995. Les chaires de mathématiques. In Les jésuites à la Renaissance. Système éducatif et production du savoir, ed. Luce Giard, 239-261. Paris: Presses Universitaires de France.
Hunt, John Dixon. 2000. Greater Perfections. The Practice of Garden Theory. Philadelphia: University of Pennsylvania Press.
Hunter, Michael. 1998. John Evelyn in the 1650s: A Virtuoso in Quest of a Role. In John Evelyn's "Elysium Britannicum" and European Gardening, eds. Therese O'Malley and Joachim Wolschke-Bulmahn, 79-106. Washington D. C.: Dumbarton Oaks Research Library and Collection.
Hyde, Elizabeth. 2005. Cultivated Power. Flowers, Culture, and Politics in the Reign of Louis XIV. Philadelphia: University of Pennsylvania Press.
Jong, Erik de. 1988. "Netherlandish Hesperides" – Garden Art in the Period of William and Mary 1650-1702. Journal of Garden History 8: 15-40.
Jong, Erik de. 2000. Nature and Art: Dutch Garden and Landscape Architecture, 1650-1740. Trans. Ann Langenakens. Philadelphia: University of Pennsylvania Press.

Jong, Erik de. 2005. Of Plants and Gardeners, Prints and Books: Reception and Exchange in Northern European Garden Culture, 1648-1725. In Baroque Garden Cultures: Emulation, Sublimation, Subversion, ed. Michel Conan, 37-84. Washington D.C.: Dumbarton Oaks Research Library and Collection.
Knobloch, Eberhard. 1989. Klassifikationen. In Maß, Zahl und Gewicht: Mathematik als Schlüssel zu Weltverständnis und Weltbeherrschung, eds. Menso Folkerts et al., 13-40. Weinheim: VCH.
Lauterbach, Christiane. 2004. Gärten der Musen und Grazien: Mensch und Natur im niederländischen Humanistengarten 1522-1655. Munich/Berlin: Deutscher Kunstverlag.
Leslie, Michael. 1998. "Bringing Ingenuity into Fashion": The "Elysium Britannicum" and the Reformation of Husbandry. In John Evelyn's "Elysium Britannicum" and European Gardening, eds. Therese O'Malley and Joachim Wolschke-Bulmahn, 131-152. Washington D. C.: Dumbarton Oaks Research Library and Collection.
Mukerji, Chandra. 1990. Reading and writing with nature: Social claims and the French formal garden. Theory and Society 19: 651-679.
Mukerji, Chandra. 1997. Territorial Ambitions and the Gardens of Versailles. Cambridge et al.: Cambridge University Press.
O'Malley, Therese, and Joachim Wolschke-Bulmahn (eds.). 1998. John Evelyn's "Elysium Britannicum" and European Gardening. Washington D. C.: Dumbarton Oaks Research Library and Collection.
Parry, Graham. 1992. John Evelyn as Hortulan Saint. In Culture and Cultivation in Early Modern England: Writing and the Land, eds. Michael Leslie and Timothy Raylor, 130-150. Leicester et al.: Continuum International Publishing.
Phibbs, John. 2006. Projective Geometry. Garden History 34: 1-21.
Remmert, Volker R. 1998. Ariadnefäden im Wissenschaftslabyrinth. Studien zu Galilei: Historiographie - Mathematik - Wirkung. Berne: Lang.
Remmert, Volker R. 2004. Hortus mathematicus: Über Querverbindungen zwischen Gartentheorie und -praxis und den mathematischen Wissenschaften in der Frühen Neuzeit. Wolfenbütteler Barock-Nachrichten 31: 3-24.
Remmert, Volker R. 2007. "Of a Gardiner, and how he is to be qualified": John Evelyn, Gartenkultur und mathematische Wissenschaften im 17. Jahrhundert. In Revolution in Arkadien, eds. Berthold Heinecke and Harald Blanke, 23-37. Hundisburg: Kultur-Landschaft Haldensleben-Hundisburg.
Remmert, Volker R. 2008. "Il faut être un peu Geometre": Die mathematischen Wissenschaften in der Gartenkunst der Frühen Neuzeit. In Wunder und Wissenschaft. Salomon de Caus und die Automatenkunst in Gärten um 1600, ed. Gabriele Uerscheln, 51-58. Düsseldorf: Grupello.
Remmert, Volker R. 2011. Picturing the Scientific Revolution: Title Engravings in Early Modern Scientific Publications. Philadelphia: Saint Joseph's University Press.
Remmert, Volker R. 2013. "and as we remember Kircher speaks of it": Schriften jesuitischer Autoren als Quellen von John Evelyns Elysium Britannicum. In Landschaften – Gärten – Literaturen: Festschrift für Hubertus Fischer, ed. Irmela von der Lühe and Joachim Wolschke-Bulmahn, 159-169. Munich: Akademische Verlagsgemeinschaft.
Romano, Antonella. 1999. La contre-réforme mathématique: Constitution et diffusion d'une culture mathématique Jésuite à la Renaissance (1540-1640). Rome: École Française de Rome.
Schweizer, Stefan. 2012. Die Erfindung der Gartenkunst: Gattungsautonomie - Diskursgeschichte – Kunstwerkanspruch. Munich: Deutscher Kunstverlag.
Seller, Vanessa Bezemer. 2001. Courtly Gardens in Holland 1600-1650: The House of Orange and the Early Hortus Batavus. Amsterdam, Woodbridge: Architectura & Natura Press, Garden Art Press.
Stafford, Barbara Maria. 1994. Artful Science: Enlightenment Entertainment and the Eclipse of Visual Education. Cambridge (Mass.), London: MIT Press.
Stehr, Nico. 1994. Arbeit, Eigentum und Wissen: zur Theorie von Wissensgesellschaften. Frankfurt a.M.: Suhrkamp.
Thomas, Keith. 2001. No High Heels in Paradise. London Review of Books 23: 21-22.

Thompson, Ian. 2006. The Sun King's Garden. Louis XIV, André Le Nôtre and the Creation of the Gardens of Versailles. London: Bloomsbury.
Valleriani, Matteo. 2011. The Garden of Pratolino: Ancient Technology Breaks Through the Barriers of Modern Iconology. In Ludi naturae. Spiele der Natur in Kunst und Wissenschaft, eds. Natascha Adamowsky, Hartmut Böhme and Robert Felfe, 121-141. Munich: Wilhelm Fink.
Vanpaemel, Geert. 2012. Jesuit Mathematicians, Military Architecture and the Transmission of Technical Knowledge. In The Jesuits of the Low Countries: Identity and Impact (1540-1773), eds. Rob Faesen and Leo Kenis, 109-128. Leuven, Paris and Walpole: Peeters.
Vérin, Hélène. 1991. Technology in the Park: Engineers and Gardeners in Seventeenth-Century France. In The History of Garden Design. The Western Tradition from the Renaissance to the Present Day, eds. Monique Mosser and Georges Teyssot, 135-146. London: Thames and Hudson.

"Without Design, or Fate, or Force": Why Couldn't John Evelyn Complete the *Elysium Britannicum*?

Michael Leslie

Abstract
Despite living for 50 years after beginning his "Elysium Britannicum" in the 1650s, John Evelyn (1620–1706) failed to complete what was meant to be a masterwork on gardens. The exceptionally complex manuscript left behind was only published in a heroic edition by John Ingram in 2000. Why did Evelyn embark on the work and why didn't he complete it? This essay suggests that the initial project developed in a uniquely unstable intellectual, religious, and political moment, in which Evelyn felt free to respond to neo-Epicurean physics and philosophy. As order returned, Evelyn found himself unable to accept the worldview implied by neo-Epicureanism but also unwilling to publish a revised version that denied conclusions his contemporaries were coming to accept. Evelyn's response is contrasted with that of the plant anatomist Nehemiah Grew (1641–1712), who was also aware of the radical implications of neo-Epicureanism, but whose social and religious circumstances were markedly different.

> How now, *Lucretius*; you an *Epicurean*, and yet against liberty of judgment among Philosophers?[1]

The mid-seventeenth century bore witness to an almost pervasive crisis in the West—famously, a "world turned upside down"[2]—not least in religion (the Second

[1] (Charleton 1657: 51). In his fictionalized discussion, Charleton (1619–1707) represents Evelyn as Lucretius.

[2] The phrase "the world turned upside down" comes from a ballad of the later 1640s, opposing the suppression of Christmas festivities and rituals by the Parliamentary government. But it has its modern resonance as the title of a book by the historian Christopher Hill (1972), which has the

M. Leslie (✉)
2274 North Drive, Memphis, TN 38112, USA
e-mail: mpmjleslie@gmail.com

H. Fischer et al. (eds.), *Gardens, Knowledge and the Sciences in the Early Modern Period*, Trends in the History of Science, DOI 10.1007/978-3-319-26342-7_3

Reformation and the Thirty Years' War), politics (the subsequent convulsions producing and resulting from the Fronde in France and the Civil Wars in Britain), and science (Galilean astronomy and Cartesian mathematics, logic, and philosophy), with the birth of a characteristic and ultimately dominant form of empiricism in England once Francis Bacon's publications concerning experiment and scientific method were embraced as enunciating foundational principles by disparate groups in the 1640s and 1650s, and by the Royal Society after 1660. These seemingly independent phenomena are profoundly interconnected; they are also inseparable from questions concerning garden and landscape design, not least the beginnings of the competition between a geometric style that emphasized a privileged proprietor, lord of all he surveyed, and a style that would ultimately seem to embody an ideology of liberty befitting a moral senatorial class. Close to the centre of many of these topics are competing ideologies and sciences of the culture and cultivation of the earth, plants, agriculture, and horticulture: the "georgical" and the "terracultural".[3]

The seventeenth century sees in Britain a startling change in gardens, landscape, and terraculture, from provincial derivativeness in style, thought, and literature in the sixteenth century to acknowledged pre-eminence in the eighteenth.[4] At its centre is the remarkable intellectual ferment of the 1650s, the decade of the Interregnum, when kings and hierarchies were overthrown, thought unfenced, and ideas "disparked".[5]

This essay focuses on John Evelyn's "Elysium Britannicum", intended by its author to be a defining contribution to debates on terraculture (the science of man's engagement with gardens, the landscape, plants, and much more) and terraculture's cultural importance in human and divine history. The "Elysium", first mentioned by Evelyn in 1658 but as having been begun substantially earlier, was anticipated by his contemporaries as just such an encyclopedic *magnum opus*. But it was never completed, the manuscript only being published in a modern edition by John E. Ingram in 2000 (Penn Studies in Landscape Architecture).[6] Evelyn's failure to complete the work despite his remarkable longevity (1620–1706) demands

subtitle, *Radical Ideas During the English Revolution*. Geoffrey Parker (2013) argues that the scope of the disruption extended not just beyond the British Isles to Europe, but indeed worldwide.

[3] For an overview of some of these topics, see (Leslie/Raylor 1992) and (Greengrass/Leslie/Raylor 1994). See also (Low 1985) for the definition of the seventeenth century as "georgical".

[4] Rebecca Bushnell makes a valiant attempt to find more depth in the workmanlike English publications of the age before Evelyn (Bushnell 2003). However, their intellectual weight seems limited, despite her advocacy.

[5] See Edmund Waller's use of this metaphor, below, in his commendatory poem introducing John Evelyn's partial translation of Lucretius's *De rerum natura*.

[6] John Ingram's successful attempt to create a modern publication of the "Elysium" was not the first: there had been at least two previous attempts. A glance at the surviving manuscript will reveal why the creation of an edition proved so difficult: sheets are full of deletions, insertions, and corrections; pages have been removed, pages added; passages and new information are on slips of paper, originally pasted into place but now floating within the folder, unsure of place. John Ingram's edition will surely be refined over time, but we should recognize his labour as heroic.

explanation; and the explanation offered here situates the author, his manuscript, and his difficulties at the centre of scientific debates of the period.

The Culture of Knowledge and Communication in Evelyn's Lifetime

John Evelyn's lifetime begins with the dramatic proposition of new approaches to and definitions of knowledge by luminaries such as Galileo, Descartes, and, most importantly for Evelyn, Francis Bacon. Once Bacon's writings had been disseminated throughout not only English intellectual life but that of the whole of Europe, an astonishing period of creativity and discovery began, claiming him as its founding father. By the time Evelyn died, knowledge of physics, chemistry, biology, botany, and a host of other intellectual realms had changed to the point of being unrecognizable. Indeed, many of the names of fields and disciplines just used are novelties from this period. Even the terminology of knowledge had changed, itself a profound phenomenon.

Equally importantly, and inevitably connected, the human organization of knowledge also changed beyond recognition. By the time Evelyn commenced work on the "Elysium Britannicum", the technology of printing with movable type had been in use for two centuries in the West, and with every decade, the volume of printed works and their dissemination had increased startlingly. Ideas were, by 1640, being disseminated with extraordinary geographical scope and remarkable speed. The scale of potential private libraries for people—usually, but not always, men, affluent but not necessarily of exceptional wealth—had grown in ways that would have defied belief a few decades earlier. The manuscript of the "Elysium Britannicum" is studded with notes and references, indicative of Evelyn's attempt to incorporate all he had read and keep up with new publications, not only those by new authors but also editions of older works available for the first time. In this and other works, he laments his inability to absorb and manage the torrent of printed knowledge.

Evelyn had access to books and other printed materials on a grand scale. In fact, he had to control a tendency to addictive acquisition, discovering that he was unable to restrain his enthusiasms and was running up against limits of space and finances. Late in life, he wrote that collecting (in this case, prints) "took up a great deal of time, and no little money; and that the thirst of still augmenting grew upon me; I at last gave it over" (Darley 2006: 149). But as well as access to printed works, Evelyn found himself in his early years part of intellectual correspondence networks that remain among the most thrilling aspects of the period. In Rome, Evelyn met the great French scholar and indefatigable correspondent Marin Mersenne (1588–1648), the French Minim friar, sometimes called the 'father of acoustics', but someone who was interested in so much more of what we now call science. In 1644, both visited the laboratory in Rome of the even more extraordinary Athanasius Kircher (*c.* 1601–1680), recently the subject of a multi-authored intellectual biography subtitled *The Last Man Who Knew Everything* (Findlen 2004:

23–25). But the "Intelligencer" most important for Evelyn was the Anglo-Prussian Samuel Hartlib (*c.* 1600–1662), who made it his life's work to foster and communicate all knowledge that tended to the betterment of mankind and the establishment of peace (irenicism, an intellectual response not least to the Thirty Years War and the Wars of Religion). Hartlib arrived in England in 1622 and spent the rest of his life almost exclusively in London, devoting himself to operating what was called an "Office of Address". In its ultimate and grandest form, this enterprise enabled Hartlib to orchestrate contact between adventurous thinkers throughout the known world, from the occasional correspondent in the Levant, through more frequent contacts in central Europe, finding its centre in Britain, and all the way to the newly established plantations on the North American seaboard. The list of his correspondents is a roll-call of many of the experimenters and writers on whom our modern world's science depends; and there is a notably large group interested in matters to do with horticulture and agriculture. Hartlib's activity can be summarized thusly: he sought out connections with anyone he thought might potentially have something to offer of intellectual value; and he recorded his meetings with them in his working logbook, which he called his "Ephemerides". He recorded in the "Ephemerides" that he first encountered Evelyn in 1653:

> Mr Evelin at Dedford a chymist hath studied a great Worke of all Trades, and wants no more to it but the description of 3 trades.[7]

Everything about that entry is fascinating: Evelyn has been introduced to the Hartlib circle as an experimental chemist. But we should note that Hartlib's real interest lies in Evelyn's claim that he has been writing a "history of trades", and that this work is close to completion. But Evelyn's "history of trades" never saw the light of the printing press, joining the "Elysium" and, indeed, a "history of religion", also begun in the 1650s, as projects worked on for decades and yet never published. Evelyn was nothing if not ambitious (even if that ambition was at odds with a countervailing desire for independence from a world of public strife and striving): all three projects addressed topics deemed central to the intellectual life of his society. We should explore the reasons for Evelyn's reluctance to print all that he had claimed to have written and to complete works upon which he had expended so much time and energy.

Evelyn was also ambitious to be accepted as a member of Samuel Hartlib's circle. Hartlib would learn of a likely and promising thinker, would seek to get a written statement of whatever new knowledge that person was thought to have, and then, using his team of scribes and even rudimentary copying technology, would send that material to the other people in his network with similar interests, asking them to write back with comments, evaluations, criticisms, and supplementary ideas and information. These responses—and Hartlib had a sort of contract with his correspondents—would then be made into a digest to be returned to the original

[7] Hartlib Papers (Sheffield University) 28/2/66B-67A.

correspondent and then circulated again, with that person's replies. When satisfied with the accumulation of material, the results might be printed, as with *The Reformed Husband-man* of 1651 (and augmented editions thereafter). The scale and ambition of Hartlib's enterprise is breathtaking.

Such activity is one of the defining features of European life in the seventeenth century: vast spiders' webs of contact between adventurous and ambitious thinkers, corresponding on an astounding range of interconnected ideas. And one of the key fields that Hartlib and his colleagues were interested in was the "georgical": everything to do with the land and, in particular, that which would render it more fertile and productive, fulfilling its potential and restoring it to the perfection of its original divine creation. Although Evelyn becomes known to Hartlib as a 'chymist', it rapidly becomes apparent that the centre of his interests is the "georgical" and "terracultural".[8] In 1659, Evelyn sent the outline of the "Elysium Britannicum" to Hartlib, as he did to a range of other leading thinkers on such matters. He surely knew that, through Hartlib's network, both he and the "Elysium" would become known; and Evelyn subsequently corresponded with many of the other major thinkers in England concerning matters agricultural and horticultural. The developments that used to be called the eighteenth-century agricultural revolution have now been securely identified as truly belonging to the previous century; and they are not least the product of the people Evelyn was associating with in the middle of the century.

Evelyn's Subordinate Position in the Pantheon

Evelyn lived to be 86 years old and he encountered many of the greatest minds of his age. Indeed, he was a contemporary of much greater men: eight years younger than John Milton, another member of the Hartlib circle; in his old age, an acquaintance of Isaac Newton, another member of the Royal Society. But Evelyn was no Milton or Newton. If we were to list all the people who turned out to be more significant than Evelyn in his period in the history of gardens and of knowledge pertaining to the plant and natural world, we might think he was effectively crowded out: the botanist John Ray exceeds Evelyn in significance in that field; Samuel Pepys is more exciting as a diarist; Evelyn's close friend Robert Boyle, the father of chemistry, is obviously more significant as what we would call a scientist; the Anglo-Prussian Samuel Hartlib is more effective and more central as a communicator of ideas; Thomas Hobbes's political theories dwarf any interventions Evelyn might make in that field.

[8] The Royal Society was formed after the Restoration, and one of its principal sections was the "Georgical Committee", of which Evelyn was the principal early member. The term "terraculture" dates from the beginning of the eighteenth century; its advantages are well described by (Otten 1985: xiii and xix, n. 1).

As a result, a brief case should be made for his significance. The point is that Evelyn knew all these men: he corresponded with them; he corresponded about them; he attempted to assimilate their ideas and their examples, wrestling with incompatibilities and struggling to adapt himself and his thought-world to developments in what is one of the most exhilarating periods in intellectual history. His works and his gardens show these struggles in a way and to a degree that are just about unmatched in England in the period; and not least of these revealing works is the unfinished "Elysium Britannicum". If we want to understand the extreme disruptions to the intellectual world of the early modern period, disruptions to ideas of gardens, landscape, the natural world, epistemology, human society, even religion—all these are interconnected—Evelyn acts as perhaps the ideal laboratory animal. What will he do in the face of this or that encounter?

As mentioned above, Evelyn benefitted from two major phenomena in the history of the communication of knowledge. He came to maturity at a time when it had become possible for a man like him to have access to printed knowledge on a scale hitherto unknown, in his own library or those of his friends, or even in new institutional libraries like Oxford University's Bodleian. And he had entry to correspondence and idea-sharing networks that were driving explicitly towards the creation of encyclopedic knowledge of the world as we experience it, culminating after the Interregnum in the Royal Society. The "Elysium Britannicum" was conceived of and understood by others as a work that would be an encyclopedia of everything that should be known about gardening, or at least a particular branch of gardening. Evelyn was without doubt participating in this ambitious programme for the creation and dissemination of knowledge.

Evelyn and Religio-political Crisis

There is a danger that, in summary, these rapid developments can seem entirely positive. But for Evelyn, there was a dark side to his times. He was born into a well-to-do family, but not within the traditional aristocracy, or even the landed gentry. The Evelyn family had risen to fortune through, in effect, ownership of extractable raw materials and industrial processes: they owned saltpeter mines in Surrey and were heavily involved in the manufacture of gunpowder. As so often in England down to the present day, the ambition of the newly-prosperous was to separate themselves from the connotations of newly-acquired wealth and its source in trade, acquiring instead the trappings of ancient gentility: land, titles, and high culture. In the case of the Evelyn family, and also perhaps because the royal government was such a good patron and customer for military materials, the result appears to have been close ideological and emotional sympathy with the traditional structures of state and society, in particular with the institutions of the monarchy, the state church as envisaged by the episcopal hierarchy, and the cultural and educational enthusiasms shared by those elites.

Evelyn came to maturity at just the point at which any such allegiance could no longer be assumed to be automatic and unquestionable. For over a decade before his twentieth birthday, there had been a simmering conflict between a king attracted by the idea of greater autocracy and a parliament that sought to defend its existing powers, to gain more, and to protect the state church from any attempt to return it to Catholicism. In 1642, that conflict erupted into open war, and people like Evelyn had to choose sides. That was traumatic in ways we can readily imagine, and the conflict forever destroyed any sense that there was a single, natural way of organizing the state and society. In the 1640s and 1650s, there was an explosion of new ideas and proposals for how human society could best be organized. Everything was open for discussion, nothing was excluded as unthinkable, even if it excited violent opposition. All societies and all individuals have framing political and religious ideas and myths which are crucial to our "imaginary", but in 1642, not one of those frames was left unchallenged. Just as the world of ideas was now being recomposed, reshaped, and reconceived without presuppositions—without what was conceived of as the stranglehold of academic Aristotelianism, the tyranny of received ideas—so too human society was suddenly free to be reconceived.

Evelyn and Bacon

We can see the effect of all this in Evelyn's career by returning to that first entry in Samuel Hartlib's "Ephemerides":

> Mr Evelin at Dedford a chymist hath studied a great Worke of all Trades, and wants no more to it but the description of 3 trades.

What does Hartlib mean by "a great work of all trades"?

The "History of Trades", the creation of a database or encyclopedia of all information about all productive activities and processes, was perhaps the corner stone proposal of the seventeenth century's scientific patron saint, Francis Bacon. Bacon was himself no great experimenter, but he described a set of attitudes to the gathering and testing of knowledge that was of unparalleled and continuing influence. In terms of Evelyn's activities, there is no better epigraph than Bacon's warm praise of practical studies as "these *Georgickes* of the Mind, concerning the husbandry and tillage thereof", "no less worthy than the heroical descriptions of virtue, duty, and felicity" (*Advancement of Learning* 2.20.3). Not long after his death, Bacon's works began to circulate throughout Europe, and it is impossible to overstate their influence. Bacon's method of scientific enquiry became the rallying point for almost everyone with intellectual ambitions; and when Evelyn identified himself as being on the verge of completion of the central Baconian project of a History of Trades, the world was going to sit up and take notice (Walter Charleton, Evelyn's friend, fellow intellectual, and almost exact contemporary, includes the historians of trades, "who enquire into the mysteries of Refiners, Belfounders, and all others who deal in Metals", etc., among the "*Merchants for light*" (Charleton 1657: 41)). The "Elysium Britannicum" is, we might say, a large, never-completed section of that infinite, unfinished History of Trades.

Fig. 1 Godfrey Kneller, Portrait of John Evelyn (*c*. 1687). ©The Royal Society

This is important because Evelyn presents himself, throughout his early career, as an heir to Francis Bacon, one who might fulfill Bacon's prophecies of encyclopedic knowledge. In his most famous portrait (Fig. 1), we see Evelyn presenting himself as the author of the first book to be published by the newly-formed Royal Society: *Sylva, Or a Discourse of Forest-Trees* (1664). Interestingly, it turns out, as was often the case in his career as a writer, that this published work is actually multi-authored, a compilation in the manner of Hartlib's publications; and much of what is truly practical and innovative, that which engages with the new intellectual culture of natural philosophy, is by authors other than Evelyn. There ought to be portraits of more people within this picture frame; but throughout his life, Evelyn tended to accept freely the contributions of others, indicate the sources in his manuscripts (if anywhere), but then remove from the eventual printed work acknowledgement and attribution, unless the contributor was of sufficiently high social status to add to Evelyn's lustre.[9]

Here, though, we should pay attention to the book's title, which clearly—even more clearly if one compares the title-pages—evokes Francis Bacon's great work,

[9] See (Yale 2011). Even in his manuscripts, Evelyn was not scrupulous, finding the need to justify himself even in unprinted work: "I do not always charge the margin with their names, which I acknowledge to be a defect in my *adversaria*" (Evelyn 1850: Postscript, xxxv).

Sylva Silvarum: or A Naturall Historie In Ten Centuries (London, 1627). We know better its frontispiece, but Evelyn's *Sylva* does not emulate that image. Rather, Evelyn's titlepage (Fig. 2) is clearly structured to remind readers of its great progenitor, Bacon's own encyclopedia of "Naturall History. In Ten Centuries" (Fig. 3).

Evelyn's association of his activities with Bacon and his intellectual tradition is equally clear at the very beginning of the "Elysium Britannicum":

> When the Almighty had exiled our Fore-Fathers out of Paradise, the memorie of the delicious place was not yet so far obliterated, but that their early attempts sufficiently discover'd how unhappy they were to live without a Garden ...

So begins Book One, Chapter One. And Chapter Two begins,

> God almighty was the first gardener we have [,] scripture said.

This may be something of a commonplace, but Evelyn expects his reader to hear a loud echo of the opening of Bacon's famous essay, "Of Gardens" (Bacon 1625: 266):

> GOD Almighty first planted a garden.

The echo is an unambiguous attachment of his project to the patron saint of the most advanced intellectual circles. Evelyn has taken on the task of creating an encyclopedic History of Trades, as Bacon urged; the title of his *Sylva* imitates Bacon's *Sylva Sylvarum*; and his "Elysium Britannicum" opens—not once, but twice, and that repetition might be an early sign of his growing difficulties—with explicit echoes of Bacon's famous essay.

The opening is bound to be important, for two reasons. First, it is the opening section of the "Elysium" that Evelyn reports as causing the most trouble and, indeed, its perfection being the objective that he never attained ("My purpose was quite to alter the philosophical part of this first booke"—see below). Second, the issue of how such a work might open is addressed by William Rowley in his prefatory letter "To the Reader" in Bacon's *Sylva Sylvarum*:

> No wher to depart from the Sense, and cleare experience; But to keep close to it, especially in the beginning. (A1v)

Rowley's letter stresses Bacon's resolution to avoid "deceiving Notions, and Theories", and to "unloose Mens minds, being bound" (A1r), an aspiration captured visually in the famous "Ne Plus Ultra" frontispiece image, with its "Mundus Intellectualis" globe seen beyond the pillars of the known, in the shoreless ocean. Conversely, Evelyn immediately applies a scriptural framework that necessarily establishes boundaries to permissible explanation. Like Evelyn, even Bacon's editor, William Rowley, seems unnerved by *Sylva Sylvarum*'s apparent lack of method and perimeters, asserting that the author deferred to no man in his "love of

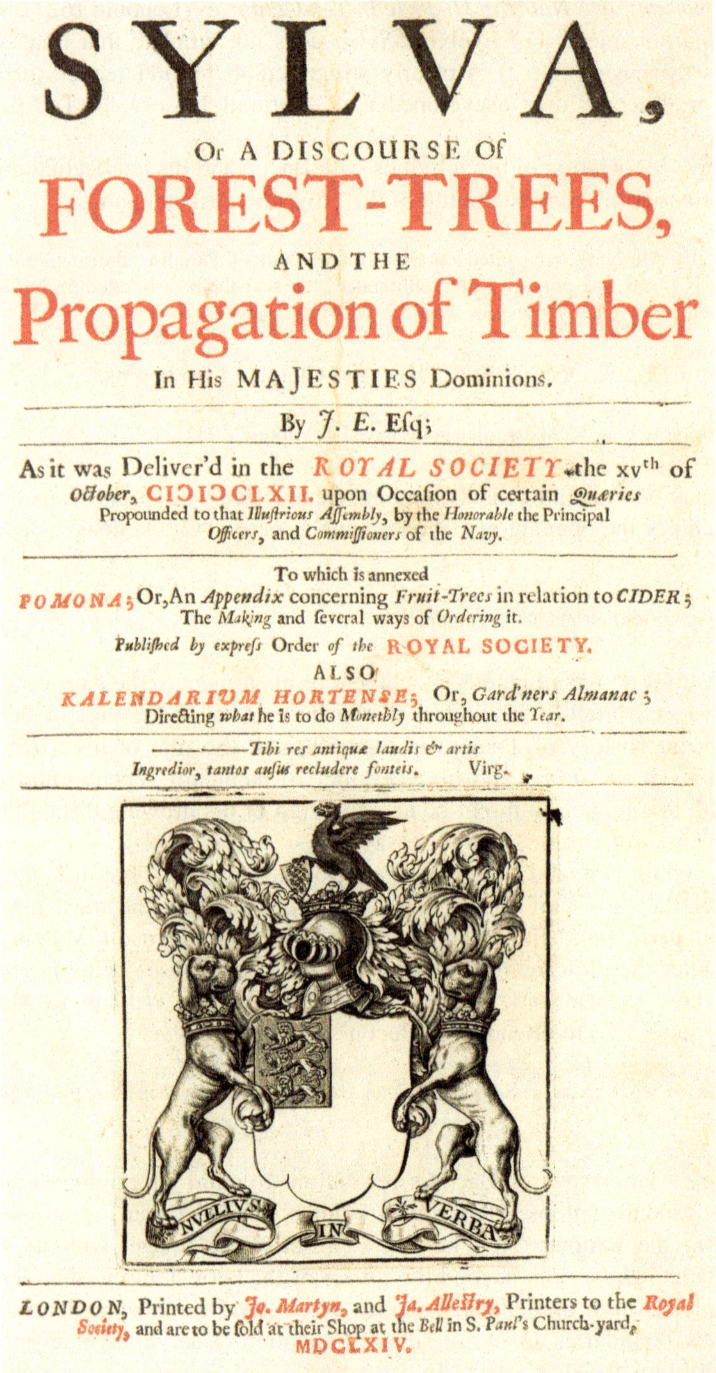

SYLVA,

Or A DISCOURSE Of

FOREST-TREES,

AND THE

Propagation of Timber

In His MAJESTIES Dominions.

By *J. E.* Esq;

As it was Deliver'd in the *ROYAL SOCIETY* the xvth of *October*, CIƆIƆCLXII. upon Occasion of certain *Quæries* Propounded to that *Illustrious Assembly*, by the *Honorable* the Principal *Officers*, and *Commissioners* of the *Navy*.

To which is annexed

POMONA; Or, An *Appendix* concerning *Fruit-Trees* in relation to *CIDER*; The *Making* and several ways of *Ordering* it.

Published by express Order *of the* ROYAL SOCIETY.

ALSO

KALENDARIUM HORTENSE; Or, *Gard'ners Almanac*; Directing *what* he is to do *Monethly* throughout the *Year*.

——*Tibi res antiquæ laudis & artis*
Ingredior, tantos ausus recludere fonteis. Virg.

LONDON, Printed by *Jo. Martyn*, and *Ja. Allestry*, Printers to the *Royal Society*, and are to be sold at their Shop at the *Bell* in S. *Paul's* Church-yard, MDCLXIV.

Fig. 2 John Evelyn, *Sylva, Or a Discourse of Forest-Trees* (London, 1664), titlepage. © The Huntington Library, San Marino, California

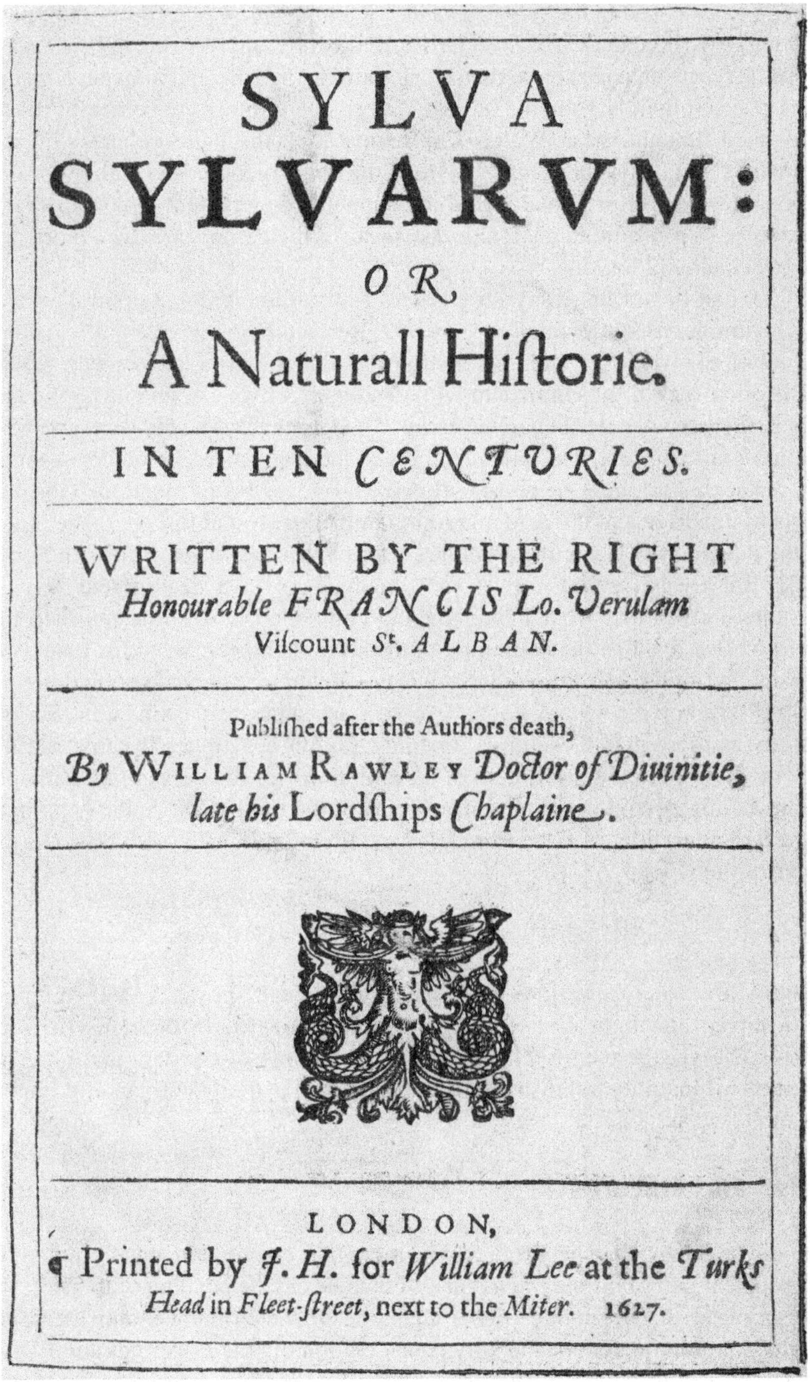
SYLVA
SYLVARVM:
OR,
A Naturall Hiſtorie.

IN TEN CENTVRIES.

WRITTEN BY THE RIGHT
Honourable FRANCIS Lo. Verulam
Viſcount *St. ALBAN.*

Publiſhed after the Authors death,
By WILLIAM RAWLEY *Doctor of Diuinitie,*
late his Lordſhips *Chaplaine.*

LONDON,
¶ Printed by *J. H.* for *William Lee* at the *Turks*
Head in *Fleet-ſtreet,* next to the *Miter.* 1627.

Fig. 3 Francis Bacon, *Sylva Silvarum: or, A Naturall Historie. In Ten Centuries* (London, 1627), titlepage. © The Huntington Library, San Marino, California

Order", as evidenced in his "*De Augmentis Scientiarum*, which (if my Judgment be any thing) is written in the Exactest Order, that I know any Writing to bee" (A2v). It is certainly true that Bacon's rhetorical training had taught him an unmatched rigour of disciplined argument. But Rowley is clearly more concerned by Bacon's ungoverned thought, what Walter Charleton called the philosopher's "liberty of judgment". None too subtly, Rowley steers the conveniently dead author back from the boundless ocean of knowledge into channeled waters. Questions of order and boundaries, so essential to contemporary ideas of the garden, are even more urgent in intellectual explorations.

Bacon had begun his essay on gardens with a thundering scriptural assertion: "GOD Almighty first planted a garden". But reading his essay alongside the opening of Evelyn's "Elysium" is instructive. Bacon was a lawyer who practiced the rhetorical arts of the courtroom with deadly effectiveness, and he applied these to his *Essayes* as well. He began the essay "Of Gardens", as he did so many, with an attention-grabbing sentence. The members of the jury—here the readers—sit up in their seats: this is going to be fun. But Bacon never again mentions God or the subject of the origins of the garden, turning immediately and forever to the practical and the detailed (a "Heap of Particulars", like *Sylva Sylvarum*, if not "Indigested" [A1r]). But Evelyn could not let such matters go: in a deep-seated way, they constitute the heart of what interested and compelled him, so he returned to the subject of God again at the beginning of the second chapter. What for Bacon was a flourish to introduce a 1000-word essay was adopted by Evelyn as something of the *raison d'etre* for the whole book. Evelyn's insistence on taking this Scriptural authority as more than a rhetorical opening gambit constitutes the first stumbling block on the way to the fulfillment of his ambitions for the "Elysium". Perhaps the most revealing note in the manuscript occurs in the margin above the beginning of the third chapter, titled "Of the Principles and Elements in generall". At some point, Evelyn wrote (Fig. 4),

> My purpose was quite to alter the philosophical part of this first booke.

There are long insertions indicated, additional passages and references, clarifications, and indications of second, third, and fourth thoughts. It is clear that, for Evelyn, these early chapters were the source of constant and unresolved difficulty; his inability to finish the work was rooted in their contradictory impulses.

Evelyn and Epicurus

What did he mean when he wrote that he thought it essential to alter completely the philosophical part of his *magnum opus*? A clue lies in the position of this note in the manuscript, at the beginning of the chapter "Of the Principles and Elements in general". In order to understand why consideration of "the Principles and Elements in generall" could be so debilitating, we have to look at Evelyn's contemporary non-gardening work, particularly from the time that he embarked on creating the

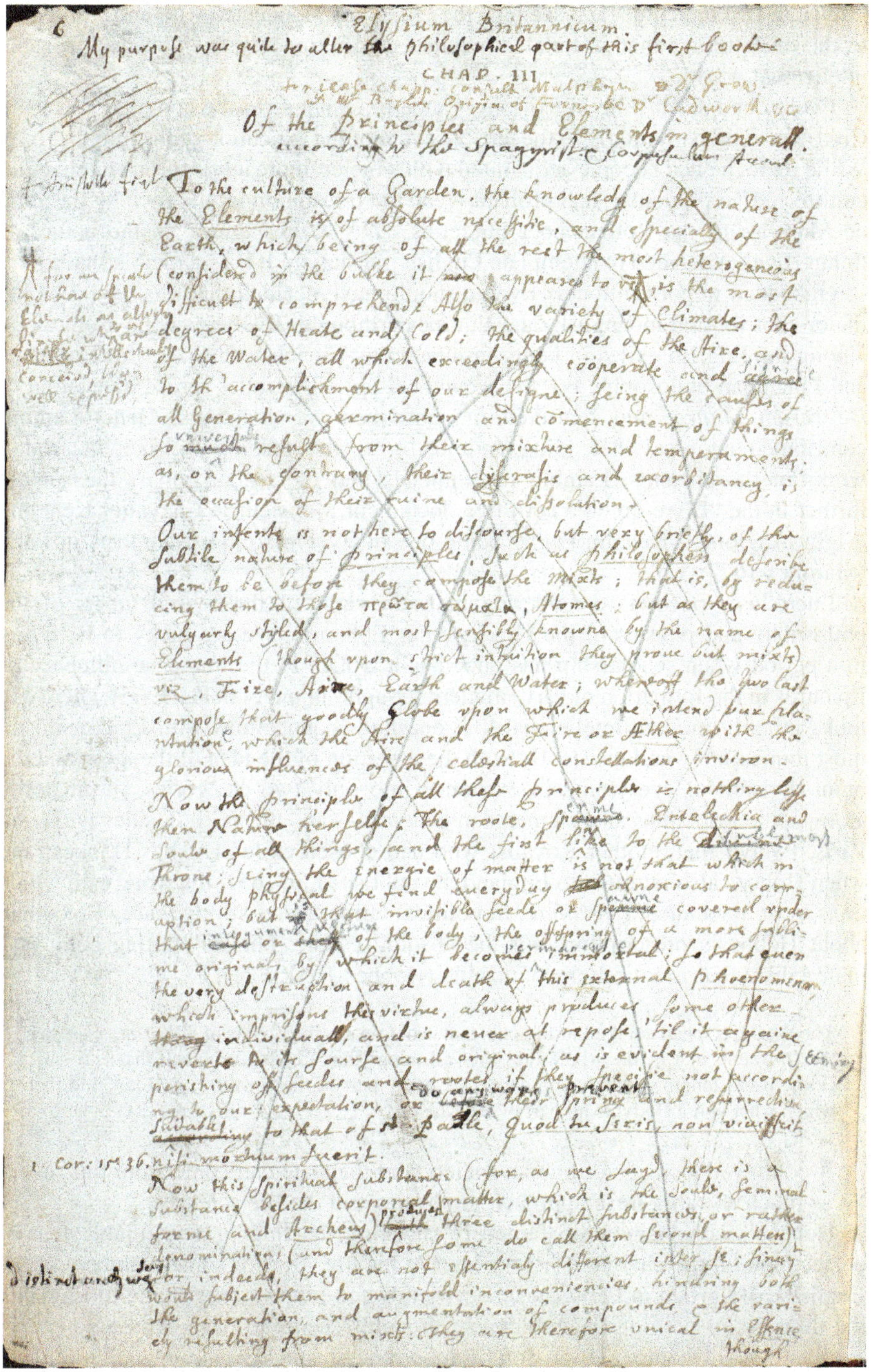
6 Elysium Britannicum.

My purpose was quite to alter the philosophical part of this first booke

CHAP. III

Of the Principles and Elements in generall.

To the culture of a Garden, the knowledge of the nature of the Elements is of absolute necessitie, and especially of the Earth, which being of all the rest the most heterogeneous (considered in the bulke it appeares to us) is the most difficult to comprehend; Also the variety of Climates; the degrees of Heate and Cold; the qualities of the Aire, and of the Water, all which exceedingly cooperate and [illegible] to the accomplishment of our designe; seing the causes of all Generation, germination and comencement of things so universally result from their mixture and temperaments; as, on the contrary, their [illegible] and discordancy, is the occasion of their ruine and dissolution.

Our intente is not here to discourse, but very briefly, of the subtile nature of Principles, such as Philosophers describe them to be before they compose the Mixts; that is, by reducing them to those πρῶτα σώματα, Atomes; but as they are vulgarly styled, and most sensibly knowne by the name of Elements (though upon strict inquisition they prove but mixts) viz. Fire, Aire, Earth and Water; whereof the two last compose that goodly Globe upon which we intend our plantation; which the Aire and the Fire or Æther with the glorious influences of the celestiall constellations inviron.

Now the principle of all these principles is nothing else then Nature herselfe, the roote, [illegible], Entelechia and Soule of all things [illegible]

[illegible]

Fig. 4 John Evelyn, "Elysium Britannicum", page 6. ©British Library Board (Add. MS 78342)

"Elysium Britannicum", and see how the "Elysium" connected with and responded to the other things that stimulated, excited, and ultimately frightened him in his intellectual world.

Evelyn had spent time in exile in France in response to the convulsions of the Civil Wars and particularly to the terrible event of the execution of the king in 1649. While there, he had become acquainted with a racier, more adventurous intellectual culture and, in particular, had probably read the translation into French by the abbé de Marolles of the Roman poet Lucretius's *De rerum natura*, on the nature of things. Evelyn decided to attempt an English translation. It is remarkable that a man so embedded in convention as Evelyn should have decided that this would be his major contribution to English, and European, thought. Evelyn was far from alone among the English exiles in being fascinated by French enthusiasm for Lucretius and Epicurean philosophy. For the second edition of Marolles' prose translation of *De rerum natura*, and reprinted in subsequent versifications, Denis Lambin contributed an essay defending and justifying the reading of such heterodox works of the ancients. Lambin recommends that the reader pursue the subject further in the "Livre qu'en a écrit tres-doctement Monsieur le Chevalier Kenelme d'Igby Anglois"—Kenelm Digby being one of Evelyn's closest contacts in Paris (Marolles 1677: 32).

Lucretius's poem is a rendering in Latin verse of the central tenets of the philosophy and physics of the Greek writer Epicurus. We are lucky to be living in a period when scholarship has once again woken to the immense influence of Epicurus in the formation of early modern thought, in many fields. Everywhere one looks in early modern intellectual life, one finds connections with Epicureanism, most immediately and insistently with the assertion of the idea of the atom and the blunt division of everything in existence into only two categories, the material atom—"the Principles and Elements in generall"—and the void. Besides Digby, in Paris, Evelyn met another Royalist who had chosen exile, Thomas Hobbes; and when Hobbes also returned to England, he published, in 1651, the work with which he is most associated, *Leviathan*. In political terms, there is little that is Epicurean about Hobbes's prescriptions for the absolutist state, but his starting-point is a materialism that is entirely recognizable (Hobbes 1651: 371):

> The World, (I mean not the Earthly one, that denominates the Lovers of it *Worldly men*, but the *Universe*, that is, the whole masse of all things that are) is Corporeal, that is to say, Body; it has length, breadth, and depth; . . . every part of the Universe, is Body, and that which what is not Body, is no part of the Universe. . .

A cold-eyed analysis of human society and the state can come from this, every bit as much as can a fresh approach to physics and the life sciences.

Hobbes's conclusions are famously shocking, leading to a justifiable fear of persecution on the charge of heresy later in his life. But for some members of the English exile group in Paris, and their friends and fellows back home in England, the thinking of previously unthinkable thoughts seems to have been a source of exhilaration—a compensation for disappointment and despair in their political, social, and religious lives. Evelyn gives every sign of having participated in that. At exactly the moment Hobbes was taking *Leviathan* through the press, Evelyn was working on his translation of Lucretius, eventually publishing the first book in 1656 (Evelyn 1656).

But despite having completed the translation of more than 50 % of the poem, he never published more: his Lucretius translation joins the "Elysium Britannicum", the History of Trades, and the "History of Religion" as great unfinished works, all having their origin in the same period of his life. Why?

The answer can be illuminated by looking at the paratext of the translation, in which were printed various commendatory poems. The most powerful is by another fascinating figure of the period, the poet Edmund Waller, at this time a close friend of Evelyn's with whom he travelled in continental Europe. Waller was a controversial figure: well-connected on both sides of the political divide in England, a Royalist who nonetheless promoted substantial political reform, he got himself into hot water with a political plot that pleased no-one and led him to imprisonment before he went into exile. There, because he was wealthy and had managed to protect his assets, he was well-known for providing the best dinners in Paris.

Waller made a specialty of writing well-turned but also very perceptive introductory poems. His praise for Evelyn's translation of Lucretius is both decorous and penetrating, and its metaphors are clearly chosen with knowledge of the translator. Waller praises Lucretius, and through him Epicurus, in these terms:

> For his immortal boundless wit
> To nature does no bounds permit;
> But boldly has removed those bars
> Of Heaven, and Earth, and Seas and Stars,
> By which they were before suppos'd
> By narrow wits to be inclos'd,
> 'Till his free Muse threw down the Pale,
> And did at once dispark them all.

Epicurus established his academy in a garden—hence the title of the essay by Sir William Temple, "On the Gardens of Epicurus" (Temple was another friend of Evelyn's, whose garden at Moor Park Evelyn described and praised). So Waller's choice of metaphors of gardening is by no means surprising. But we should notice what he is saying about the garden and through it the world of ideas, as both are revealed by the work Evelyn has been translating: the boundaries have been removed, boundaries that were like prison bars that kept ideas in their place and we thinkers limited in our freedom to imagine and explore. The Muse throws down the Pale, the fence, and "disparks" everything. It was well known to contemporaries that much of the vocabulary of gardens in many languages goes back to ideas of enclosure, definition of a space by boundaries, separating a cultivated and ordered inside and a wild, disordered outside. Evelyn certainly knew this and Waller's bracing lines elicited a marginal comment that dissociated the translator from them and their implications:

> According to the Institution of *Epicurus* in relation to *whom* these *Verses* were intended, and not that the *Interpreter* doth justifie this irreligion of the *Poet*, whose Arguments he afterwards refutes. (p. 3)

That Evelyn should react in this manner and yet leave the poem unaltered in the printed work should make us wary, however: is this "plausible deniability"? There seems no doubt that Evelyn, at this time, responded to the prospect of the dissolution of intellectual barriers in exactly the way that Waller proposes: with an exhilarating sense of freedom from the tyranny and limitations of the dogmas of the past. Evelyn was so closely associated with the Royal Society that he gave it its name, and when it came time to give the Society a motto, the phrase chosen was *Nullis in verba*: take no one's word for anything, a notion Baconian and Epicurean in spirit.

Edmund Waller was surely aware of Evelyn's devotion to gardening and so chose his metaphors knowingly. But that aptness is not sufficient to conceal from the translator the awful truth about his endeavor, which comes out all-too-clearly in Waller's opening:

> Lucretius with a Stork-like fate,
> Born and translated in a state,
> Comes to proclaim in English Verse
> No monarch rules the Universe;
> But chance and Atomes make this All,
> In Order Democratical,
> Where Bodies freely run their course,
> Without design, or Fate, or Force.

No design; no fate. All chance. All "democratical": no hierarchy. "No monarch rules the Universe". Responding to views such as Waller's, and to his own unnerving temptation to agree, Evelyn vividly depicts this atomization of thought in the Preface to his "History of Religion": "[T]here was no king in Israel, but everyone did what was right in his own eyes" (p. xix). In the Epicurean and Lucretian universe, everything is a product of chance, not design or control. There is no God, no universal monarch. With that comes the consequence wryly noted by James I: "[N]o bishops, no king". Take God out of the universe and the structures of order and hierarchy in all areas of life become challengeable at best, arbitrary and unsustainable at worst. It is impossible to imagine a less welcome conclusion for a man of Evelyn's stamp; and yet, through his ambition to translate Lucretius, he manifests a giddy fascination with the prospect of universes both material and intellectual without rulers or predetermined paths and patterns.

This might seem little more than the usual conflict between the ancients and the moderns, and Evelyn does have a role in that debate, particularly towards the end of the century. That conflict is often represented as a binary choice: you're either with us or against us. But Evelyn was in a far more interesting position in the 1650s: he had to repudiate the stance of the moderns, but not because of a reflex conservatism. The problem for Evelyn was that he was only too aware of the enormous attraction of the modern position, in this case, the position of the atomists, the corpuscularians, the Epicureans. By virtue of his translation project, we might say that he was there, if not at the birth of English materialist philosophy, certainly in its infancy, and acting to usher it into prominence. Famously, Evelyn proposed a scholarly community, a retreat from the world's turmoil, where like-minded studious people could live amid gardens to foster plants, knowledge, and ideas. His plans are usually associated with the structure of Carthusian monasteries, of which Evelyn certainly knew. But the community he proposes, with women as well as men, sounds equally, if not more, like one of the Epicurean societies recorded from the ancient world. It is of a piece with the future he proposed in a letter to his relative William Prettyman:

> I shall therefore bring over with me no ambitions to be a statesman, or meddle with the unlucky Interests of Kingdomes, but shall contentedly submit to the losse of my education, by which I might have one day hoped to have bin considerable in my Country. A Friend, a Booke, and a Garden shall for the future, perfectly circumscribe my utmost designes. [2 December 1651; Letterbook #36]

This is Epicurus's ideal life: free from anxiety, removed from public affairs, concentrating on the immediate sources of personal contentment (a friend, a book, a garden). Evelyn knew how attractive the Epicurean example was, across a wide intellectual range and in terms, as here, of an alternative self without social or political ambitions. For the introductory essay to his translation, he even adopted a strikingly different writing style, unlike anything else he ever wrote: light, volatile, playful, and ironic. He adopted, very recognizably, the style of the Lucretian French sceptics he had been reading in exile.[10]

For Evelyn, Epicureanism was most attractive in the period immediately before he began composing the "Elysium". Walter Charleton knew of Evelyn's immersion in Epicureanism's bracing scepticism, and recognised that this enthusiasm suited a period in which apparent certainties of politics and religion had been overturned.[11] Evelyn commented with a carefully calculated Epicurean playfulness and no serious resistance to the atheist implications; and in response, Charleton represented Evelyn by the name of Lucretius, spokesman for Epicureanism, in his *Immortality of the Human Soul Demonstrated by the Light of Nature* (1657).

[10] Evelyn was not alone in adapting his style to suit the likely reception of his ideas. See Anthony Grafton's comments on Athanasius Kircher's "deliberate ambiguity and complex wit" when publishing views on historical time that conflicted with Biblical chronology (Findlen 2004: 183).

[11] Charleton had probably visited the exiled Stuart court in Paris, perhaps meeting Evelyn there and with him many of the leading English and French pioneers of new intellectual and scientific approaches. Like Evelyn, he felt the force of Epicurus's natural philosophy; like Evelyn, he was conscious of the challenge Epicurean ideas mounted to traditional Christianity; and he set out to render it compatible in works such as *The Darkness of Atheism Dispelled by the Light of Nature* (1652), *Physiologia Epicuro-Gassendo-Charltoniana* (1654), and *Epicurus's Morals* (1656). He was the perfectly-placed foil for Evelyn's own ideas.

As the epigraph to this essay shows, Charleton shrewdly recognized his friend's dilemma. Modern Epicurean physics was thrillingly persuasive in its atomic materialism; Epicurus's life philosophy of friendship, the community of like-minds, and detachment offered a thoroughly plausible alternative to a hostile world. But Evelyn was unable to suppress his understanding that acceptance of Epicureanism brought in its train, irresistibly for a mind like his, a set of conclusions that would undermine the whole of his world view, his religion, and the structures that gave his life meaning, purpose, and context. For Evelyn, living in a world turned upside down by the execution of the king, Epicureanism was an all-too-obvious fit for the times. But as an emotional royalist, adherent and beneficiary of traditional social and political structures, the opening vistas of a world "without design, or Fate, or Force" also threatened everything. The scientific historian Michael Hunter has written of the fear experienced among many that Epicurean atomism would lead to the draining of the universe of "spiritual forces and moral absolutism" (Hunter 1995: 14). Evelyn first responded with excitement to that boundless world; but he then recognized that, if it really is boundless, it is without respect for any inherited structures.

Evelyn's predicament, and the nature of his eventual choice, can be seen clearly in a comparison with a different mind grappling with the same issues, but with a different outcome: the medical doctor Nehemiah Grew (1641–1712), Evelyn's younger contemporary. In his notes for the revision of the "Elysium", Evelyn sometimes invokes Grew, whose *Anatomy of Plants. With an Idea of a Philosophical History of Plants* was presented as lectures to the Royal Society and published in 1682. Grew's work is an absolute treat, depending on wonderfully precise engravings of plants derived from painstaking microscope studies (Fig. 5), engravings and observations worthy of being ranked with those of Robert Hooke's more famous work *Micrographia*. But we are concerned with what Grew calls the "philosophical history of plants". In his opening pages, Grew is careful to stress that he received the encouragement of a senior figure in the state church, the Bishop of Chester; he also praises Evelyn and the latter's close collaborator John Beale (1608–1683) for their assiduousness in "ordering [plants] with respect to their *Alimental* and *Mechanick* Uses" (p. 1). But Grew clearly thought such ordering to be of secondary value; it was not really doing the essential job of addressing the philosophical nature of vegetable life. In this, he looks back to the lack of order that so unnerved William Rowley in Francis Bacon's *Sylva Sylvarum*; and Grew surely self-consciously picks up the Baconian image of the natural philosopher boldly sailing beyond the pillars of the known world into the trackless ocean. This is how Grew understands his own enterprise, the exploration of "*Terrae Incognitae* in Philosophy", equivalent to those in "Geography" (dedicatory epistle "To his most Sacred Majesty Charles II"). Grew concludes the dedication by expanding on the metaphor:

> In sum, Your Majesty will find, that we are come ashore into a new World, whereof we see no end.

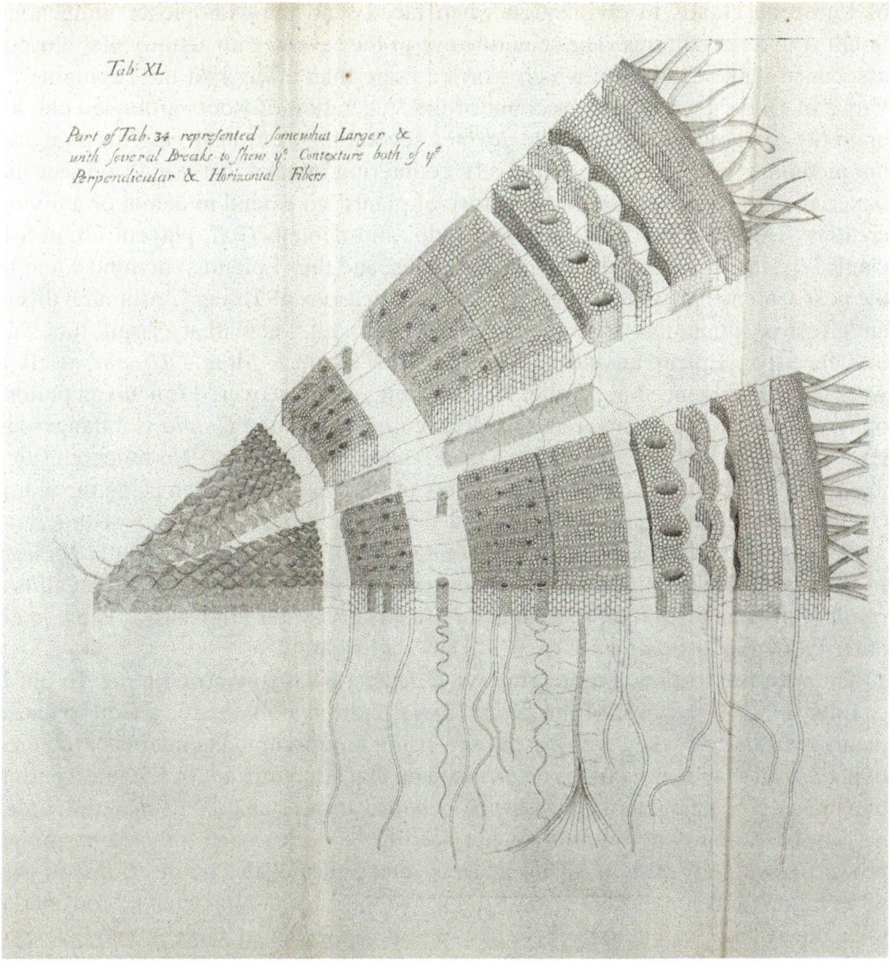

Fig. 5 Nehemiah Grew, *Anatomy of Plants. With an Idea of a Philosophical History of Plants* (London, 1682), Table 40. © The Lilly Library, Indiana University, Bloomington, Indiana

> It may be, that some will say, into another *Utopia*. Yet not I, but Nature speaketh these things: the only true *Pallas*, wherewith it is treasonable for the most curiously handed Arachne to compare.

Grew is finding himself drawn ineluctably into analogies between the studies of plants he has undertaken and the discoveries made with geographical exploration, the discovery of hitherto unknown worlds and societies, and the inevitable challenges to received opinions and established order and structures that those discoveries present. (The most famous example of this scepticism, again shot through with knowledge of Epicurus and well-known in seventeenth-century England, is Michel de Montaigne's questioning in his essay "Of the Cannibals" of European claims to civilization when faced with the example of indigenous South American cultures. The scientific eye of the seventeenth century was closely associated with scepticism, which worried more than a few.) At the beginning of Part 2 of Grew's work, "An Account of the Vegetation of Roots grounded chiefly upon the foregoing Anatomy", he devotes several pages to justifying his own and his method's orthodoxy, pre-emptively countering attacks that would project the materialism of his "philosophical history of plants" to extend to denial of a divine creator: "He, therefore, that philosophiseth, and denieth *God*, playeth a childish Game" (p. 79). It is clear that he has Lucretius, and thus Epicurus, in mind when in the next sentence he writes of "the *Causes* and *Reasons* of Things". Although this is an effective quotation from Virgil's *Georgics*, all knew that Virgil too was influenced by Epicurean ideas. Grew invokes Thomas More's *Utopia*, itself a work clearly influenced in parallel by early news of rediscovered Epicurean philosophy and the newly discovered American cultures. But the *Utopia* is a dangerous text. To stretch a point, we might adapt Edmund Waller's line: No monarch rules that Universe. As he recognises the shadow of the treasonous, even using the word, Grew again deflects criticism by saying that he is merely a reporter of the great works of nature: Nature speaketh these things, not me. And besides, rightly pursued from cause to preceding cause, the philosopher will eventually perceive that all is a result of divine creation, of a "*Universal Monarchy*, as it is eminently Visible in all other Particular *Oeconomies*" (p. 81). That's all right, then.

Grew's introduction is a pre-emptive defence. What so worried him? To put it simply, in his new world of the life sciences, there is no need for a God to cause plants to be the way they are. We can see, using microscopy, the internal structures of plants and that their nature is determined by that structure, not by the intervention of a maker. The characteristics of the plant world are a product of its material; order has not been imposed; nature does not need to be policed and directed by divine power. Grew felt the need to address the anti-theistic implications of his observations and conclusions; but he did so without abandoning what he considered to be the truths apparent to his eyes and reason. A mechanist, Grew steered clear of non-mechanical explanations that depended on non-material "vital forces", the doctrine of signatures, and such like. Divine intervention was given a place, but only at the moment of creation; thereafter, the nature of the vegetable world was clearly the result of its own forms and components, the causes being immediately evident in the structure of plants. Grew points out with relief that there is some Scriptural authority for banishing God: the Bible tells us that, after six days' labour, God withdrew: "He *resteth from*" further intervention (p. 80). It is striking that, in his index to the volume, Grew's only entry not concerning the details and characteristics of plants is

> Philosophy, *begins and ends with Theology* (Sig. Zz1r).

The entry is surely there just to be sure that no-one could accuse him of atheism. Nonetheless, beyond such assertions of orthodoxy, Grew is emphatically on the atomist, or corpuscularian, or Lucretian, or Epicurean side of the debate, precisely the side that Evelyn found it necessary, for political, religious, and social reasons, to abandon. As Evelyn put it in his unfinished and unpublished "History of Religion",

> Atheists are properly those who derive all from senseless matter. These are called our new philosophers, men of high thought, *Esprits forts*, who yet ascend no farther than this dunghill earth; so they will have God contribute nothing more to the fabric of the universe than his whirling about the *vortices* of matter, globose and striate particles, from whose casual motions, according to certain catholic laws of nature and matter, all things, animate and inanimate, proceed; without the conduct, forsooth, of any mind, wisdom, or providence whatsoever (p. 288).

Nehemiah Grew can have been in no doubt that there were those in the Royal Society, Evelyn not least, who would judge his *Anatomy of Plants* to be the work of a dangerous *esprit fort*, "deifying the power of matter" ("History of Religion", Preface, p. xxvii).

Nonetheless, things were perhaps easier for Grew because of his position within English society. He was the son of a distinguished Dissenting minister, and his religious non-conformity contrasts with Evelyn's intense commitment to the established Church of England. In terms of position and opportunity determined by religious affiliation, and this mattered a great deal in later seventeenth-century England, Grew had little or nothing to lose; he would never be on the inside track. But it was the reverse for Evelyn, for whom doors opened by virtue of conformity and adherence to the doctrines and authority structures of the English church and its sponsoring state.

The contrast with Nehemiah Grew sheds light directly on Evelyn's problems with the "Elysium". And there is an associated phenomenon, both cause and consequence. Whereas an outsider like Nehemiah Grew saw no reason to inhibit the circulation of ideas and the communication of knowledge—such openness could only benefit his kind—Evelyn is striking for his growing sense that knowledge, especially knowledge that offered intellectual power, needed to be restricted to an approved group, people of his own station in life and with similar outlooks. Again, we see a reason why he had a history of failing to bring works to wider publication through the printing press. The more inventive, the more stimulating the work, the more careful one must be about dissemination, especially in a period he labels "this Catalysis of piety amongst us, and Hydra of Errors". With the translation of Lucretius, we can see this most starkly: faced with the challenge of Lucretius's enthusiastic, lyrical, and frank discussion of sex and procreation in the fourth book of *De rerum natura*, Evelyn simply omitted it. In the context of Evelyn's successive failures to complete major works, it is significant that he wrote,

> [T]his age is now [so] deprav'd I had much rather all the poems in the world should perish, then that any thing of mine should contribute & minister to vice.

Evelyn's friend and correspondent John Beale, who also had close connections with Samuel Hartlib and his circle, had a similar response to dangerous ideas, most notably when responding to publications by Milton after the Restoration (Milton too was a correspondent of Hartlib). Having read and admired *Paradise Lost*, early on recognizing its great power, Beale, in 1670, nonetheless wrote to Evelyn saying it should never have been published in print:

> Milton is abroad againe, in Prose, & in Verse, Epic, & Dramatic. I have not yet seen his History . . . He hath great faults in his *Paradyse lost* in his plea for our Original right, & in ye long *blasphemies* of Devils; For w*hi*ch he hath no Authority, & they beget a bad, and afflict a good spirit.[12]

From his correspondence, we can see that Evelyn became ever more worried as he translated Lucretius—what had he done by embarking on this project? He discussed his predicament with his spiritual guide, Jeremy Taylor (1613–1667); characteristically, he sought a remedy in writing, channeling himself into conformity and orthodox belief ("*to settle and satisfy myself*") in "A History of Religion", "begun in the year 1657, when the Church of England was in persecution", but only first published in 1850 (Evelyn 1850: Postscript, xxxiii). Like the "Elysium Britannicum", Evelyn seems to have been dissatisfied with this work, leaving notes for emendations and radical rewritings. It is close-to-contemporary with the "Elysium", and strikingly similar in that, though he worked on it for decades, Evelyn could not (or would not) bring it to completion or to publication that would result in critical scrutiny.

And as he turned away from Lucretius, I think he also saw the "Elysium Britannicum" as a safer way to contribute to knowledge, particularly if he chose a restricted version of his potentially universal topic. Though Evelyn understood better than us the close connection, the essential interpenetration of the world of gardens, the world of knowledge (science), and the world of politics and society in the period, he declared at the outset that his chosen masterwork's treatment of his subject would be within firm boundaries:

> ELYSIUM BRITANNICUM
> OR THE
> ROYAL GARDENS
> IN
> THREE BOOKS

Where Edmund Waller had giddily pointed out that in Evelyn's translation of Lucretius "No monarch rules the Universe", in the "Elysium Britannicum", Evelyn pins his colours to the royalist mast in all stages of his work on the project. The kind of garden culture he is depicting and the kind of knowledge culture he is participating in are inseparable from political and social beliefs and attitudes.

But he did not complete it. Surely it could not have been so very difficult to produce a final version of this kind of work over all those decades? Except, for Evelyn, it was. He found himself in a succession of irresolvable dilemmas: he

---

[12] British Library Add. MS 78313, letter 108 (24 December 1670). See (Leslie 1992: 161; Poole 2004).

would not complete his translation of *De rerum natura* because its implications were too revolutionary; he would not complete his History of Trades because it would disseminate knowledge too widely; he would not complete the "Elysium" because he knew its self-limiting approach was inadequate as a contribution to the knowledge-world of which he wanted to be a part. He wanted to demonstrate his encyclopedic command of all aspects of the discourse of gardens and plants, and to provide a new, comprehensive "philosophy" of the vegetable world, like Nehemiah Grew. Doing so would enable him to claim his place as the heir to Francis Bacon, fulfilling the potential of Bacon's essay "Of Gardens". But he knew that the limits to knowledge that he must maintain for reasons to do with religion, society, and politics rendered his text implausible in the world of the Royal Society that he was so instrumental in founding. And he had a powerful example in Francis Bacon, the founding father of empirical science. At just the moment Evelyn was translating Lucretius's poem and becoming ever more queasy about doing so, Bacon's juvenilia were published (Bacon 1658), showing an early, strong affiliation with Lucretian and Epicurean ideas. But Bacon had back-pedalled hard in later works, as Evelyn knew well.

Evelyn was far from alone is his disquiet. His great friend and contemporary Robert Boyle, having read people like Thomas Hobbes early in his life, certainly understood the threat to traditional religions posed by the Epicurean and Lucretian direction being taken by science in all its forms. But some, among them Boyle, proved capable of separating the compelling truths of Epicurean physics from the unpalatable implications for religion and politics. Partly, Boyle does this through linguistic sleights of hand: it is easier to get away with a theory based on 'corpuscles' than it is one that proposes atoms, an unconcealed borrowing from Lucretius and Epicurus. But Evelyn turns out to have a particular kind of honesty, a particular kind of truthfulness: he could not obfuscate. In this way, somewhat paradoxically, he identifies himself as anything but a dilettante or *virtuoso* playing at science and ideas. Evelyn knew that, taken seriously, the truths of Epicurean materialism were incompatible with those aspects of his existence that he held dearer than anything else. That kind of knowledge would burn away the things that comforted him. But at the same time, he knew that an account of his subject with pretentions to encyclopedism, pretentions to be the last word, would have to include these very elements. The "Elysium Britannicum" is a mirror image of the aborted project to publish an Englished Lucretius, which remained in the dust of his study because he did not wish to render public the vision it contained, knowledge that he felt challenged and overturned all that he held dear; the "Elysium Britannicum" remained next to it in the study's dust because, having to limit it in the way he had chosen, Evelyn couldn't bring himself to reveal the omissions, the gaps: "[M]y purpose was quite to alter the philosophical part".

Bibliography

Sources

Bacon, Francis. 1605. The twoo bookes of Francis Bacon. Of the proficience and aduancement of learning, diuine and humane To the King. London.
Bacon, Francis. 1625. The essayes or counsels, ciuill and morall. London.
Bacon, Francis. 1627. Sylua syluarum: or A naturall historie In ten centuries. Ed. William Rawley. London.
Bacon, Francis. 1658. Opuscula Varia Posthuma, Philosophica, Civilia, et Theologica, Francisci Baconi.... Ed. William Rawley. London.
Beale, John. Letters. British Library Additional Manuscripts 78313.
Charleton, Walter. 1652. The Darkness of Atheism Dispelled by the Light of Nature. London.
Charleton, Walter. 1654. Physiologia Epicuro-Gassendo-Charltoniana. London.
Charleton, Walter. 1656. Epicurus's Morals. London.
Charleton, Walter. 1657. The Immortality of the Human Soul, Demonstrated by the Light of Nature. London.
Evelyn, John. 1624. Sylva, or A Discourse of Forest-Trees. London.
Evelyn, John. 1656. An Essay on the First Book of T. Lucretius Carus: De Rerum Natura. London.
Evelyn, John. 1850. The history of religion. A rational account of the true religion. Ed. R. M. Evanston. London.
Evelyn, John. 2001. Elysium Britannicum, or The Royal Gardens in three books. Ed. John E. Ingram. Philadelphia.
Evelyn, John. Letterbooks. 2 vols. British Library Additional Manuscripts 78298-9.
Grew, Nehemiah. 1682. Anatomy of Plants. With an Idea of a Philosophical History of Plants. London.
Hartlib, Samuel. 1651. The Reformed Husband-man. London.
Hartlib, Samuel. 2013. The Hartlib Papers. Eds. M. Greengrass, M. Leslie and M. Hannon. Sheffield HRI Online Publications. http://www.hrionline.ac.uk/hartlib.
Hobbes, Thomas. 1651. Leviathan, or, The matter, form, and power of a common-wealth ecclesiastical and civil. London.
Marolles, abbé Michel de (tr.). 1677. Les Six Livres de Lucrece De La Nature des Choses. 3rd edition. Paris.

Secondary Literature

Bushnell, Rebecca. 2003. Green Desire: Imagining Early Modern English Gardens. Ithaca, NY: Cornell University Press.
Darley, Gillian. 2006. John Evelyn: Living for Ingenuity. New Haven and London: Yale University Press.
Findlen, Paula (ed.). 2004. Athanasius Kircher, The Last Man Who Knew Everything. New York: Routledge.
Greengrass, Mark, Michael Leslie, and Timothy Raylor (eds.). 1994. Samuel Hartlib and Universal Reformation: Studies in Intellectual History. Cambridge: Cambridge University Press.
Hill, Christopher. 1972. The World Turned Upside Down: Radical Ideas During the English Revolution. London: Temple Smith.
Hunter, Michael. 1995. Science and the Shape of Orthodoxy: Intellectual Change in Late Seventeenth-Century Britain. Woodbridge: Boydell.
Leslie, Michael. 1992. The Spiritual Husbandry of John Beale. In Culture and Cultivation in Early Modern England: Writing and the Land, eds. Michael Leslie and Timothy Raylor, 151-172. Leicester: Leicester University Press.

Leslie, Michael, and Timothy Raylor (eds.). Culture and Cultivation in Early Modern England: Writing and the Land. Leicester: Leicester University Press.
Low, Antony. 1985. The Georgic Revolution. Princeton: Princeton University Press.
Otten, Charlotte F. 1985. Environ'd with Eternity: God, Poems, and Plants in Sixteenth and Seventeenth Century England. Lawrence, KS: Coronado Press.
Parker, Geoffrey. 2013. Global Crisis: War, Climate Change and Catastrophe in the Seventeenth Century. New Haven: Yale University Press.
Poole, William. 2004. The Early Reception of Paradise Lost. Literature Compass 1: 1-13.
Yale, Elizabeth. 2011. Marginalia, commonplaces, and correspondence: Scribal exchange in early modern science. Studies in History and Philosophy of Science 42: 193-202.

The Power of the Sun-King at the Potager du Roi

Chandra Mukerji

> *"Les Jardins Fruitier set Potagers m' ont été trop favorables, pour cacher l'extréme reconnoissance des biens que je leur dois: Je leur suis obligé de l'honneur que VOTRE MAJESTE´ m' a fait d'avoir augementé en ma personne le nombre des Officiers de sa Maison. Une telle obligation merite bien au moins que je la publie."*[1] *(Jean de la Quintinie 1692, epistre, p ij)*

Abstract

This chapter explores how natural knowledge was used as a tool of politics in the *potager du roi* at Versailles. Under the Sun King, Louis XIV, the royal gardens gained particular importance as a means of demonstrating the Sun King's natural authority over the earth and his power to bring fertility and felicity to his lands. Jean de la Quintinie, director of the *potager,* the kitchen gardens at Versailles, contributed to this program with both his gardening and a book he wrote on raising fruit. In the book, he described the three different logics of governance he used in the garden to manifest royal power: physical control, land stewardship, and control of the effects of the sun on fruit. He drew the analogy between sun and Sun King as forces governing the earth, naturalizing royal power. La Quintinie described absolutism in France as a natural system of authority that paralleled the sun's power in nature. But he also showed by analogy that absolutism could be dangerous and corrupting like the sun, providing a surprisingly critical portrait of absolutism even while celebrating the Sun King.

[1] The fruit and vegetable gardens have been too favorable to me, for me to hide the extreme debt I owe them. I am obliged to them for the honor Your Majesty has given me by elevating me to be a member of the king's household. Such an obligation merits at least that I publish.

C. Mukerji (✉)
Department of Communication 0503, UCSD, 9500 Gilman Drive, La Jolla, CA 92093, USA
e-mail: Cmukerji@gmail.com

H. Fischer et al. (eds.), *Gardens, Knowledge and the Sciences in the Early Modern Period*, Trends in the History of Science, DOI 10.1007/978-3-319-26342-7_4

Steven Shapin and Simon Schaffer (1985) famously assert that solutions to problems of knowledge have also historically been solutions to problems of politics, both resolving questions of authority. In science, the political stakes of an epistemic system (Knorr-Cetina 1999) may be formally hidden, and in politics, the problems of knowledge may be disputed or disguised (Leibler and Breslau 2005), but natural knowledge has had many uses in addressing problems of political authority.

Nowhere would conceptions of nature be more obviously important to regime power than at the court of the Sun King, where understandings of the sun were used both to represent and extend the power of Louis XIV. At Versailles, the authority of the monarch was equated with the power of the sun in bringing life and abundance to the earth. The natural world was Apollo's to govern, and the virtue of his governance was manifested in the beauty and fertility of the earth. To make the point, the art in the gardens of Versailles presented the king as Apollo and the plantings as a model of order, suggesting with the beauty and abundance of the result nature's willingness to subordinate itself to the Sun King.

Jean de la Quintinie, director of the kitchen garden or the *potager du roi*, brought his botanical and horticultural knowledge to the task of demonstrating Louis XIV's power by managing and displaying the effects of the sun on fruits and vegetables. La Quintinie provided exquisite produce throughout the year to Louis XIV's table, and made the *potager* as a garden an exquisite display of techniques of control (La Quintinie 1692; Mukerji 1997; Thacker 1972). For La Quintinie, solutions to problems of horticultural knowledge were also solutions to problems of politics. As a member of the king's household, his good standing depended on his successful use of natural knowledge to please and celebrate his patron. He had to bring foods worthy of a Sun King to the royal table, and create an Eden in the *potager* so perfect that it provided tangible evidence of Louis XIV's efficacy in ruling his lands.

Natural Knowledge and State Power

The displays of natural knowledge in the microcosm of the royal gardens were particularly important because they paralleled techniques of territorial governance being used by the administration to rule the macrocosm of France (Mukerji 2009). The minister of finance and director of the king's household, Jean-Baptiste Colbert, deployed three distinct methods of territorial governance to augment the king's power: dominating territory physically, making land improvements to demonstrate stewardship, and exploring the analogy of Louis XIV to Apollo to naturalize the king's powers and associate his reign with an imperial destiny.

With these methods, Colbert crafted an approach to territorial politics that was startlingly effective (Mukerji 2009, 2012). Evading patrimonial practices of networked power, he reconfigured the relationship between the state and land. (Mukerji 2009, 2010a; cf. Appuhn 2009; Carroll 2006). Elites in both the clergy and nobility had political networks that could rival those of the king, but the king had territorial authority and access to natural knowledge that his rivals did not and used them to pursue dreams of empire (Kettering 1986; Mukerji 2009; Soll 2009). Louis XIV could justify territorial interventions as part of his moral duty as a Christian to act as steward of his lands. And as the Sun King, he was culturally defined as a natural force capable of making the earth fertile.

The administration used natural knowledge to work out these logics of power, and pursued territorial politics, combining techniques of physical domination, stewardship politics, and the image of Louis XIV as the Sun King. The results were impressive. The monarchy in France that had been notoriously weak when Louis XIV ascended the throne became instead notoriously strong (Anderson 1974; Machiavelli 1966). Solutions to problems of natural knowledge had become answers to problems of monarchical power, and helped make the king a potent and legitimate territorial ruler (Mukerji 2009).

The Gardens of Versailles

The territorial politics pursued by Colbert and the vision of the Sun King as a natural ruler were embedded in the great landscape gardens built at royal residences, most notably Versailles, testifying to the king's capacity to rule the earth. The gardens were models of logistical control with complex infrastructures of roads, battlement-style walls, and waterworks—demonstrating the physical methods available to the administration to extend royal power (Mukerji 1997; Weiss 1995). The great parks with their peaceful beauty were also models of stewardship, versions of a restored Garden of Eden. They were peaceful, productive, verdant and lush. And they featured both overt and indirect allusions to the Apollo and the power of the sun. The royal gardens, in other words, used natural knowledge both symbolically and practically to display this regime's politics of territorial domination and legitimation (Mukerji 1994, 1997, 2012).

Peter Burke (1992) has argued that the propaganda associating Louis XIV with Apollo failed miserably, never convincing anyone that the king had godlike powers or that France was heir to Rome's classical past. But Burke did not address the demonstrations in royal gardens of Louis XIV's ability to govern like a Sun King. French formal gardens were a great success as models of territorial governance, and widely imitated throughout Europe in spite of their critics (Goldstein 2008; Thacker 1979). These stunningly large and imposing landscapes demonstrated what strong control of the land could do, while indexing with their designs and art the equation of Louis XIV's power to that of Apollo (Givry et al. 2000; Hazlehurst 1980; Hunt et al. 2002). As Burke has suggested, whatever was asserted in print about the Sun King could be met with counter-narratives of derision in a kind of discursive standoff (Burke 1992), but the royal gardens provided silent and tangible evidence of the king's powers over his territory—the power of an Apollo on earth (Goldstein 2008; Hazlehurst 1980; Hunt et al. 2002; Mukerji 2012).

The equation of Louis XIV with Apollo was most famously elaborated at Versailles in the grotto and fountain of Apollo (Apostolidès 1981; Félibien 1672; Néraudeau 1986). The grotto, sitting on the terrace next to the chateau, silently equated the residence of the king with the underground home where Apollo and his horses rested after a day of pulling the sun through the skies. The statues depicted Apollo and his horses cared for by attendants in a world with sparkling fountains and walls encrusted with gems and shells that twinkled in the dim light. Directly down the hill from the chateau, the Apollo Fountain depicted the god rising in the morning on his chariot drawn by his fierce and powerful horses—their powers underscored by massive sprays of water. This Apollo was a beautiful and muscular deity, and emerged from a large circular pool that mirrored the sun and sky, bringing the sun to earth and into the royal garden. The effects of the grotto and fountain were technical as well as aesthetic; they were part of an elaborate hydraulic system making water serve the Sun King, and suggesting that Apollo naturally dominated the forces of nature around him (Apostolidès 1981; Félibien 1672; Weiss 1995; Mukerji 2012; Néraudau 1986).

Building a *petit parc* worthy of Apollo was an act of territorial domination that made use of the most advanced natural knowledge of the period. It was a product of experiments in botany, horticulture, hydraulic engineering, metal casting, and materials science, using architecture, pumps, reservoirs and collections of bulbs, trees, minerals and shells to make a "second nature" (Cronon 1991) that stood for royal power. It was quite an effort to bring water where it was needed, collect rare and valuable gems and shells, cast large bronze statues at the arsenal, import trees, bushes and bulbs from around the world, maintain formal parterres of flowers, and raise fresh fruit for the royal table every month of the year in a cold, grey climate (Givry et al. 2000; Mukerji 1997; Weiss 1995; Tomasi 1983). The garden at Versailles became a showcase of powers that could be derived from nature to serve the king, managing natural objects and forces to display what a Sun King was and could do.

The *potager du roi* did not have great fountains or statues to represent the Sun King or his powers, but it nonetheless demonstrated most directly the power of the Sun King to bring abundance to his lands and provide food for his people. La Quintinie developed the most refined, showy and effective horticultural methods to show how the power of the sun could be deployed to manage the fertility of the earth, providing fruit and vegetables during each month of the year. The *potager* became a mute but rhetorically powerful model of good governance, promoting the image of the Sun King as a force of nature capable of making the world more fertile and human life more felicitous.

Horticulture and Collection

The writings of Jean de la Quintinie in his *Instructions pour les jardins fruitiers et potagers*, 1692, provide interesting evidence of the natural knowledge that this influential gardener assembled and put to work in the royal kitchen garden in order to further the king's standing as a political leader. Although its title did not imply this, the *Instructions* focused mainly on raising fruits rather than vegetables, since the latter had already been extensively discussed in garden books by other French authors, most notably by Olivier de Serres (1600). La Quintinie's great contribution lay in his extensive knowledge of fruit trees and vines, and his systemic approach to their cultivation. Like other garden writers before him, the director of the *potager* could describe the exact horticultural needs of particular varieties of fruit, but he approached the *potager* as a system and a collection—a model of territorial governance of land and fruiting plants worthy of a patron designated as the Sun King.

His book was not a philosophical treatise, but La Quintinie nonetheless was concerned about the natural forces at work in plants. It was his understanding of nature as a system that was the foundation for his system of gardening. So, for example, he reflected on the question of how plants got their nutrients:

> [F]or example, whether there is in plants a circulation of sap as much as in animals there is a circulation blood, or if the roots attract the fluid for the nourishment of each plant through positive action, or without any action on their part.[2]

This "philosophical knowledge" about circulations in plants was not just of abstract interest; it was a basis for understanding how to intervene in the growth of plants and use the power of the sun. Heating the earth stimulated the flow of sap, making it possible for trees to set, mature, and ripen fruit. Knowing nature's dynamics provided the intellectual foundation for developing horticultural solutions to problems of territorial politics in the microcosm of the *potager* (La Quintinie 1692: I, 20–22 and 46–47).

Many of the techniques and ideas used by La Quintinie were familiar ones that he learned from well-known gardeners and/or their books. He studied classical works, such as Columella's *De re rustica* (1537) on estate management and Pliny (1989) on natural history. He also knew literature on French and Dutch gardening methods, particularly the French *mesnagement* literature that emphasized rational land use, moral stewardship, and political leadership through gardening and agriculture. He tried to synthesize a vast repertoire of ideas, developing a book not simply of practices, but principles of practice to guide thinking about gardens.

La Quintinie wrote in the style of *mesnagement* writers, producing both a practical how-to guide and a reflection on good conduct, using Christian ideas about stewardship of the earth and classical ideas about rational land use, botany and horticulture (Hazlehurst 1966; Hunt et al. 2002; Mukerji 2009: 209–213). While echoing *mesnagement* ideas about the importance of natural knowledge to soil preparation, land use, and the design of microclimates, La Quintinie also repeated ideas about moral discipline, diligent labor and the reasoning ability needed by good gardeners (La Quintinie 1692: I, 33–37, II, 202–204). La Quintinie added, however, that it was not enough for a gardener to be skilled if his master had

[2] "[P]our savoir, par exemple, s'il y a dans les Plantes une circulation de seve, aussi bien que dans les animaux il y a une cirdulation de sang. Pour sçavoir si les recines attirent par un action effective le suc qui sert de nourriture à chaque Plantes, ou si simplement elles reçeoivent ce suc sans aucune action de leur part." (La Quintinie 1692: I, 20)

ignorant ideas about land and its use. Good gardening was a form of governance that depended on the master's ability to direct the work. This logic obviously allowed La Quintinie to credit Louis XIV for his successes, but it also elaborated his view of gardens as systems of governance that intersected relations of power (La Quintinie 1692: I, 66).

In seeking ways to bring the best produce to table, La Quintinie was in competition with commercial gardeners who served the elites of Paris, trying to raise novel and tender imported plants with experimental methods (Dunerin 1983; Mukerji 1997: 167–171). La Quintinie synthesized a massive amount of their practical knowledge with his more philosophical knowledge, developing his own ways of experimenting in horticulture around principles he developed about collecting, planting, pruning, and cultivation as techniques of garden governance. Following his understanding of how to govern a garden, he developed a *potager* worthy of his patron, providing a surprising array of delicious foods for the royal table (La Quintinie 1692: I, 9–12).

The point of this system was to defy the limits of the seasons, pushing plants to their limits and yet keeping them alive and productive even if they were rare imports. To assure fresh produce year-round and produce delicacies that others could not get, La Quintinie assembled hundreds of varieties of a given type of fruit, most notably pears and apples, and allocated space for the different varieties according to principles of continual productivity. The number of each type to grow and proportions of space allocated to them were crucial (La Quintinie: I, 22–23, 30 and 40–47).

> The change of seasons could sometimes make me prefer in a garden plan a less good fruit to a better one, & this because the better comes at a time when I could have sufficient others which are admirable, & the less good comes in a season when one is happy to have at least mediocre ones the absence of more excellent ones being great.[3]

La Quintinie had the material advantages of a royal gardener in making the finest kitchen garden in France. He had few limits on what he could spend on trees, walls, trellises, glass bells, hot-houses and screens—the tools of forcing plants to act in unnatural ways—and he had abundant space for his collection. He famously used massive amounts of glass to cover bedding plants even though glass was very expensive at the time (Mukerji 1997). He could also employ as many gardeners as he needed to make the *potager du roi* a paradise of plenty (La Quintinie 1692: II, 136). But he insisted in the *Instructions* that it was not resources that made a good garden. Gardeners with smaller plots and fewer resources could do something similar with a limited collection of trees. It was knowledge that was needed to

[3] "[L]a suite des Saisons pourra bien quelquefois me faire preferer dans les plans un moins bon Fruit à un autre meilleur, & cela, parce que ce meilleur vient dans un temps où j'en puisse avoir suffisemment de ces autres, qui son admirables, & que le moins bon vient dans une Saison, où la disette des plus excellens étant trés-grandes on est trop heureux d'en avoir au moins de mediocres." (La Quintinie 1692: I, 30)

use any available space wisely and govern a collection of fruit trees and vines to advantage (La Quintinie 1692: I, 36). But his advantages did help him make the garden of the Sun King appropriately exceptional.

The *Instructions* (1697) was certainly not a political treatise, and not meant to refer directly to Colbert's programs of political territoriality. But it nonetheless showed how natural knowledge could be used to further territorial governance in the microcosm of the *potager* just as Colbert was using natural knowledge to govern the macrocosm of France. That is why the *potager* could indeed demonstrate how Louis XIV was exercising territorial power: dominating nature by physical force, exercising stewardship, and governing the earth like the sun to make it fertile. In his book, La Quintinie described horticultural knowledge as singular, but employed three different logics of governance in the garden. He furthered the Sun King's power by exploring in his practices how control over nature could work as a tool of empire, a mode of caring for Creation, and a way of bringing the power of the sun to earth (Dunerin 1983; La Quintinie 1692; Mukerji 1997: 167–171).

Gardening by Force

Gardening by physically altering the soil, irrigating land, grafting trees, and forcing plants was the most traditional approach to using natural knowledge for territorial governance in gardens, and the one most vigorously pursued by La Quintinie. Governance by force was an extension of military efforts at land domination—simply one that used natural knowledge to make land control more beneficial to human life. It had ancient provenance in Roman gardens, and gained particular importance in the *potager* as a way to make the king's garden a model of control.

Most of the techniques for gardening by force had provenance in the ancient world and were elements of Roman estate management. Working an estate was seen as a war against soil depletion, garden pests, wind and weather that worked against good harvests. The keys to success were physical: developing irrigation, adding amendments to the soil, grafting new fruit slips to rootstock or setting up windbreaks with rows of fast-growing trees. Landholders could take advantage of and battle against natural forces and tendencies of plants by learning about nature and being observant in their practices. The point was to make plants and the places where they grew serve human interests by engaging natural forces in a system of governance.

This vision of gardening as part of estate management was articulated most clearly by the classical author and Roman aristocrat, Columella (1537), but was also widely reproduced in France in *mesnagement* texts (Estienne et Libault 1564; Serres 1600). Writers in this tradition presented gardening as a form of local governance, requiring leadership as well as rational analytic ability. Columella (1537) explicitly linked estate management to politics, too, by arguing that a decline in noble knowledge of agriculture was contributing to the decline in the empire. He advocated systematic study and use of the landscape, and emphasized how important it was for governing an empire to know what varieties of useful plants would grow in a particular climate and type of land. He treated gardening as a necessary means for assuring political dominance, maintaining control of the land, and addressing the needs of people in the empire. And the *mesnagement* writer Olivier de Serres used Columella's argument to advocate the political development of gardening in France.

La Quintinie picked up these themes in his book, and repeated the idea that successful empires depended on natural knowledge. He also echoed the idea that true leaders had a propensity for struggling with nature in the garden. It was a form of land control that demonstrated their wisdom as rulers. "If one believes one of the ancients about this [Pliny], the same virtues that bring felicity to their people also brought fertility to their lands."[4] Good governance in gardens, as much as empires and states, entailed beneficial control of land. Gaining this kind of control had been a battle of importance throughout history: "[. . .] to taste the pleasures of our first fathers as well as to surpass the glory of the most illustrious monarchs in overturning every day the ambitions of an infinity of enemies with new victories."[5] Gardening was a way of exercising force to gain pleasure and solidify power.

The view of gardening as an exercise of power equivalent to the military use of force was physically embodied in the architecture of the *potager du roi*. The garden was built like a fortress surrounded with battlement-style walls and crisscrossed with ramparts. The outer defenses had utility as well as symbolic significance; these walls protected the king's food supply from poachers and animals. In contrast, the inner ramparts were primarily for controlling the growth and productivity of plants. They were covered with espalier trees and vines, managing the effects of wind, sun, and climate on the plants and soil. In winter, they kept the garden warmer at night; in summer, they provided shade; and throughout the year, they protected fragile plants from strong winds and bad weather. The walls also allowed visitors to walk the ramparts to survey the plants, using a military practice for viewing troops to appreciate the beauty and control of the monarch's lands (Mukerji 1997: 167–171).

La Quintinie himself had battled like an ancient estate manager against the natural features of the site to make the *potager* a model of beneficial governance. He was given an unpromising parcel of land at Versailles in a lowland area of muddy clay near the chateau that was convenient to the kitchen, but not good for plants. To make the site serve its intended purpose, he had workers move massive amounts of soil into the garden, erect walls, drain land, dig irrigation ditches, and add soil amendments. Only this dramatic show of force and will could produce an environment that would sustain plant life, and serve the king's table.

Building the collection of fruit trees was also a significant act of power, some of which was surprisingly political. A large set of fruit trees for the *potager* at Versailles came from a collection that had belonged to Nicolas Fouquet, the former

---

[4] "Si on en croit un Ancien [Pliny], les mêmes vertus qui faisoient la felicité de leurs Peuples, faisoient aussi la fertilité de leur Terres." (La Quintinie 1692: I, iij)

[5] "[. . .] goûter les plaisirs de nos premier Peres, aussi-bien que surpasser la gloire des plus illustres Monarques, en renversants tous les jours l'ambition d'une infinité d'Ennemis par de nouvelles Victoires." (La Quintinie 1692: I, a iij)

treasury minister who had been put in jail for graft. They were booty from a political war waged by the king to take control of his kingdom at the start of his reign. The rare specimens were so poorly known in France that it was hard to identify them, but once they were identified, the valuable assets were put on display at Versailles (Goldstein 2008). The citrus trees went to the Orangerie, and the other fruit trees came to the *potager du roi*. The rest (and bulk) of the plants at the *potager* had less dramatic and political provenance. They came from other gardeners or plant traders who imported and experimented with horticulture themselves (Mukerji 1997: 108 and 167–174). Not fruits of battle, these specimens nonetheless testified to the geographical reach of France in a global system of trade and knowledge.

The plants, once collected, were submitted to various disciplinary regimes. La Quintinie experimented with all the known means of physically controlling plants: grafting, trimming, pruning, stretching, bending, covering, mulching, and defoliating. These acts of physical domination were ways of "fighting" against natural forces and tendencies in plants. The techniques were used so widely that most of the trees in the *potager* were artificial constructs in substance and form; trees made from grafts on rootstock were arrayed among the walls of the *potager* in espalier, their branches stretched along trellises attached to the walls to give them geometrical form. These plants had what La Quintinie described as pleasing qualities, yielding beautiful fruit on manicured plants arrayed artfully but with clear force along the battlement walls.

La Quintinie lauded espalier methods for controlling the growth of fruit on trees and vines so as to produce notable effects on the size and beauty of fruit. Espalier techniques were particularly effective in raising tender plants or ripening fruit earlier in the season. But with less demanding trees like pears, La Quintinie said, these techniques were also useful for increasing the fruit's size and production (La Quintinie 1692: I, 274–275).

La Quintinie acknowledged that espalier forms were so unnatural that they also could be dangerous to the plants. A good gardener had to know how to lay them against the walls with care. Some trees were flexible enough to be stretched along wires. But for many varieties, this would not work.

> The methods of using wire seem suitable enough, but they are hardly good, as much because they are not solid enough and large branches that you need to force can break them or stretch them until the wire injures or scratches the branches which are young & as a consequence tender, & thus can make them exude their gum which makes them die, as well young branches can slip too easily behind the wires, where it is not easy to retrieve them without spoiling them.[6]

[6] "[Les] manieres [d'employer un fil] paroissent assez propres, mais elles ne sont guere bonnes, tant parce qu'elles ne sont pas assez solides, si bien que les grosses branches, qu'il faut quelquefois forcer, les rompent, ou les allongent, que parce que ce fil est subject à blesser, & écorcher les branches, qui sont jeunes, & par consequent tendres, & ainsi leur font venir de la gomme qui les fait perir, joint que les jeunes branches se glissent trop facilement derriere ces fils, d'ou il n'est pas aisé de les retirer sans les gâter." (La Quintinie 1692: I, 266)

The better solution for espalier trees was to build wooden trellises, but La Quintinie admitted not everyone could afford it. Leadership in the garden required not only intelligence and careful observation, but also thoughtful use of available resources.

Physical governance of fruiting plants in the *potager* was not limited to the work on espaliers. What La Quintinie called bushy fruit trees were also set out in geometrical order within measured spaces, and trimmed to take geometrical forms. Some were planted in small walled enclosures. Ground plants like strawberries and melons were protected and forced under glass, and some trees were protected with netting. The gardeners regulated exposure to sun, water and air, creating disciplinary regimes with horticultural knowledge in order to make plants behave in extraordinary ways in the garden of the Sun King.

The *potager* became such a successful model of logistical governance—a beautiful showcase of land control—that it was included in some of Louis XIV's itineraries for promenades (Thacker 1972; Babelon and Hoog 1992). These formal tours for distinguished visitors were designed personally by the king to introduce them to his court and the character of his reign. Clearly, the king thought the *potager* was important to see even though it lacked the artworks and waterworks of the *petit parc*. It was a well-governed corner of the French landscape protected by military means and displaying an exquisite, rational order that testified to Louis XIV's ability to exercise power over his lands to good effect.

Exercising Stewardship

La Quintinie also made the *potager* an Edenic garden, an enclosed paradise of plenty, displaying Louis XIV's stewardship as well as control of his lands. Stewardship legitimated the king's territorial ambitions, linking political territoriality to the moral duty of Christians to exercise dominion over the creatures of the earth. A Christian king's stewardship of God's gift of Creation was meant to enhance the natural productivity of the earth and serve the well-being of people; the *potager* demonstrated precisely Louis XIV's ability to do this.

Good stewardship also meant restoring Eden, making up for the sins of Adam and Eve, and this had two components. One was collecting species from around the world that had been dispersed during the Fall. The other was developing the horticultural techniques to allow them to exist together in one garden, using the intelligence given men by God to exercise dominion (Palissy 1988; Rivet 1990). Both of these activities were central to La Quintinie's experiments in the *potager du roi*, and described that way in the *Instructions*.

At the beginning of his book, La Quintinie referred to gardeners as descendants of Adam burdened with the labor of gardening and the responsibility of managing Creation to make it flourish and sustain life.

> My thought in this regard is, that the first Man having been created in a Garden, & after his sin received the order to cultivate the earth, to take from it his nourishment from the sweat of his brow, it follows that one of his principle functions, as well as that of his direct descendants was to devote themselves to the cultivation of Fruits and Vegetables, since it is only that which provides to human beings all the necessities of life.[7]

La Quintinie followed stewardship principles, too, as he assembled a wide range of fruit trees and vines from around the world and experimented with techniques for acclimatizing and nurturing imported plants with microclimates, providing the necessities for human life. He maintained a garden that was beautiful, useful, abundant and testified with its Edenic qualities to the moral capacity of the king to govern his lands according to God's wishes.

La Quintinie developed many of his ideas about stewardship and the horticultural means for practicing it from Olivier de Serres (Serres and Geffe 1971). Serres had developed methods for raising rare, imported plants, and promoted stewardship as a political philosophy, celebrating his patron, Henri IV, for improving his lands with plantations of trees and gardens. Calling the state a great estate, Serres advocated in particular the cultivation of useful plants like mulberries that could be used to raise silkworms. He also wrote about pharmacopia, and acclimatizing new fruits and vegetables not only for the purpose of furthering human wellbeing with food but also stimulating commerce and reducing human suffering due to poverty. To make it possible to raise the rare imports that had the greatest market value, Serres experimented with microclimates, using hot houses and glass bell jars that permitted plants from warmer climates to grow in France. The point was to care for the land by using it with knowledgeable methods; to tend to the needs of the creatures made by God to serve men; and to govern the kingdom by providing the necessities for human life.

La Quintinie extended the experiments started by Serres to raise vegetables and bedding plants, applying the same principles to fruit trees. Serres had focused on creating microclimates, using glass jars and hot houses that came to be called *serres*. Building microclimates for fruit trees was trickier, but the wide walls and different pockets of the *potager* helped to create artificial climates inside the walled garden. La Quintinie used the exposures of different parts of the structure, scaffolding and fabrics to create microclimates, planting delicate, heat-loving trees or vines along south- and west-facing walls and covering them to protect them from pests and keep them warm at night. His work was particularly successful with Muscat grapes.

[7] "Ma pensée à cet égarde est, que le premier Homme ayant été creé dans un Jardin, & y ayant aprés son peché réçu ordre de cultiver la terre, pour en tirer sa nourriture à la sueur de son front, il s'ensuit qu'une de ses fonctions principales, aussi-bien que celle de les permiers descendans fut de s'addonner à la culture des Fruits, et des Legumes, puisque c'étoit elle seule qui produisoit au genre humain tout le necessaire pour la vie." (La Quintinie 1692: I, 2)

> Whether a fruit is large or small does not matter, provided that it is sweet, firm, yellow, hard, and crisp, and that the juice in it is sweet, perfumed; the Muscat grape, I tell you, what pleasure does it not give to one that eats it? Could one see a more excellent fruit during the months of September, October, and sometimes until the end of November? ... we have clear need of espaliers facing east or south. In 1676, we had particularly good production of the most delicious in the world at these exposures.[8]

La Quintinie's ideas about the systematic development of a collection of fruit trees in the *potager* were derived from principles espoused by Serres. Collecting was a form of systematic governance designed to care for fruit trees so that the collection would stay intact. He understood that if all the trees were forced to grow in unnatural ways, they could be damaged or die, endangering the collection as a whole. La Quintinie needed grafting stock for plants that he grew in espalier so he allowed some specimens to grow large across allées or in bushy form in areas of the garden where they would not shade other trees. This increased the likelihood that the collection would remain viable even if some espalier trees died, and the garden's Edenic diversity would be maintained (La Quintinie: I, 363–367). This was the ultimate demonstration of stewardship.

The Effects of the Sun

In the *potager* of the Sun King, La Quintinie also cultivated natural knowledge of the effects of the sun on the production of fruit, and used it to develop his own understanding of the natural authority of the Sun King over the earth. He saw the sun acting as the primary element in nature, providing the context in which other elements could exercise their powers. He also understood the sun to be the fundamental source of fertility on earth. In his view, the sun's power was necessary, but he also noted the danger and corruption that that kind of unrestricted power could produce. In acquiring natural knowledge about the power of the sun in the *potager*, La Quintinie developed means not only to naturalize absolutism as inevitable and necessary, but also to depict it as a system of power with palpable dangers.

Nature, La Quintinie argued, was itself a system of governance that made it possible for plants to grow and affected how they did so. It was a system (family?) of interacting forces that gardeners had to study, understand and use to increase the productivity of the garden and sustain a collection of plants. He described nature this way:

[8] "Soit que le grain en soit gros, soit qu'il en soit menu . . .il n'import, pourvu qu'il soit clair, ferme, jeune, dur, & croquant, & que l'eau en soit douce, sucrée, & parfumée; telle grape de Muscat, dis-je, quel plaisir ne donne-t'elle pas à celuy qui la mange? peut-on voir un plus excellent fruit pendant les mois de Septembre & d'Octobre, & quelquefois jusqu'à la fin de Novembre?. . .nous avons necessairement besoin des Espalier du Levant, ou du midy; l'année 1676, nous en particulierement produit du plus delicieux du monde à ces expositions." (La Quintinie 1692: I, 409–410)

> She is a wise and knowing mother who having regarded all parts of the earth like so many children that would belong to her equally also she desired to divide equally among them the gifts and favors she wanted them to have in this way to build a union, and to see good intelligence eternally reign among them, she had regulated all things well that each indicated by its products what made it singular.[9]

La Quintinie's garden with its microclimates was an effort to intervene in nature's system of self-governance, using and displaying the power of the sun. It was the sun's heat that was captured in microclimates, and used in forcing espalier trees that stretched along walls. La Quintinie witnessed the regulatory powers of the sun over the creatures of the earth, and understood the power of the sun as a natural form of absolutism. So, when he reflected on the power of the sun, La Quintinie also obliquely addressed the character of absolutism.

La Quintinie best understood the importance of the sun's warmth from failures in the garden. The problem in cold climates was that the sun did not have enough power to keep plants from warmer climates alive.

> They transplant into quite cold climates a plant of a vine that was destined only for warmer lands... It seems that [nature] goes so far in creating losses that all the Grapes of Egypt, Africa, Italy, etc. die in the neighboring lands to the North; we truly try by means of walls with the best exposures to provide enough warmth... & there are some years and some locations where we succeed not too badly with some, but there are also many in which we need to find ways to console ourselves for our lost labors.[10]

This story of horticultural failure might have seemed an embarrassment to the king's kitchen gardener, but La Quintinie understood the story as revealing the power of nature itself, and the distinctive power of the sun over the life and death of plants.

Nature was a system of forces outside of human control, so trying to intervene in that system was not always going to work. There was a natural hierarchy of powers, and gardeners had only a limited capacity to exercise their own will. But the power of the sun was different. And it was in these terms that La Quintinie described the Sun King himself as different.

---

[9] "C'est une mere sage, & bien entenduë, qui ayant regardé toutes les parties de la terre, comme autant d'enfans qui luy appartenoient également, aussi leur a-t-elle volue également partager les biens & les faveurs qu'elle avoit à leur faire de maniere que pour entretenir l'union, & la bonne intelligence, qu'elle vouloit voir éternellement regner entre-elles, elle a si bien reglé toutes choses, que chacune a de se signaler par des production qui luy font singulieres." (La Quintinie de la Quintinie 1692: I, 408)

[10] "Ils transplantent en des climats assez froids le plan[t] de Vigne, qu'elle n'avoit destiné que pour les pays les plus chauds.... Il semble qu'elle [nature] ait creu, qu'il y iroit de son honneur, si elle se laissoit aller jusqu'à souffrir que tous les Raisins d'Egypte, d'Afrique, d'Ialie, &c, meurissent dans des pays du voisinage du Nord; nous essayons à la verité par le moyen de nos murs bien exposez de procurer autant de chaleur,.. [I]l est de certaines années, & de certains terroirs, où nous ne reüssissons pas mal en quelques-uns, mais aussi il y en a beaucoup, où nous avons plus besoins de chercher à nous consoler de nos peines perdues...." (La Quintinie 1692: I, 408)

La Quintinie said that the Sun King was like the sun, in having a natural capacity to govern the earth and exercise independent powers over the fertility of land. In the dedication of the *Instructions* to Louis XIV, the royal gardener argued that his patron was not like ordinary men or even like other great leaders. Nature herself showed particular favor to *this* king, endowing him with insights about governing the earth that she had not revealed to his predecessors.

> Nature which (it seems) takes pleasure in refusing nothing to your majesty, and regards [your majesty] in effect as the most perfect of its works, has revealed in your reign what the earth has kept hidden in all the past centuries. [We] who have the honor of cultivating your gardens, can see what to the glory of our monarch and the advantage of humankind, has been unknown to antiquity will never be so again to anyone.[11]

La Quintinie, as a member of the king's household dedicating his book to Louis XIV, was certainly expected to say that all good things emanated from the monarch, but how he said it was significant. He identified the king as part of nature, and a Sun King because of capabilities endowed by nature. He had a natural propensity for governance and gardening that was demonstrated on his lands.

La Quintinie went on to argue that just as the sun was the primary force in nature, bringing fertility to the land, the Sun King should be accepted as the primary force in France. "This Land [of France] that seems so stubborn in the view of all the world, will finally subordinate itself and also to respond with some joy to even the least commandments of a grand Prince, which all the other elements gain glory to obey."[12] The power of Louis XIV, according to La Quintinie, should be treated like the power of the sun, something superhuman, necessary, and inevitable from which other powerful forces can gain glory.

La Quintinie treated the garden as a system in part because he saw governance as natural. If gardening was a matter of engaging nature, this was best done by building a system for aligning the natural tendencies of things. That was why he thought a collection of fruit trees should reproduce some of the diversity of nature's "children", and contain a range of different specimens with distinctive attributes, creating a healthy redundancy of the sort found in nature (La Quintinie 1692: I, 22–31, II, 3–8).

---

[11] "Nature qui (ce semble) prend plaisir à ne rien refuser à VOTRE MAJESTE´, & qui la regarde en effet comme le plus parfait de ses Ouvrages, a sans doute reservé pour son auguste Regne, ce que la terre a caché à tous les siecles passez....[Nous] qui ont l'honneur de la cultiver dans ses Jardins, nous verrons à la Gloire de Nôtre Monarque, & à l'avantage du Genre humain, que ce qui a été inconnu à tout l'Antiquité ne le sera plus pour personne."

[12] "Cette Terre qui paroît si opiniâtre à l'égard de tout le monde, cedera enfin, & mesme, pour ainsi dire, avec quelque joye aux moindres commandemens d'un grand Prince, à qui tous les autres élemens font gloire d'obeir." (La Quintinie: I, a ij–iij)

Although La Quintinie had drawn direct analogies between the sun and the Sun King in the early pages of the *Instructions*, in Volume II, in which he described how the sun worked on fruit for better or worse, he did not draw political analogies and praise the king. In this part of the book, which was full of horticultural detail, he seemed to address practicing gardeners like himself. In this section, he provided a surprisingly detailed and vivid portrait of absolutism as a system of governance that was not only necessary but also dangerous.

When La Quintinie reflected on the positive role of the sun in the garden, he spoke not only about the sun's direct effects, but also about its capacity to organize the forces around it. The sun was the dominant element in nature, but more than that, it was the heart of the system in nature that brought life to the land. As well as generative, he saw the sun as dangerous, but also unstoppable because it was necessary to nature as a whole.

> La Quintinie argued that all four of the elements were necessary to the generativity of the earth, but the other elements were only effective in the presence of the sun, warming the earth and air. "[The] function of the Sun is thus to warm the air, as much as the winds permit it, and the function of the warm air is to take part in the warming of the earth, & to all Plants; this warmed earth takes action and the source of life . . .[the sap] is set in motion.[13]

The earth was the medium for gardening, and provided the salts needed by plants (Palissy 1988). But the earth could do nothing for plants without the primary action of the heat of the sun to warm the air and earth, and the contribution of water to the soil:

> [T]he earth (considered as one of the four elements) has no primary and natural disposition for vegetation [but] once she is succored by the rays of the Sun, and the underground fires which warm her, and some water wets her, she changes so to speak her nature as if to obey an absolute commandment of a sovereign Master, she seems to resemble a living and animated being.[14]

According to this passage, the earth became fertile on command, as though obeying a sovereign master. Even the language La Quintinie chose treated absolutism as a natural system. La Quintinie has turned a model of natural elements into a model of governance, but he never made this argument directly in relation to practical gardening. He was not being a propagandist as he had been in the

[13] "La fonction du Soleil est donc d'échauffer l'air, autant que les vents le luy permettent, & la fonction de l'air échauffé est de faire part de sa chaleur à la terre, & à toutes les Plantes; cette terre échauffée fait d'abord agir & le principe de vie . . .[la seve] est mise en action." (La Quintinie 1692: II, 202)

[14] "Je pourrois bien avancer icy premierement que la terre (à la considerer en soy comme un des quatres élemens) n'a veritablement aucune disposition premiere & naturelle pour la vegetation. . . [mais] qu'ensuite elle est secourruë tant des rayons du Soleil, & des feux soûterrans que l'échauffent, que de quelques eaux qui l'humectent, elle change pour ainsi dire de nature si bien que pour obéïr à un commandement si absolu du souverain Maître, elle paroit ce semble un être vivant & animé." (La Quintinie 1692: I, 140)

introduction. La Quintinie seemed instead to be translating political observations into the language of gardeners to describe the nature of power.

La Quintinie was surprisingly candid about the limits of the natural powers of the sun, pointing to situations where this power attenuated. He noted what qualities of earth and air made plants fail to respond to or resist the powers of the sun. He observed that, in some places, the air would be warmed by the sun as usual, but the earth would stay cool. In other words, the other elements did not always join forces with the sun. He attributed the differences to subtle vapors in the earth, but still noted that the difference had consequences for stimulating the roots and thus setting fruit (La Quintinie 1692: II, 202–203). La Quintinie also noted that fruit trees set and opened buds at the same time of year no matter how warm the weather was, so it did not depend on the warmth of the sun—although the direct rays of the sun might affect when buds opened (La Quintinie 1692: II, 202).

Most strikingly, La Quintinie warned about dangers of the sun to fruit, and prescribed ways to protect against them. He argued, for example, that sunlight, although necessary for giving mature fruit good color, could burn the outside of tender young fruit. To protect against this mortal danger, La Quintinie developed a system of managing the growth of leaves on fruit trees. When fruit first set, he said, the leaves naturally shaded them, and protected the young fruit from the harsh rays of the sun. The leaves should be left in place until the fruit was almost mature. Then, the trees should be stripped of these leaves to expose the fruit, allowing them to acquire a good color before ripening. Only when they reached nearly full size were the young fruit strong enough to face the power of the sun (La Quintinie 1692: II, 184–186).

Killing young fruit in gardens this way resulted from ignorant misunderstanding of systematic power. Gardeners who did not see nature as a system were tempted to take away the leaves. They knew that the power of sun could make the fruit grow faster and acquire good color, but they did not recognize when that power became dangerous, so they ended up killing the young fruit. The sun had the power of death as well as life, and gardeners had to recognize the vulnerability of the young to this ultimate power, learning how to protect them (La Quintinie 1692: II, 184–186).

La Quintinie also pointed to the danger of the sun as a source of corruption. Ripening of fruit itself, he argued, was a form of corruption that developed as the sun warmed the soil and affected the circulation inside plants. (La Quintinie: II, 202). This explained why fruit so often rotted from the inside out. The fire of the sun, he noted, worked on fruit in the opposite way to the action of fire on roasting meats. Rather than changing the outside, the warmth around the roots of trees made fruit become soft and exude juice, affecting the interior of fruit faster than the exterior, and making fruit rot from the pit toward the skin (La Quintinie 1692: II, 190–202).

This kind of corruption could be hard to detect, and was certainly hard to avoid altogether if one wanted soft, sweet fruit. It was an effect of the sun that was natural and generally beneficial, but the danger had to be recognized and managed wisely (La Quintinie 1692: II, 190–204). The power of the sun could not safely be left unchecked.

It is not surprising that La Quintinie would reflect on and describe the power of the sun as he supervised the *potager*, and promote absolutism as natural. He was, after all, an officer of the king's household charged with extending and fueling the glory of the Sun King. But it is surprising that he would describe the dangers of the sun as a threat to the young and an inevitable source of corruption. Perhaps this explains why he left the political analogy to lie dormant in the practical part of the text.

In the end, La Quintinie treated systematic governance as necessary, and celebrated Louis XIV as a singular ruler with a distinct capacity for ruling his land. He made the case that the greatness of France was contingent on the king's successes in governing his lands. He suggested, too, that just as plants in the garden could not choose to be governed by the sun's warmth, so those who were governed by the Sun King could only resist his power by denying its nature. They could not stop the Sun King from animating the earth and making his territory more glorious any more than they could stop the sun from having its effects in the garden. But still, he seemed to confide to other gardeners, there were dangers to absolutism, pointing to natural limits of the sun's power and the corrupting force of the sun.

Conclusions

In three distinct ways, then, La Quintinie linked natural knowledge to politics, and furthered the power of his patron by cultivating and learning from the *potager du roi*. His solutions to problems of natural knowledge in the garden were also solutions to problems of politics. He used horticultural knowledge to naturalize state absolutism; he nurtured plants in the garden to present the king as a good steward; and he used force over natural things to make nature submit to the will of the monarch. Above all, the royal gardener brought beautiful and sweet fruit to the king's table, making life at court more felicitous and the power of the Sun King more palpable and seductive.

The equation of Louis XIV to Apollo may have remained problematic and dismissed as propaganda, but no one could doubt that the French monarch could rule his lands effectively if they visited the gardens at Versailles. The park and its gardens were a marvel of beauty, abundance, and control. Louis XIV acted like a Sun King whether he was Apollo or not. La Quintinie, along with the other gardeners at Versailles, helped draw attention to and justify political territoriality as a technique of power, too. Applying rational land use practices to dominate the landscape seemed an effective method for making France more like Rome—an empire controlled by both military and horticultural means. This vision of power/knowledge was most clearly articulated in the *potager* where fortress walls evoked military domination, and the gardening displayed exquisite horticultural finesse.

All the heroic feats of masterful gardening in the *potager* and the analogies drawn between the king and Apollo in the *Instructions* used natural knowledge to promote the vision of Louis XIV as a force of nature capable of bringing to his lands an extraordinary abundance and beauty only made possible by the sun. La Quintinie seemed to promote absolutism as a natural and effective form of power in the way he described the power of the sun over other elements, but he also suggested to fellow gardeners that there were limits and dangers to this system of power. He seemed to speak in horticultural terms as to what he had learned in the hothouse of the *potager* about the king's household, royal power, and systematic governance of nature, people, and territories. The sun's dominance in nature provided a model of control that the Sun King could imitate to govern a kingdom with equally good effect.

La Quintinie told a story of power in nature that any gardener could recognize, having struggled with varying degrees of success to make things grow where they did not naturally live, and feeling the relative powerlessness of gardeners in relation to the powers in nature like the sun. In this context, La Quintinie championed the Sun King's capacity to bring felicity and fecundity to the earth. But the royal gardener who had studied the sun so mindfully to use it well in the garden also knew that the power of the sun/Sun King carried mortal danger and had to be managed with carefully developed techniques. In this way, La Quintinie used natural knowledge not only to address problems of politics, but also to understand them.

Bibliography

Sources

Columella, Lucius Junius Moderatus. 1537. De re rustica L. Iunii Moderati Columell libri XIII. Lugduni: Apud Seb. Gryphium.

Estienne, Charles. 1564. L' Agriculture et maison rustique de M. Charles Estienne,... en laquelle est contenu tout ce qui peut estre requis pour bastir maison champestre, nourrir et médeciner bestail et volaille... Plus un bref recueil de la chasse et de la fauconnerie. Paris: J. Du Puis.

Félibien, André. 1672. Description de la grotte de Versailles. Paris: S. Mabre-Cramoisy.

Machiavelli, N., and D. Donno. 1966. The Prince. New York: Bantam.

Palissy, Bernard. 1988. Recepte véritable. Ed. K. Cameron. Génève: Libr. Droz.

Pliny. 1989. Natural History. Book XXIII. Harvard: Loeb Library.

Quintinie, Jean de la. 1692. Instructions pour les jardins fruitiers et potagers. Amsterdam: Henri Desbordes.

Serres, Olivier de. 1600. Le theatre d'agricvltvre et mesnage des champs. Paris: I. Métayer imprimeur ordinaire du roy.

Serres, Olivier de, and Nicholas Geffe. 1971. The Perfect Use of Silkworms. New York: Da Capo Press.

Secondary Literature

Anderson, Perry. 1974. Lineages of the Absolutist State. London: NLB.

Apostolidès, Jean-Marie. 1981. Le Roi-machine: spectacle et politique au temps de Louis XIV. Paris: Editions de Minuit.

Appuhn, Karl Richard. 2009. A forest on the sea: environmental expertise in Renaissance Venice. Baltimore: Johns Hopkins University Press.

Babelon, Jean-Pierre, and Simone Hoog. 1992. Manière de montrer les jardins de Versailles par Louis XIV. Paris: Réunion des Musées Nationaux.
Burke, Peter. 1992. The fabrication of Louis XIV. New Haven: Yale University Press.
Carroll, Patrick. 2006. Science, culture, and modern state formation. Berkeley, Calif.: University of California Press.
Cronon, William. 1991. Nature's Metropolis: Chicago and the Great West. New York: W. W. Norton.
Dunerin, A. 1983. "Le Potager du Roi 1683-1983." ENSP.
Givry, Jacques de, Pascal Lobegeois, and Michael Tournier. 2000. Versailles: the fountains of the Sun King. Paris: Mémoire & Lumières.
Goldstein, Claire. 2008. Vaux and Versailles. Philadelphia: University of Pennsylvania Press.
Hazlehurst, F. Hamilton. 1966. Jacques Boyceau and the French formal garden. Athens: University of Georgia Press.
Hazlehurst, F. Hamilton. 1980. Gardens of illusion: the genius of André Le Nostre. Nashville, Tenn.: Vanderbilt University Press.
Hunt, John Dixon, Michel Conan, and Claire Goldstein. 2002. Tradition and innovation in French garden art: chapters of a new history. Philadelphia: University of Pennsylvania Press.
Kettering, Sharon. 1986. Patrons, brokers, and clients in seventeenth-century France. New York: Oxford University Press.
Knorr-Cetina, Karin. 1999. Epistemic Cultures. Cambridge: Harvard University Press.
Leibler, Anat, and Daniel Breslau. 2005. The Uncounted: Citizenship and Exclusion in the Israeli Census of 1948. Ethnic and Racial Studies 28: 880-902.
Léry, E. 1926. Les travaux du potager de Versailles: la machine à transporter les terres. Conferences des Sociétés Savantes de Seine et Oise 8: 18-20.
Lestringant, Frank. 1990. L'Eden et les tenèbres extérieures. In Actes du Colloque Bernard Palissy, 1510-1590, 119-130. Albineana 4, Paru en 1992, réimpression à Niort 2010.
Mukerji, Chandra. 1994. The Political Mobilization of Nature in French Formal Gardens. Theory and Society 23: 651-677.
Mukerji, Chandra. 1997. Territorial ambitions and the gardens of Versailles. Cambridge; New York: Cambridge University Press.
Mukerji, Chandra. 2002. Bourgeois Culture and French Gardening in the 16th and 17th Centuries. In Bourgeois and aristocratic cultural encounters in garden art, 1550-1850, ed. M. Conan, 173-188. Washington, D.C.: Dumbarton Oaks.
Mukerji, Chandra. 2009. Impossible Engineering: Technology and territoriality on the Canal du Midi. Princeton, N.J.: Princeton University Press.
Mukerji, Chandra. 2010a. The Territorial State as a Figured World of Power: Strategics, Logistics and Impersonal Rule. Sociological Theory 28: 402-425.
Mukerji, Chandra. 2010b. The Unintended State. In Material Powers: Cultural Studies, History and the Material Turn, eds. T. Bennett and P. Joyce, 81-101. London: Routledge.
Mukerji, Chandra. 2012. Space and Political Pedagogy at the Gardens of Versailles. Public Culture 24: 515-540.
Néraudau, Jean-Pierre. 1986. L'Olympe du Roi-Soleil: mythologie et ideologie royale au grand Siècle. Paris: Les Belles Lettres.
Rivet, Bernard. 1990. "Aspects Économiques de l'Oeuvre de B. Palissy" et Frank Lestringant, "L'Eden et les Tenèbres Extérieures". In Bernard Palissy 1510-1590, l'écrivain, le réforme, le céramiste, ed. Frank Lestringant, 167-180. Coédition Associacion Internationale des Amis d'Agrippa d'Aubigné- Éditions SPEC.
Rivet, Bernard. 1990. Réflexions sur quelques aspects économiques de l'oeuvre de Bernard Palissy. In Actes du Colloque Bernard Palissy, 1510-1590, 167-180. Albineana 4, Paru en 1992, réimpression à Niort 2010.
Shapin, Steven, Simon Schaffer, and Thomas Hobbes. 1985. Leviathan and the air-pump: Hobbes, Boyle, and the experimental life: including a translation of Thomas Hobbes, Dialogus physicus de natura aeris by Simon Schaffer. Princeton, N.J.: Princeton University Press.

Soll, Jacob. 2009. The information master: Jean-Baptiste Colbert's secret state intelligence system. Ann Arbor: University of Michigan Press.
Thacker, Christopher. 1972. Le manière de montrer les jardins de Versailles. Journal of Garden History 1: 49-69.
Thacker, Christopher. 1979. The History of Gardens. Berkeley: University of California Press.
Tomasi, Lucia. 1983. "Projects for Botanical and Other Gardens: a 16th-Century Manual." Journal of Garden History 2: 1-34.
Weiss, Allen S. 1995. Mirrors of infinity: the French formal garden and 17th-century metaphysics. Princeton: Princeton Architectural Press.

Part II

Mathematical Sciences and the Art of Gardening

The Organ of the Villa d'Este in Tivoli and the Standards of Pneumatic Engineering in the Renaissance

Simone M. Kaiser and Matteo Valleriani

Abstract

The Villa d'Este in Tivoli near Rome was the last building project of Cardinal Ippolito II d'Este and the one that secured him lasting fame. One of its main attractions is the hydraulic organ integrated into a fountain system, called the *Fountain of the Organ*. This paper points to a source describing the organ's mechanism that has been ignored until now: Oreste Vannocci Biringucci's translation of Hero's *Pneumatics*, which was commissioned in the course of building pneumatic devices in another famous villa garden, the Medici garden in Pratolino. Hero's text is followed by a description of the technical apparatus and functioning of Tivoli's organ, which is published and translated here for the first time. The research is based on two different descriptions of the organ of Tivoli and on a virtual reconstruction of the hydraulic organ built within the scenic reproduction of the legendary *Mount Parnassus* at the garden of Pratolino. By means of a comparative analysis, it will be shown that the practical experience of Renaissance pneumatic engineers is superior to the knowledge codified in the ancient texts. As a consequence, this work also shows how Renaissance gardens represented the knowledge platform for theoretical and practical knowledge.

S.M. Kaiser (✉)
Abteilung Kunst und Kulturgeschichte, Hessisches Landesmuseum Darmstadt, Friedensplatz 1, 64283 Darmstadt, Germany
e-mail: simone.kaiser@hlmd.de

M. Valleriani
Max Planck Institute for the History of Science, Boltzmann Str. 22, 14195 Berlin, Germany
e-mail: valleriani@mpiwg-berlin.mpg.de

H. Fischer et al. (eds.), *Gardens, Knowledge and the Sciences in the Early Modern Period*, Trends in the History of Science, DOI 10.1007/978-3-319-26342-7_5

The Garden of Tivoli and Its Organ

The Italian Renaissance was a period in which technology flourished. Hydraulics, pneumatics together with architecture and metallurgy comprised some of the most cultivated practical activities of the period. Highly prestigious villa garden projects were used increasingly as a platform of expertise, connecting various disciplines in an ideal work of garden art where the latest state of knowledge could be displayed and added to the magnificence of its patron (Kaiser 2012, 2014).

Such a villa garden was envisioned by Cardinal Ippolito II d'Este in Tivoli from the 1550s onwards, with construction beginning in the 1560s[1] (Coffin 1960; Lamb 1966). This project set new standards in the art of fountain building, deliberately displaying the peak of prowess in the hydraulic engineering of its time. The masses of water that could be moved were particularly unequalled. It posed a challenge and served as a model for many subsequent garden projects. The height of magnificence was reached on the occasion of the pope's visit in September 1572. When the cardinal died in the same year only three months later, the garden—still unfinished—was further maintained and altered by several successive cardinals of the Este family. Concerning the conception and appearance of the organ as preserved up to today (Figs. 1 and 2), the periods of Ippolito II and Alessandro (1599–1624) are particularly relevant.

The organ is just one part of a myriad of water devices all drawing from the same supply system. Actually, two aqueducts had to be built in order to feed all the fountains of Villa d'Este. The first one (1560–1561) was designed to bring the clear water of the Rivellese source from Monte Sant' Angelo to the church square of Santa Maria Maggiore, supplying the populace of Tivoli as well as the palace and upper terraces of Villa d'Este. The second one (1564–1565) was built on the subterranean level. By means of pipes positioned at a natural waterfall, abundant waters from the River Aniene were diverted in order to satisfy the much larger needs of the elaborate water games on the lower terraces of the garden—the organ included (Coffin 1960: 9; Lamb 1966: 37–40).

Unsurprisingly, this ensemble, one of the most impressive early modern achievements in the field of water technology, wasn't the work of a single engineer but was brought about by the collaborative effort of many experts and their laborers. The master plan of the villa garden was conceived by Pirro Ligorio, who had been employed as the cardinal's antiquarian in 1550 to explore the villas and water systems of old Tibur (Tivoli).[2] The construction works were carried out under the supervision of the Ferrarese architect Giovan Alberto Galvani. Several fountain engineers participated in the design and construction of the manifold water games,

[1] Ippolito II d'Este became governor of Tivoli in 1550, and again—after 4 years of exile—in 1559, serving until his death in 1572.

[2] Pirro Ligorio also became papal architect in Rome (1558–1567). After having been entangled in a scandal he finally left Rome in 1569 to become the court antiquarian of Duke Alfonso II d'Este in Ferrara until his death in 1583.

Fig. 1 The *Fountain of the Organ* at the Villa d'Este, Tivoli. Lamb 1966, table 16

such as the Italians Curzio Maccarone and Tommaso Ghinucci, who are associated with several villa building sites around Rome. But the hydraulic organ was the *œuvre* of two Frenchmen, identified in Italian sources as Luca Clerico and Claudio Venardo. These two hydraulic and pneumatic engineers seem to have specialized in Heronian devices. Otherwise, not much is known about their biographies. Clerico made another Heronian device at the Villa d'Este, the *Fountain of the Owl* (1566). When Clerico (Luc Leclerc) died around 1568–1569, the work was completed by Claudio Venardo (Claude Venard), who seems to have been his pupil and nephew (Coffin 1960: 17; Lamb 1966: 55).

Initially, the *Fountain of the Organ* was also called the *Fountain of Nature*, or the *Fountain of the Deluge* (Figs. 1 and 2). These denominations allude to a narrative interrelating various features of the garden to an overarching iconological system (Kaiser 2014). In achieving the realization of the iconological program, the role of water in such gardens was not purely decorative. In Renaissance conception, the 'garden' is a place where nature is observed, imitated (by means of art) and even

Fig. 2 Earliest preserved depiction of Tivoli's organ, showing it enclosed with an open circular screen which is not mentioned in any of the known descriptions. Fresco in the Salotto of Villa d'Este, (1568–1569) and Dernie (1996: 71)

improved. This conception is often addressed within the iconology of the gardens in a self-reflexive manner. The *Fountain of the Organ* was, for instance, part of a large water cross axis that was intended to illustrate the course of life in analogy to the course of water—a topic which Ligorio considered ancient and most suitable for fountain programs *all'antica*. The fountain was located on a slope at the northeastern edge of the garden. In its central niche, in front of the organ, a statue of *Nature* originally stood upon a rock, a rustic copy of the multi-breasted Diana of Ephesus, spilling water from her numerous breasts (Appendix 1). On the lower terrace, beneath the slope, the water was led on into rectangular basins full of fish and aquatic birds. Two of these basins had water jets directed from herm pillars at the rim toward the center of the basin—the ballistic curves and pressure of the latter were calculated so that they created a sort of vaulted water ceiling and a rainbow effect when the sun was shining (Lamb 1966: 47). At the Western end of this cross axis, a *Fountain of the Sea* was planned to complete the idea of this cycle. On its way along that cross axis, water was fashioned into many different shapes imitating natural phenomena, such as rain, stormy weather and rainbows. These effects were aesthetically pleasing and scientifically interesting, and also had a moral overtone insofar as they invited association with principal elements of salvific history, such as the Deluge and the subsequent covenant with God indicated by a rainbow (Gen.: 9 and 1–17). Above all, in the frame of operative pneumatics, water is a natural element which lends itself most readily to integration of natural and artificial elements, imitating nature, channeling its forces and putting them to good use in ways both practical and pleasurable. This is certainly true of the hydraulic organ at Tivoli as well.

The organ had no need of a player, but was able to produce a musical piece automatically. This was a novelty compared to the antique art of hydraulic organ-building as passed on by Heron of Alexandria and also by Vitruvius (Degering 1989; Markovits 2003). But similar devices had already been further developed during the late Middle Ages. Famous for its automata was, for instance, the late medieval pleasure park in Hesdin, Northern France, established at the end of the thirteenth century. Robert II, Count of Artois, built this park from 1295 to 1302. The Italian knight Renaud Coignet, whom Robert had met on his way back from the Crusades, supervised the construction. He had been in Sicily and Naples where he may have been inspired by the example of famous Saracen and Norman pleasure parks. He also possibly came across the expertise of Islamic engineers or Arabic sources who had preserved the mechanical knowledge of Hero's time (like the ninth century *Book of Ingenious Devices* by Banū Mūsa or al-Jazarī's later *Book of Knowledge*). The building project at Hesdin was further maintained and extended in the fourteenth and fifteenth centuries (Van Buren 1986: 115–134; Miller 1986: 143f; Franke 1997: 135–158). Among other things, there was a big clock that presumably possessed a bellows-operated organ and a chamber where rain could be imitated, as well as thunder—mechanisms that were likely inspired by Hero's writings (Franke 1997: 136 and 148). In this respect, it may be no coincidence that the masters of Tivoli's Heronian devices were Frenchmen, assuming that this art was popular and well developed in France. Ippolito II lived at the French court in the late 1530s and often returned there afterward, though it is still unclear how he got acquainted with his French engineers. His diplomatic correspondence proves that he had visited Hesdin in 1537 (Occhipinti 2001: 17) when the French king had just conquered the town and before it was utterly destroyed by Emperor Charles V in 1553. He certainly saw the state of the park at that time and heard of its former fame, which could have inspired a desire for automatic devices in a garden of his own. It also remains speculative who first brought Hero's writings to the cardinal's attention—Greek manuscripts of the text in Italian libraries had already aroused the interest of humanists, but still no Latin or vernacular translations had been printed at the time (Boas 1949; Valleriani 2007)—and what role this knowledge really played in the creation of Heronian devices at Tivoli. A remark in Ligorio's treatise on the *Nobility of the Ancient Arts*, written between 1568/9–1573, indicates, at least, that the author had also engaged with Hero's knowledge and that it must have been a favorite topic in the academic circle of Ippolito II. Given that he complains, somewhat indignantly and snobbishly, about the ignorance of fellow architects, implying, among other things, that one shouldn't make hydraulic machines without having studied Hero's automates, it seems obvious that he himself did so[3] (Ligorio, fol. 21v; Schreurs 2000: 421f).

Tivoli's organ was, anyhow, the first automatic one in a Roman villa garden (Dernie 1996: note 52). The sensation it caused was such that similar devices immediately became an indispensable feature of other important villas. The organ was integrated into a fountain façade (a mannerist, *alla rustica* version of the antique water castles Ligorio had studied thoroughly; Fig. 2) and designed to perform a madrigal for five voices. The performance started with a blare of trumpets. During the whole length of the madrigal, the sound of birds accompanied the composition. When the piece was finished, the amount of water that had activated the organ was released, swooshing out all at once—this effect was called the 'deluge'. Moreover, this caused the sound of a blowing horn, which was positioned in the water basin in front of the organ (Appendices 1 and 2).

From a technical point of view, the peculiarities of Tivoli's organ are threefold: its dimensions, its position and the multiple functionality of the technology that activated it. According to the archeological findings and the ensuing studies, the apparatus of the organ, situated beyond the wall of the *Fountain of the Organ*, appears to be considerably larger than any other known Renaissance hydraulic organ. This could be related to its multiple functionality, because the water and the mechanisms served not only to make the organ play, but also to power the mentioned set of trumpets and to cause a number of artificial birds to sing in harmony (the latter is nowadays called a typical Heronian device). Finally, the position of the organ was peculiar too, for it was situated outside and, originally, important parts of its apparatus were not covered—particularly the water reservoir behind the fascia of the fountain's façade where the Aniene water arrived before being led on into the organ's chamber (Figs. 2 and 3). According to Carl Lamb

[3] "[. . .] et quelli che vogliono fare delli istrumenti da tirare acqua non hanno studiato che forza ha l'acqua con le machine Aftomauto [sic], che per accidenza di ruote si muovono, come le cose di Herone, et di Philostrato, et di quegli che scrivono di Horioli: Vogliono essere ogni muratore Architetto et ogni Misuratore di Terreni; cosa da vomitare, che tanta ignorantia sia caduta tra gli huomini! Essi per parere eccellenti come è stato Michelangelo Buonarroti si metteno il capello et li stivaletti in gamba con le scarpe di sopra et tengono le ciglia sotto il capello peloso per imitare la scienza di quello, come quella stesse nel capello et nelli stivali et come pratticando con uno dottore se impari nell'andar vestito senza havere lingua o metodo da potere studiare, cosi costoro per andare cosi vestiti, vogliono parere dottori et Michelagnoli [. . .]".

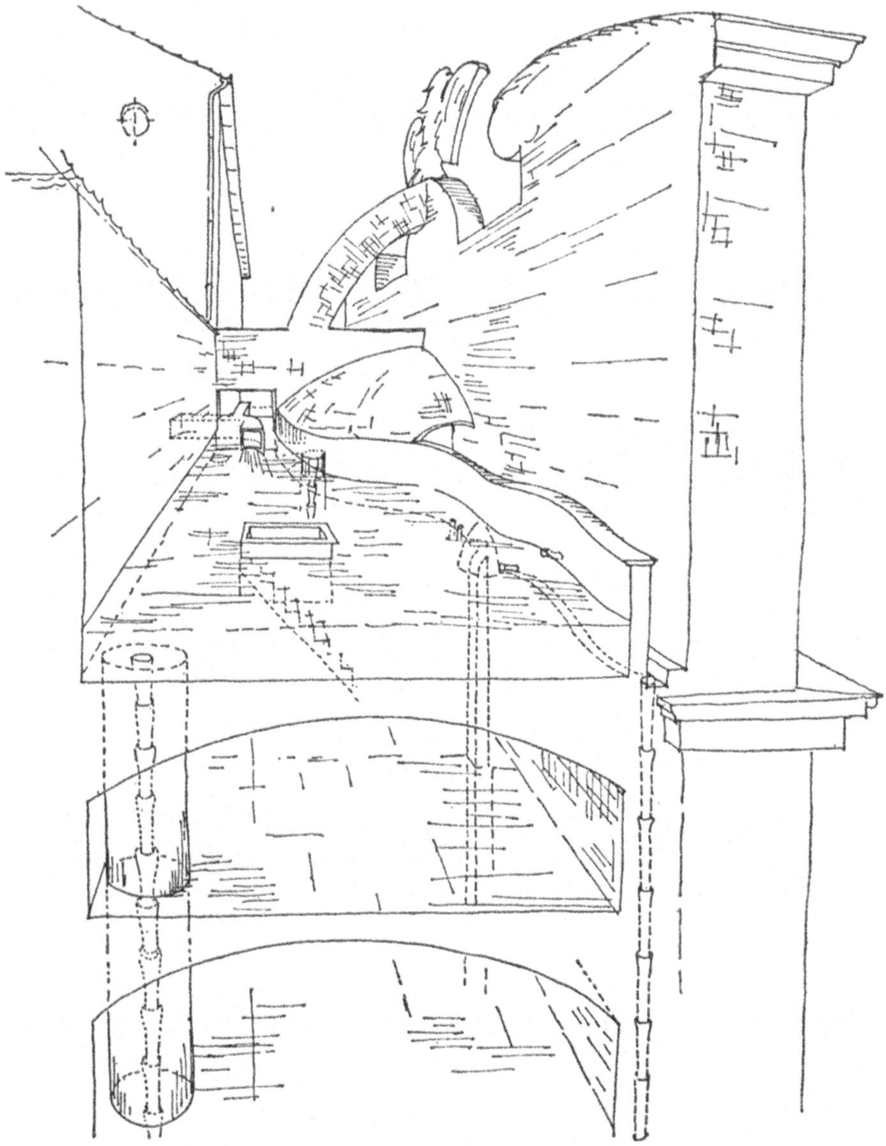

Fig. 3 Sketch of the hydro pneumatic apparatus of the organ at Tivoli Lamb (1966: 54)

(1966: 56), this is the only hydraulic organ conceived to diffuse sound in the open and, after this experience, all other hydraulic organs were built in closed spaces to better appreciate the musical performance. This, however, is not true. In the garden of Pratolino, for instance, built only a few years later, two hydraulic organs were constructed: one inside underneath the mansion, the other outside at *Mount*

Fig. 4 *Mount Parnassus* at Pratolino. Pen Drawing, Giovanni Guerra, 1598. Architektur, graphische Sammlung, Albertina, Vienna, 37222r

Parnassus (Valleriani 2009, 2010a, b). The latter device was also able to diffuse sound in the open (Fig. 4).

The very first version of Tivoli's organ must have been completed and functioning before September 1572, as it is conveyed that Pope Gregory XIII attended a musical performance played on the instrument during his visit to Villa d'Este and, deeply impressed, asked to be introduced to Claudio Venardo (Zappi 1920: 62; Coffin 1960; Lamb 1966: 55). It was destroyed in early May 1582: someone had damaged its conducts and broken the organ pipes at night (Lamb 1966: 55). Besides this attack, the *Fountain of the Organ* also suffered serious maintenance problems, especially after the death of Cardinal Luigi d'Este in 1586, when the villa garden passed by reason of testamentary clauses into the hands of the College of Cardinals and was rather neglected. When Cardinal Alessandro d'Este took over, he finally ordered the construction of an architectural element to cover the organ pipes (Lamb 1966: 52–54; Fig. 1).

A most elaborate description of the technical apparatus and functioning of the first version of Tivoli's organ (prior to the attack of 1582), known and published for the first time in 1964, is included in the travel account of the young Frenchmen Nicolas Audebert. He visited Villa d'Este late in 1576 or early in 1577, and finished the manuscript on his *Voyage d'Italie* by 1582, after his return to France (Lightbown 1964; Audebert, Olivero 1981; Appendix 1). It is the description of an amateur to whom everything had been explained, supposedly by his fellow countryman, Claude Venard himself. The other detailed description, presented below for the first time, was also finished in 1582, but written in the common technical language of contemporary engineers (Appendix 2). The information that can be found therein, in comparison to Audebert and together with the analysis of the archeological findings, offers a unique chance not only to complete the description of the functioning of the first version of the organ of Tivoli but also to learn more about the standards of the theoretical and practical knowledge of Renaissance engineers.

Oreste Vannocci Biringucci's Text

In 1575, 3 years after Tivoli's organ had performed in honor of the pope, Federico Commandino's Latin translation of Hero's *Pneumatics* was published posthumously. It was the first printed edition of Hero's text. This edition had not received any final text editing before publication but was nevertheless enriched by a series of new engravings that greatly improved its comprehensibility. This most likely led the chief engineer of the Medici family in Florence, Bernardo Buontalenti, to commission an Italian translation of Hero's *Pneumatics* while supervising the construction of all pneumatic devices—including the two hydraulic organs—installed at the garden of Pratolino. The translation, made by Oreste Vannocci Biringucci, nephew of the famous Vannocci Biringucci who published *De la pirotechnia* in 1540, was eventually finished in 1582 (Figs. 5 and 6). It was ready for publication but then remained in manuscript form because of the premature death of the author. Unfortunately, the drawings, for which empty spaces were left in the manuscript, were never found. Although this document does not seem to be linked to the garden of Tivoli, it is unexpectedly followed by a detailed description of the pneumatic mechanisms of Tivoli's organ.

As is well known, Hero's *Pneumatics* ends with a description of the construction and functioning of the Aeolian and hydraulic organs. What has been overlooked in modern times is that engineers of the sixteenth century had already noticed that Hero's description of the hydraulic organ was incorrect and attributed this problem to the corruption that the text most certainly underwent. The same Oreste Vannocci Biringucci complains about the lack of clarity in Hero's text concerning the hydraulic organ. This is his explicit motive for adding a description of an actually functioning practical application to the original text, that of the hydraulic organ built in the garden of Tivoli. This addition was meant to improve the description of the devices as they appeared at that time in Hero's text.

A further motivation may have been the issue of competition. The famous travel account by Michel de Montaigne, who visited both sites in 1580–1581, indicates that the garden at Pratolino was built with the aim of competing with and outmatching the highly prestigious wonders of its predecessor in Tivoli (Montaigne

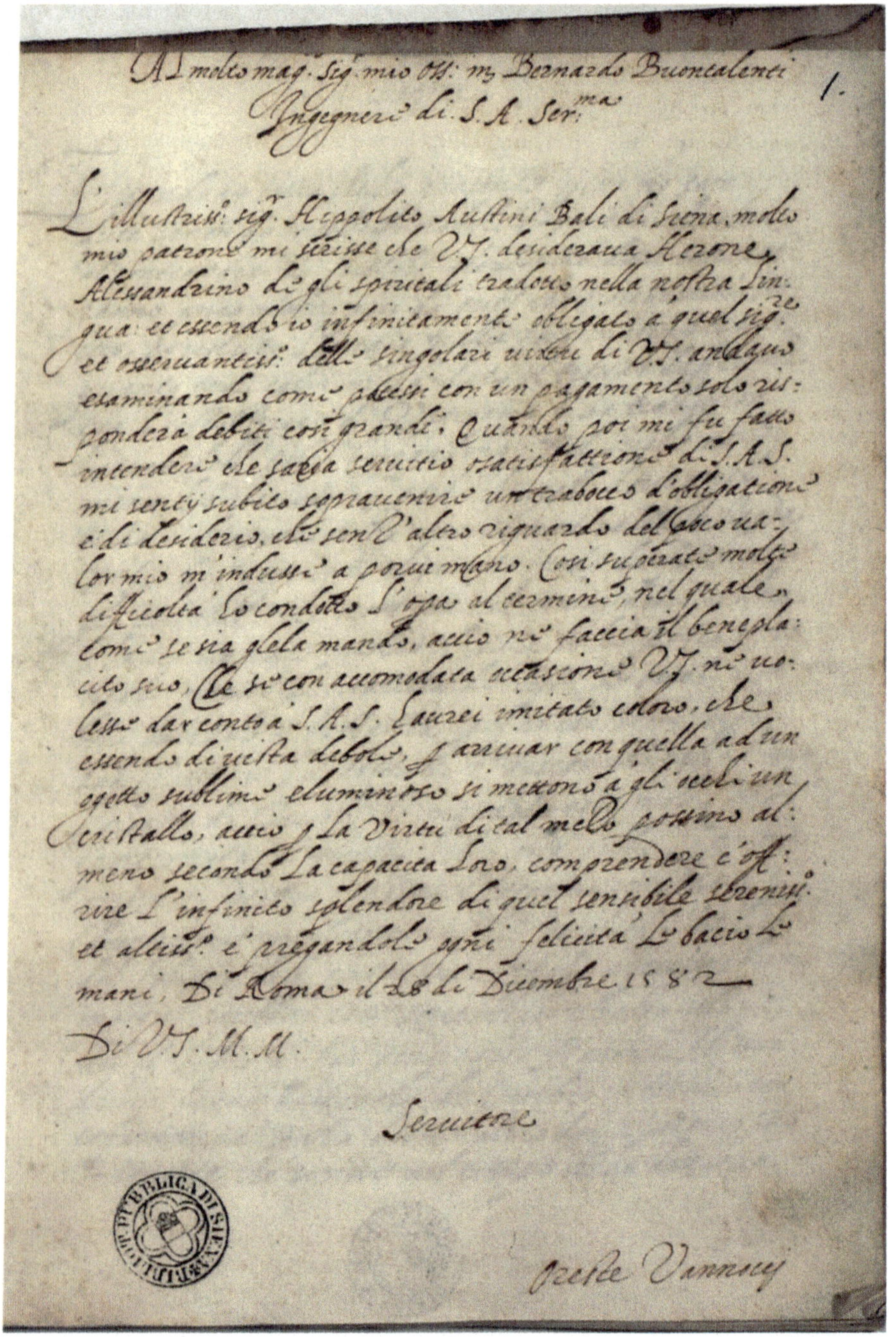

Al molto mag.co Sig.r mio oss.mo m. Bernardo Buontalenti
Ingegnero di S.A. Ser.ma

1.

L'Illustriss.o Sig.r Hippolito Agostini Balì di Siena, molto
mio patrone mi scrisse che V.S. desiderava Herone
Alessandrino de gli spiritali tradotto nella nostra lin-
gua: et essendo io infinitamente obligato a quel sig.re
et osservantiss.o delle singolari virtù di V.S. andavo
esaminando come potessi con un pagamento solo ris-
pondere a debiti così grandi. Quando poi mi fu fatto
intendere che sarà servitio o satisfattione di S.A.S.
mi sentij subito sopravenire un trabocco d'obligatione
e di desiderio, che senz'altro riguardo del poco va-
lor mio mi indusse a porvi mano. Così superato molte
difficoltà ho condotto l'opra al termine, nel quale
come se sia gliela mando, acciò ne faccia il benepla-
cito suo, che se con accomodata occasione V.S. ne vo-
lesse dar conto a S.A.S. haverei imitato coloro, che
essendo di vista debole, per arrivar con quella ad un
oggetto sublime e luminoso si mettono a gli occhi un
cristallo, acciò per la virtù di tal mezzo possino al-
meno secondo la capacità loro comprendere e ser-
vire l'infinito splendore di quel sensibile sereniss.
et altiss.o e pregandoli ogni felicità le bacio le
mani, Di Roma il 28 di Dicembre 1582

Di V.S. M.M.

Servitore

Oreste Vannocci

Fig. 5 Dedication letter of Oreste Vannocci Biringucci to Bernardo Buontalenti, December 28, 1582. Hero, Vannocci, 1582, Biblioteca degli Intronati, Siena, Ms. L. VI. 44, 1v

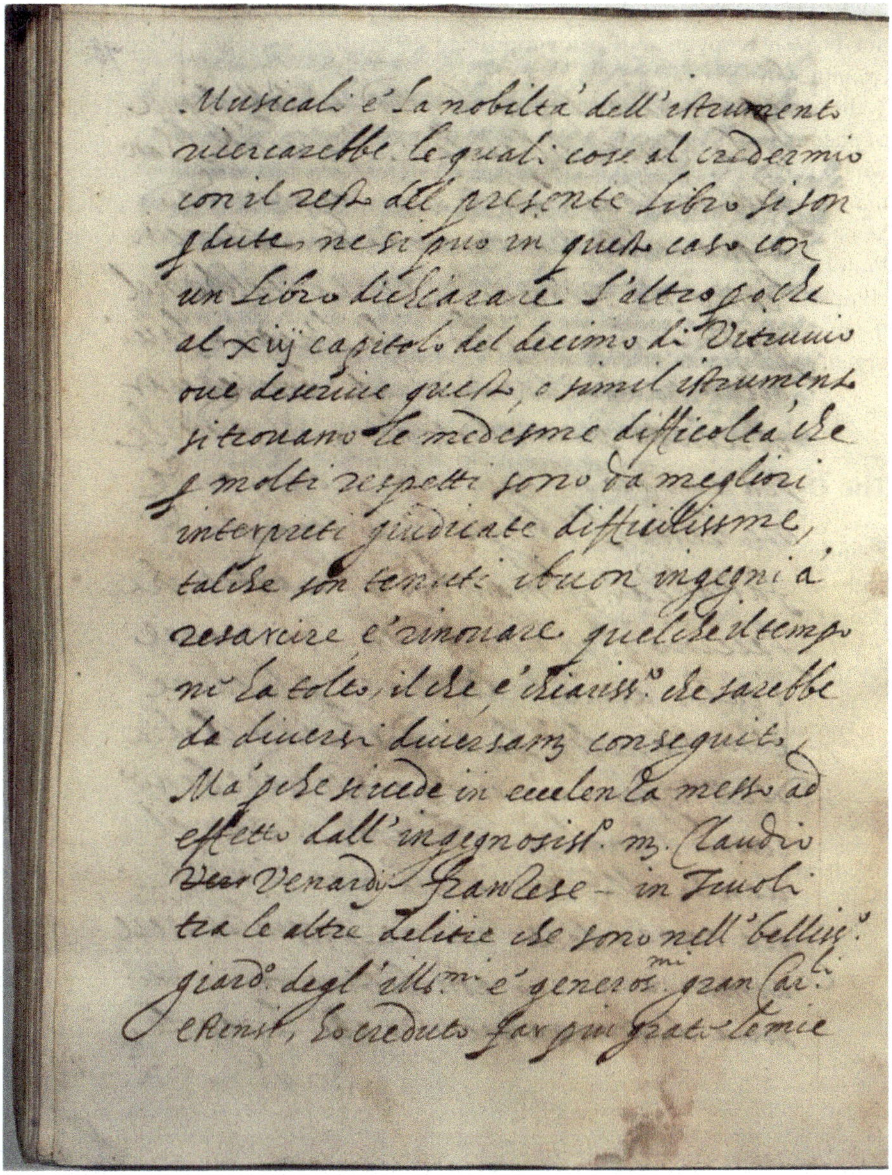

Fig. 6 On this folio of Biringucci's text, he quotes the name of the engineer Claudio Venardo. Hero, Vannocci, 1582, Biblioteca degli Intronati, Siena, Ms. L. VI. 44, 70v

1983: 235).[4] If this was the case, Buontalenti and his crew must have had an interest not only in copying but also improving its hydraulic organs compared to the examples of Hero, as well as of Tivoli.

In contrast to the textual tradition related to pneumatics, the practical activities of pneumatic engineers show an impressive continuity throughout the Middle Ages, which is particularly well-documented in reference to the Latin West starting from the eleventh century on (Butzer and Lohrmann 1993). Since the translation of Vannocci Biringucci deeply influenced the construction plans of the garden of Pratolino, and since Hero's description of the hydraulic organ is corrupted, a comparison is made in the present work between the sources at disposal concerning the hydraulic organ built at *Mount Parnassus* in the garden of Pratolino and Hero's organ (see also Valleriani 2010b).[5]

The Organ of the Garden of Pratolino

Heinrich Schickhardt, the court architect and engineer of the duke of Württemberg with whom he traveled to Rome in 1599–1600, explains the workings of the pneumatic music machine at Pratolino as follows:

> This work plays only two pieces [compositions]. One, when the wheel has completed a half-turn, the other when it is found in the other half. The notes that the organ lets out are the Principal and an octave [higher]. [. . .] 35 wood blocks [. . .]. (Schickhardt 1600: b–53r)[6]

[4] Montaigne compares Pratolino and Tivoli as follows: "Pratolino est fait justement à l'envie de ce lieu. En richesse et beauté de grottes, Florence surpasse infiniment; en abondance d'eau, Ferrare [Villa d'Este of the cardinal of Ferrara]; en diversité de jeux et de mouvements plaisants tirés de l'eau, ils sont pareils, si le Florentin n'a quelque peu plus de mignardise en la disposition et ordre de tout le corps du lieu; Ferrare en statue antiques, et en palais, Florence; en assiette du lieu, beauté du prospect infiniment Ferrare [. . .]". He also describes Tivoli's organ briefly (233): "La musique des orgues, qui est une vraie musique et d'orgues naturelles, sonnant toujours toutefois une même chose, se fait par le moyen de l'eau qui tombe avec grande violence dans une cave ronde, voûtée, et agite l'air qui y est, et le contraint de gagner pour sortir les tuyaux des orgues et lui fournir du vent. Une autre eau, poussant une roue à tout certaines dents, fait battre par certain ordre le clavier des orgues; on y oit aussi le son de trompettes contrefait. Ailleurs, on oit le chant des oiseaux, qui sont des petites flûtes de bronze qu'on voit aux régales, et rendent le son pareil à ces petits pots de terre pleins d'eau que les petits enfants soufflent par le bec, cela par artifice pareil aux orgues [. . .]". And he mentions Pratolino's organ beneath the mansion as being its biggest wonder (175).

[5] A further source exists that describes the organ of the garden of Tivoli; Giambattista della Porta included such a description in his book on pneumatics first published in Latin in 1601 and then in its own Italian translation (Della Porta 1606: 88–89). This source has not been given a translation here because Della Porta does not describe the pneumatic apparatus but merely the mechanics of the organ. In particular, his description clearly shows that he did not understand the pneumatic functioning of the instrument.

[6] "Es schlecht die Werckh nur zwai Stuckh. Das ein, wan die Roller halb herum und die ander in dem uberigen halben Tail. Die Noten, so das Clavier trukhen, ein Princibal und ein Octaf".

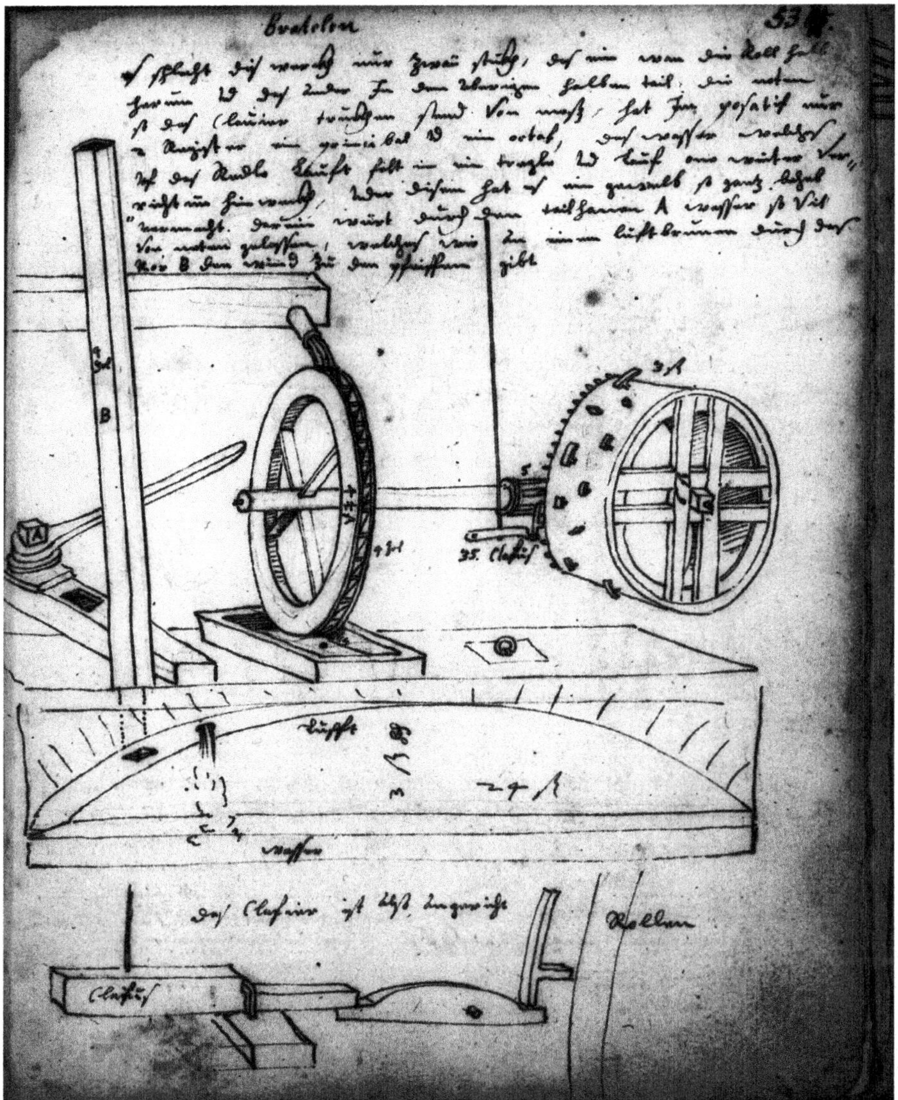

Fig. 7 Sketch of the mechanical apparatus of the hydraulic organ built in the *Mount Parnassus* at the garden of Pratolino. Pen Drawing, Heinrich Schickhardt, 1600. Cod. Hist. 4, ms. 148, Württembergische Landesbibliothek, Stuttgart, b-53r

The melody was very short, being composed of only 17 notes, and was repeated an octave higher. Adding a probable anacrusis, we arrive at the number of 35 blocks counted by Schickhardt on the organ roller (Fig. 7). Once the melody and registers had been determined, the Renaissance hydraulic machine had to have a pneumatic apparatus that could play them with the desired intensity. After this was assured, however, there was no need for any other mechanism that would allow different compositions, possibly with different intensity or volume, to be performed. It is probably for this reason that none of the known Renaissance hydraulic organs was equipped with pumps that could further increase the pressure in the air system.

An analysis of Schickhardt's notes and drawings shows that the pneumatic apparatus of the Pratolino water organ is fairly simple. The hydraulic systems that activated the pneumatic and mechanical apparatuses seem to be separate, even if there was probably a key in common to control the whole system. When this first key was opened, the mechanical apparatus started up immediately: the water wheel set the transmission mechanisms into motion; they then turned the roller with the blocks. This opened and closed the organ pipes, which were located in the mount and whose extremities were visible from outside of it (Fig. 4), through the mechanism shown in Schickhardt's drawing at the bottom of the page (Fig. 7). The pneumatic apparatus was started up by another key, visible to the left of the water wheel. Opening this allowed water to flow into the basin drawn at the bottom. Next, the water that had been introduced forced the air to escape through the tube drawn by Schickhardt near the key that started up the pneumatic apparatus. Finally, this tube brought the air to the organ pipes. Schickhardt describes the pneumatic apparatus as follows:

> The water that runs over the wheel arrives in a basin and flows away without other mechanisms. Below it, there is a very solid and closed vault. Opening key A, a great amount of water is introduced into this vault, as much water as is needed. The water then provides wind for the pipes through tube B as in a fountain of air. (Schickhardt 1600, b–53r)[7]

It follows from this that the pneumatic apparatus of Pratolino's organ is, in fact, made up of a cylinder and a reversed vessel which, transferred into the real dimensions observed by Schickhardt, become a small, probably cylindrical basin and a 'very solid' vault that seals it off perfectly. The fact that a reversed vessel is used represents the real peculiarity of this organ if compared with the descriptions offered in the ancient sources, notably in Hero's *Pneumatics*. All ancient descriptions of pneumatic apparatuses, and in particular of those intended to perform music, describe pneumatic cylinders, which are closed by flat surfaces. As will be shown in the following, this new early modern solution was possibly developed to deal with a peculiar problem of musical pneumatic instruments.

[7] "Das Wasser, welches uf das Redle lauft, felt in ein Tregle und lauf/t/ one weiter Verrichtun/g/ hinweckh. Under disem hat es ein Gewelb, so ganz beheb vermacht. Darein würt durch den Tailhanen A Wasser sovil vonneten gelassen. welches wie an einem Luftbrunen durch das Ror B den Wind zu den Pfeifen gib".

The Technical Solution of the Choking Vessel

Pratolino's organ did not envision the use of a mechanism for introducing air into the system in order to raise the pressure, but its dimensions were such—we need only think, for example, of the distance between the pneumatic apparatus at the bottom of the artificial mount and the pipes set above it—that it very likely required very high pressure of emission for it to work well. In the case of Pratolino, this condition was satisfied through the construction of a hydraulic system that could introduce large quantities of water into the system in a short time, for example, by taking advantage of the great height of the fall, and by having an intake tube with a wide perimeter, which was slightly narrowed at the end, that is, at the point where the tube is inserted into the pneumatic device. If we imagine, then, a very rapid process of increase in the water level, the hypothesis can be formulated that air turbulences in the upper part of the basin took place, and that Buontalenti intended to eliminate their effects with the use of the choking vessel. In fact, in phases in which an elevated air pressure is reached in the cylinder, the degree of viscosity of the air increases, and some turbulence in its escape becomes possible. In a hydraulic organ, this consequence would clearly create sudden changes in the intensity of the sound emitted, and this would make for a rather unpleasant, or even completely mistaken, musical execution. It is therefore possible to postulate that the choking vessel was an additional device for stabilizing the flow of the escape of air when pressure was high, as modern fluid mechanics is able to demonstrate.

If we follow this interpretation, the choking vessel becomes the device of most interest, from the point of view of both the history of technology and the history of science. The specific use of such a vessel in Pratolino should, actually, be traced back to a physical intuition supported by everyday experience in the use of the funnel, above all in the presence of high-viscosity substances. The accumulated experimentation and experience of technicians led to the establishment of the use of the technology of the choking vessel for rapid pneumatic processes, or in any case, processes that occur in conditions of high air pressure (Valleriani 2010b).

From the point of view of later theoretical developments, it is clear that the qualitative knowledge that led Renaissance technicians and engineers to use the choking vessel for the purpose of stabilizing air flow in cases of high air pressure could not but represent a challenge to theoretical understanding. This challenge would not begin to be solved until the eighteenth century.

The Control of Air Turbulences in the Organ of Tivoli

On the basis of the hitherto unknown description of the organ of Tivoli (Appendix 2), it is possible to infer that the technology of the choking vessel was already in use before the organ of Pratolino was built.

First of all, it is clear from Audebert's description that the cylindric apparatus of the organ of Tivoli had the shape of a "cavern [...] vaulted inside." Moreover, the pneumatic cylinder also displays a technical solution that might have been applied to obtain the same effect as the curved vessel in Pratolino.

Vannocci Biringucci describes the organ with a richness of detail. Particularly impressive is his description of the apparatus that joined the small blocks of the wheel to the keys, which in turn opened the holes from inside the pneumatic cylinder and toward the pipes. Such holes and openings were particularly elaborated. The openings were indeed shaped in the form of cones of considerable dimensions (Vannocci 73v). This means that the effects of the possible turbulences of the air exiting toward the pipes could be notably decreased thanks to these funnel-shaped constructions at the openings of the vault. The solution of the conical openings was added to that of the vaulted vessel probably because of the impressive number of pipes, 22, with which the organ of Tivoli was provided.

Conclusion

From the analysis of the technology employed in building both of the organs considered in this work, the organ of the Mount Parnassus in Pratolino and the organ of Tivoli, it turns out, first, that the organ of the Mount Parnassus in Pratolino was not a unique piece of technology. The analysis of the technology of such an organ is relevant in showing that engineers, who, at that time, were directly inspired by the reading of Hero's Pneumatics, were nevertheless aware of the power and efficiency of their own technology, since they did not literally follow him when they knew they could build better devices. The analysis of the technology of the organ of Tivoli, in turn, clearly shows that such technology was the result of the continuous efforts of hydraulic engineers over centuries rather than the result of their re-interpretation and appropriation of ancient technology.

These aspects offer a clear indication of the different transmission paths followed in history by practical knowledge on one side and codified theoretical and practical knowledge on the other. In particular, it is evident that processes of accumulating practical knowledge, embedded in the technological artifacts and transmitted from one generation to the other by means of learning-by-doing, experienced an impressive continuity and were particularly efficient methods up to the early modern period. Hence, the remark of the garden's designer Ligorio that no one should build hydraulic machines without having studied Hero may be, after all, primarily assessed in terms of a coat of education enhancing the prestige of engineers and as an indicator for the growing interest in the codification of technological knowledge for which Hero served as an important model.

Ultimately, it becomes apparent that not only did scientific knowledge have an impact on gardens but also vice versa. The growing socio-cultural relevance of early modern gardens transformed them into knowledge laboratories where codified theoretical and practical ancient knowledge was appropriated and unified with the working tradition of early modern engineers. Gardens and their building sites, moreover, became the place where different skills, traditions of practical knowledge and experience met and merged with each other, so that technology underwent further intense development, while transforming itself into a mathematical discipline on the basis of ancient models and increasing the social status of its practitioners at the same time.

Appendix 1

Description of the fountain of the organ by Nicolas Audebert (1576–1577)

334v

Passant à coste & derrière ceste muraille on entre par un petit portail en une place carree, clause de petits murs, à un bout de laquelle est *la fontaine des Orgues* qui surpasse du tout les aultres precedentes, non pour les enrichissement d'icelle, mais pour l'artifice et secrets ingenieux qui y sont. Ceste fontaine est plus rustique que bien ornee, ayant le bas comme un rocher, et au pied d'iceluy une fosse faicte en forme de troys demyz cercles & plus hault y a une statue moderne qui n'est que de pierre & en facon d'un Terme, qui est pour representer Nature, et aussy se nomme aultrement *la fontaine de la Nature*, le hault de laquelle est une teste de femme, sans aulcuns bras, laquelle a le corps tout couuert de grosses mammel	Approaching and passing this wall, you enter a small portal leading to a rectangular square, enclosed by low walls, at the end of which is situated the *Fountain of the Organ*. This fountain outmatches all its predecessors, not because of its decorative splendor but through its technical refinement and the ingenious secrets contained therein. This fountain is rustic rather than elaborately decorated. The lower part is built in the shape of a rock. At its base, there is a water basin in the shape of three half circles. Further up stands a modern statue, simply made of stone and shaped as a herm. This statue is meant to represent nature—which is why they also call the place *Fountain of Nature*—topped by a female head with no arms, and a body entirely covered with big breasts

335r

les tombantes l'une sur l'aultre, et le reste au lieu de jambes est un pillier quarré qui vient en pointe, & diminuant contrebas. Derriere ceste statue passent 22 tuyaulx d'Orgue qui vont au hault respondre dedans une petite fenestre voustee. Ces orgues sans ayde de personne sonnent une chanson de musique auec toutes ses partyes, non moins bien & melodieusement auec ses mesures et fredons, que pourroit faire le plus excellent joueur : Mais pour rendre cela plus admirable on desnye a la plus grand part de ceux qui y vont de leur faire veoir l'artifice, toutesfoys J'euz bien ceste faueur que tout me fut communiqué & allay aux lieux plus secretz ou Je remarquay curieusement tous les mouuements & moyens de ceste ingenieuse & admirable inuention, que je descriray icy selon ce que Jay veu. Dessoubs ceste forme de rocher & structure de fontaine il y a une petite chambre soubz terre comme une cauerne laquelle est voustee par dedans. Et ny a ouuerture alcune sinon au hault, a un coin de la vouste qui est persee par dessus n'y ayant lieu que pour passer un homme & pour y entrer il faut leuer une grosse	hanging one upon the other' the lower part has the shape of a pointed pillar instead of legs, declining in diameter from top to bottom. Behind this statue are 22 organ pipes, made to fit into a little vaulted niche above. Without any human intervention, the organ plays a musical piece with everything that comes with it, with beat and trills, by no means any less pleasing and melodic as the most gifted musician would perform. However, to make the spectacle appear even more admirable, most visitors are denied view of the underlying mechanical construction. In contrast, I have been granted the honour of a detailed explanation and was allowed to access the most secret places, where I curiously observed all movements and means of this adorable invention. This is what I am going to describe here as I saw it. Beneath the rock-shaped structure of the fountain is a little subterranean chamber similar to a cavern, vaulted inside. There is no aperture whatsoever except above in one corner of the vault which is built in such a way that there is just enough space to let one man pass. In order to enter you have to lift a large

(continued)

335v

pierre quarree qui ferme ceste ouuerture si iustement que l'air ne peult y entrer ny en sortir aulcunement. Par la nous deualasmes auec une eschelle en ceste chambrette; a un des bouts de laquelle y a un gros canal de plomb qui vient de hault; & a son isseüe des la vouste Jettant l'eau en grande abondance et de force, laquelle est cause que se versant en la chambre, ceste agitation vehemente engendre du vent qui la dedans demeure enclaus et bien enserré quand on a bouché le lieu par ou on ÿ entre. Dauantage l'eau qui est tombee ne trouuant isseue pour s'escouler, que un canal de beaucoup moindre grosseur que celuy du hault, & partant n'est assez suffisant pour mettre hors l'abondance d'eau qui cherche passage pour sortir : il aduient qu'une bonne partie d'icelle demeure en ceste chambrette, & peu a peu se haulsant & augmentant rend le vent plus enserré. Qui plus est le canal du hault par lequel tombe l'eau respond sur une roue de fer qui ayant de petites ailles et platines en la facon d'une roue de moulin a eau, & venant a tourner, elle augmente encores le vent par son mouuement. Ceste roue tournant, en faict aussy tourner une aultre qui est proche, et de fer comme la precedente, estant aussy de pareille haulteur qui reuient a quattre	rectangular stone sealing this aperture so closely that absolutely no air can get in or out. Through it, via a ladder, we descended into this little chamber. At one side, overhead, there is a thick tube of plumb. Water spouts from the vault through its outlet abundantly and with great force. And while the water is pouring into this chamber, this violent movement generates wind which stays within, enclosed and fairly compressed when the aperture is sealed. The increasing amount of water coming in cannot find any exit except through a tube much thinner than the one above, and the outward flow insufficiently drains the abundance of water trying to blaze a trail outside. Therefore, it remains in large parts in this little chamber, little by little rising more and more, pressing the wind ever more strongly. What is more, the upper conduit aims at an iron wheel that has little wings and leaves in the manner of a hydraulic millwheel; as soon as the latter starts to turn, its movement continues to increase the wind. Once the wheel is turning, it subsequently puts another one next to it into motion. As with the first one, it is of iron and of similar height, i.e., some four

336r

ou cinq pieds de diametre; mais ceste cy est differente en ce qu'elle est fort large, asauoir d'enuiron troys pieds, qui est comme une longue lame de fer qui circuit et couure la roue : au dessus de laquelle y a grande quantité de petites pieces de fer esleuees & soudees pour demeurer droictes, dont les unes sont plus grandes, plus petites, ou moyennes; seruant pour les mesures, poses, longues, breues, demy breues, crocheues, souspirs, et aultres choses requises & necessaires en la musique. Au hault de la vouste ÿ a 22 petits trous ausquels respondent chascun un tuyau d'orgue : lequel morceau pertuis demeure tousJours bouché par le moyen d'un petit morceau de fer blanc appliqué contre chascun trou & tuyau d'orgue : lequel morceau de fer blanc n'est plus grand qu'une piece d'argent, & couppé en quarré, estant un peu caue par le milieu affin que la force du vent enclaus e presse dauantage et tienne le trou mieux bouché. Dauantage chascune piece de fer blanc est	or five feet in diameter. But this one is different insofar as it is much larger in width, i.e., some three feet; and there is something similar to a long iron blade surrounding and covering the wheel. At its surface, a large number small iron pieces stick up, brazed in such a manner that they stay upright, some of which are bigger, some smaller and some medium-sized. They serve as measures, indicate pauses, long, short, half, eighth notes, crotchet rests and other things required for music making. Above, in the vault, there are 22 little holes, each connected to an organ pipe. Every such aperture remains always closed by means of a little piece of tin foil attached to every organ pipe. Such a piece of tin foil isn't any bigger than a silver coin, square-cut and somewhat curved in the middle in order that the power of the wind closes and presses a fortiori and keeps the hole better sealed. Furthermore, every piece of tin foil is brazed to the end of a long iron bar, the other

(continued)

336r

soudee au bout d'une longue verge de fer laquelle de l'aultre bout vient respondre sur ceste roue large, a laquelle il ny a qu'un doigt a dire que ceste queüe de fer ne touche, ayant un coste contre bas et l'aultre contre mont par le moyen d'une aultre verge de fer attachee au hault de la vouste qui tient ceste plus longue verge	end of which aims at this big wheel, just the width of a finger far from touching it. One end points downward, the other upward by means of another iron bar attached to the vault above, keeping this longer suspended bar

336v

suspendue comme en equilibre, ayant toutesfoys un costé quelque plus poisant que l'aultre pour tirer tousjours contre bas vers la roue, & l'aultre ou est attacheé la piece de fer blanc, pour tenir tousJours les pertuis bouchez. Tout cela estant ainsy disposé, et la premiere roue venant à tourner elle faict par mesme moyen tourner sa prochaine laquelle est cause de toute la musique, car ayant (comme a este dict cy dessus) ces mesures, breues, longue, & aultres partyes de musique; En ce tour se rencontre quelqu'une de ces mesures (qui sont esleuees sur la roue plus d'un bon doigt) contre le bout d'une des queües de fer suspendues, laquelle piece et mesure la faict un peu leuer, & ainsy tant qu'elle met à passer le bout du hault se tient baisse & le pertuis demeure ouuert, dedans lequel le vent entre & s'entonne, faisant sonner le tuyau d'orgue : & des aultres en semblable allant ainsy consecutiuement en l'ordre que chascune piece de fer est appliquee sur la roue : par ce moyen le son du tuyau dure tant qu'il y entre du vent, c'est asauoir plus ou moins selon la piece qui touche le bout de la verge de fer, soit une breue ou une longue : Qui est en somme tout le secret & artifice de ceste fontaine.	in a sort of balance. One end, however, outweighs the other a little in order to create a steady pull down toward the wheel and so that the other one where the piece of tin foil is attached always keeps the holes closed. As everything is arranged in this manner, the first wheel starting to turn puts the one next to it into motion which generates the whole music because it has (as mentioned above) these measures, short ones, long ones and other musical components. When one of the measures (sticking out of the wheel more than a finger) hits the end of one of these suspended iron bars during the turn of the wheel, this measure piece lifts the bar a little and, as long as it passes, the upper end remains lowered and the hole open. Therein the wind enters and enriches with sound waves because it makes the organ pipe sound. The same applies to the other pipes continuously, according to the order of the measures on the wheel. Thus, every organ pipe sounds as long as the wind enters therein, that is to say, for a longer or shorter while depending on the long or short size of the piece touching the end of the iron bar. This is, in summary, all the secret and technical refinement of this fountain.

337r

Pendant que ceste musique dure, il ÿ a un *Rossignol* qui ne cesse de chanter par le moyen d'un tuyau faict en subleau lequel par un bout recoit du vent de ceste cauerne, & par l'aultre trempe en un petit vase plein d'eau. Ces orgues ne sonnent qu'une seule chançon de musique, qui est a cinq partyes, fort longue, et bien musicale : mais celuy qui les feit sauoit le moyen de leur faire sonner toutes chancons qu'il luy plaisoit, accommodant premierement toutes les verges de fer suspendues pour les	While this music is playing, there is also a *Nightingale* singing ceaselessly by means of an underwater [*faict en subleau*] tube receiving wind from one end of this cavern, its other end dipping into a little pot full of water. These organ pipes play only a single piece of music, which is for five voices, quite long and very musical. Its creator, however, knew how to make them play all the songs that he pleased by arranging at first all the suspended iron bars in order to make them communicate with the

(continued)

337r	
faire rapporter sur telle mesure qu'il estoit besoing. & ne veux aussy oublier a remarquer que celuy qui feit toutes ces fontaines, estois un Francoys, lequel avoit lors un nepueu auec luy qui y est demeuré apres la mort de son oncle, & gouuerne encore toutes ces fontaines. La musique estant finye, on tire une petite corde qui debousche quelques conduictz, et lors commence en un instant a sortir tout au tour de la roche une si grande quantité d'eau (qui estoit enclause et serree dedans ceste cauerne & es enuirons) qu'il semble vrayement d'une tempeste et orage pour le grand bruit qui se	necessary measures. Item, I don't want to forget to mention that the creator of all these fountains was a Frenchmen who back then brought a nephew with him who stayed on after the death of his uncle and still superintends all these fountains. When the music has finished, you pull a little cord that opens a number of conduits whereupon such a large amount of water (enclosed and compressed in this cavern and its surroundings) exits all around the rock at once. This, indeed, resembles a tempestuous storm because of the deafening sound
337v	
faict par la force & impetuosité dont elle est poussee, laquelle sourdant du bas de la terre, s'eslance en l'air de la haulteur de deux lances : puis sur la fin elle vient a se rappaiser & cousler tout doulcement selon son ordinaire; ce qu'ilz nomment *le Deluge*.	created through the aggressive power that pushes the water. The latter pours out from underground and rises up into the air two lances high. Toward the end it calms down, flowing most gently as usual. This is what they call *the Deluge*.
& durant ce deluge ỹ a au milieu de ceste eau un *Triton* de pierre qui tient une trombe laquelle il faict raisonner d'un son enroué, qui continue tant que la tempeste & orage durent : Et tout ainsy qu'il commence premierement en murmurant, et peu a peu s'augmente le son tant qu'il soit en sa force; Aussy apres il s'appaise peu à peu, & sur la fin ne faict plus qu'un gargoullement et murmure qui semble venir de fort loing, & que le Triton s'en soit allé.	And during the deluge there is also a *Triton* of stone in the middle of this water holding a horn, making it sound in a hoarse tone for as long as this tempestuous storm lasts. And this happens in such a way that the tone starts murmuring at first and then increases gradually in volume. Also, afterward it calms down bit by bit until, toward the end, it makes nothing but a gargling murmure appearing to come from very far away as if the Triton had left the scene.

Audebert, Nicolas. 1576/77. Le Palais, Jardin, & Fontaines, de Tyuoly. Excerpt from id., Voyage d'Italie, Landsdowne Ms. 720, fols. 334v–337v. London: British Museum. Transcribed in Ronald William Lightbown. 1964. Nicolas Audebert and the Villa d'Este. Journal of the Warburg and Courtauld Institutes 27: 164–190. See critical edition and text normalization in Nicolas Audebert. 1981. Voyage d'Italie. Ed. Adalberto Olivero, II, 81–84. Rome: Luciano Lucarini. See partial translation into English in Dernie 1996, 64–68. This is a new and complete translation by Simone M. Kaiser

Appendix 2

Description of the fountain of the organ by Oreste Vannoci Biringucci (1582)

70v

Ma' perche si vede in eccelenza messo ad effetto dall'ingegnosiss.o m. Claudio Verr Venardij franzese in Tivoli tra le altre delitie che sono nell' belliss.o giard.o degl' ill.mi e' generos.mi gran Car.li estensi, ho creduto far piu grate le mie	Because one can observe this in its best shape in the construction realized by the very skillful Frenchman Claudio Verr. Venardij at Tivoli among the other tasty things that are in the beautiful garden of the Illustrious and Generous great Cardinals Estensi, I think to improve my

71r

fatiche a' quel nobile intelletto che se vedra con l'aggiungervi una descrittione di quel'organo, il qual per forza d'acqua senza servitu humana rende armonia suaviss.a, Ma per non haver ad interromper la descrittion dell'organo è necessario dar notitia, o, rinovarla almeno d'un effetto maraviglioso dell'acqua che in quest'istrumento fa mirabil operatione Pero che introducendo l' acqua per un canale conduce seco insieme unito il vento che poi separati l'un dall'altro vanno a' far diversi ministerij che concorgano a' far sonare l'organo, e, di questo senz'altre prove basterà dirne l'esperienza, quando dunque si conduce l'acqua per un canale elevato cioe drizzato a' piombo, o poco inclinato	respect to that noble intellect [Hero of Alexandria] by adding a description of that organ, which, powered by water and without human work, emits a very pleasant harmony. Thus, in order to proceed with the description of the organ without interrupting it, it is convenient to first speak about a marvelous effect of the water, an operation produced in this instrument. When the water is introduced through a channel, it carries wind together with it. These then are separated from each other and accomplish different functions, both useful for allowing the organ to play. Without many proofs, it will be sufficient to describe how it works. The water moves within a channel which is perpendicularly positioned or a little inclined.

71v

che sia al quanto piu largo, che non ricercarebbe la quantità dell'acqua cioe capace di piu acqua che non è quella che vi passa, purche sia tanta che entrata nella' bocca del condotto con certo dilatamente che fa' ordenariamente si venghi a congiugnere, e' con la larghezza sua quasi ad empire il condotto, in quel ingresso facendo un certo vortice, o giramento con il flusso continua rapisce seco gran quantità d'aria e' la porta e spinge dentro, mentre d'ogn'intorno s'affatica d'entrare e' con l'impeto procede verso il fondo del canale e' cotal aria non puo poi tornar indietro per che ne rincontra sempre dell'altra che insieme con l'acqua e' tra le parti dell' acqua se ne siende a' basso, Hor di questo canale propongo la presente figura	The channel is a little bigger than the quantity of water would require, that is, it could contain a greater quantity of water. Such quantity has to be enough to create a sort of vortex or circular motion when it enters the mouth of the channel and, as usual, it dilates a little bit so that the dimension of the water, joint together, almost fills the channel. With its continuous stream, it captures a great quantity of air and carries and pushes it into [the instrument] while it enters and, with impetus, proceeds toward the end of the channel. Such air cannot move backward, because more air, together with water, continuously comes in. In this way, it proceeds to the bottom. Of this channel I show the present figure.

(continued)

72r	
nella quale l'acqua CD si vede entrar d'ogn'intorno nel condotto AB ma' non è in tanta copia che lo venghi ad empir totalmente et alla bocca A s'accomoda un coperchio che a' benefitio beneplacito la chiuda che non v'entri l'acqua, et è segnato F, è chiaro dunque che gli sbalzi e' disunioni dell'acqua che d'ogn'intorno concorgono rapiscono e' per il AB introducan l'aria dove fa bisogno, la qual notitia non pare ci servira all'istrumento che siam per descrivere ma' ancora ad infinite altre cose che non è a' proposito il raccontarle, sara dunque tempo con il modo che di Herone, di passare alla promessa descrittion del organo di Tivoli. Organo tiburtino LXXVIII Intendiamo dimostrare come senza fatica humana si possin far sonare gli organi con vario contento	in which one sees the water CD while it enters the channel AB but not in such a quantity as to fill it [the channel] completely. At the mouth A, there is a lid, marked F. that can be closed to stop the water stream from entering. It is therefore clear that the jumps and disjunctions of the water all around capture and push the air into the channel AB sufficiently. The description of this contrivance will not be useful for the description of the organ but for innumerable other things, which however should not be described here. It is therefore the moment to move on with the promised description of the organ of Tivoli according to the method of Hero. The organ of Tivoli LXXVIII We want to demonstrate how organs can be made to play various performances but without human assistance.

72v	
Sia il basamento AB chiuso d'ogn'intorno e' di buona capacita nel qual entri il condotto dell'acqua CD accomodato nel modo che s'è detto di sopra che sia dalla parte D alquanto piu stretto e' l'acqua E non sia tanta che l'empi ma solo ne occupi per esempio i due terzi o la meta, e' sia sopra al CD accomodato il coperchio F di sorte che tirando la cordicella FG si venghi ad inalzare e' dar il passo libero all' acqua, sieno poi le canne dell' organo HI disposte e' proportionate con armonica ragione nel' modo che si fanno ordenariamente, ma dalla parte H sieno ciascuna d' esse congiunte con le fistole HL in modo che sien tante fistole quante canne, Le quali fistole entrino nel basamento AB saldate insieme con lui onde portino	Suppose the base AB is sealed all around and is of good dimensions. The water channel CD enters it as described above, that is, a little narrower at the end D. The quantity of water E does not fill it completely but only to two thirds or one half. At the top of CD, the lid F is positioned so that, when the rope FG is pulled, it moves up and lets the water stream. Then, suppose the organ pipes are located and conceived according to the rules of harmony in the way they are usually built. At the side H, let each of them be joined together with the small poles HL so that there are as many poles as there are pipes. The poles enter the base AB, to which they are welded. They carry

73r	
il vento alle canne, Ma acciò non suonino tutte in un tratto e' non vada il vento se non a' quelle che vogliamo si ponghino sotto alle fistole HL i tasti LN che le aprino e' serrino secondo il girar della rota OP nel dorso della quale son le coste o rilievi che apreno i tasti, et essa gira perche ha dentro di se la rota dentata ER nel medesimo asse che è girata' dal rocchetto S che ha i denti inseriti, et interposti	the air to the pipes. To avoid activating them all together, and ensuring that the air moves only toward those we want to perform, the keys LN are applied under the poles so that they open and close the pipes according to the rotation of the wheel OP. On this wheel, there are small wood blocks that open the keys. This wheel turns because it also contains the gear wheel ER along the same axis, which is turned

(continued)

73r	
con quelli della rota RE il qual rocchetto gira per il moto della rota a' cassette T che vien mossa dall' acqua che cade dall'CD la qual insieme conduce il vento che fa successivamente sonar le canne secondo l'aprir de tasti, L'acqua poi sen esce per il diabete XV che si fa di tal grandezza che non capisca tutta l'acqua che entra per il CD accio il vento non se ne possi uscire, e' per il multiplicar dell'acqua	by the sprocket S. Its gears are alternatively inserted into those of the wheel RE. The sprocket turns because of the motion of the buckets on water wheel T, which is in turn moved by the water falling from CD. This water carries the air that then lets the pipe perform according to the opening of the keys. The water then exits through the funnel XV built to such a size that it does not contain all the water so that the wind cannot exit. Moreover it increases the water
73v	
nel basamento diventa maggiore Ma' accio l'ordine el modo di quei tasti s'intenda meglio ne descriveremo separatamente uno in maggiore forma a' similitudine del quale s'intenderan tutti gli altri sia dunque una canna dell' organo HI, e' la sua fistola correspondente HL che di sotto al luogo L termini in un cono concavo a' guisa di campanello e' sia di rame ben polito con un buso piccolo nel somo della concavità che riesca nella fistola et annodisi il tasto o vero stile di ferro AB che stia e' si volga sopra al CD nel luogo D il qual CD stia fermo nell'tetto del basamento al luogo CE' nella somita segnata' A sia affisso ma non fermo un cono solido di rame ben polito segnato K che serri e' se ne stia per l'ordinario dentro al cono concavo L in modo	in the base. In order to better understand the construction and the use of those keys, we describe one of them with more detail. According to this, it is then possible to understand all the other keys which are similar. Suppose HI is a pipe of the organ and HL its corresponding pole that, under L, ends with a shape of a concave cone like a ring. Let it be of copper, smooth and with a small hole at the top of the concavity that exits into the pole. The key or iron bar is tied so that it stands over CD at the place D. CD has to be over the seal of the base at the place CE without moving, and at the top of A, a mobile solid cone of smooth copper is joined and marked as K. This closes the concave cone L and is within the same L. In this way,
74r	
che lo sigilli e non vi lasci entrar il vento et accio vi stia piu fermo e' non se ne possi allontanre se non per forza della rota vi si pone il filo di rame avvolto FG che congiunga al tetto del basamento il tasto e' quando pur dalla rota il cono K è allontanato dall' L girata la rota è restato il tasto libero il filo FG tornando nell'esser suo ritira il cono K nell' L e serra l'esito al vento Ma l'estremità del tasto segnata B sia incorvata e' termini nella penna over paletta E alla qual si deven rincontrar le coste, o rilievi della rota OP Le quali coste sieno come che girando la rota esse sole possin percotere nella paletta E e' alzando quella faccino declinare l' [unclear] per A e separar il cono K dall'L onde il vento entra nella canna, e' la fa' risonare per quanto è lunga La costa	it seals it and does not let the wind enter. To make it more stable and to avoid it from moving away, when the force of the wheel does not apply, one applies the coiled copper line FG that joins the key to the lid of the base. And when the cone K is moved away from L by the wheel, when the wheel has turned and the key is free, the line FG operates again and pushes the cone K into L and closes the air passage. The end of the key marked B is curved and ends at the small palette E which meets the small wood blocks of the wheel OP. These blocks have to be made in such a way that they can only touch the palette E and, when this happens, the pole moves toward A, the cone K moves away from L, and the air enters the pipe, allowing it to perform according to the length of the wood block,

(continued)

74v

che percote nell'E e' cosi dispone molte coste nel dorso della rota OP in modo che vi sia compartito il madrigale, o la musica che ne piace, et ordinati tanti tasti quante son canne nel modo che stà l'AB facendo la rota OP tanto largo e' con tante coste nel dorso che vi capino tutti i tasti, si sentiran sonar le canne a' quattro e' sei per volta secondo che staranno le coste e' soneranno gli organi ogni volta che tirata la cordicella si darà l'acqua che muova le rote e' oncisi il vento il qual per necessità uscira delle canne che li son' aperte perche di sopra non puo uscire rincontrando nuovo vento e' acqua che sopraviene, ne meno di sotto per che l' acqua escede e' ricopre sempre il diabete, tal che fara sonare diversamente secondo la dispositione delle coste che son nella rota OP nella quale a' Tivoli è disposto	that touches E. Many wood blocks are positioned on the wheel OP so that a madrigal results from it, or some other music if preferred. Once as many keys as pipes are built, as shown in AB, if the wheel OP is large enough and has enough blocks to touch all the keys, one will hear all pipes performing, four of six of them at once according to the positions of the blocks. And the organ performs each time the rope is pulled, opening the water stream, that moves the wheels and with it the wind that necessarily exits through the pipes which are open. This is because it cannot exit at the top as further wind and water are entering and cannot exit at the bottom either because the water completely covers the funnel. In this way, the air is used to perfom according to the disposition of the blocks, which are on the wheel OP, and at Tivoli,

75r

un bel madrigale assai lungo et è di dorso si largo che basta a XXIII tasti e' soprabonda il vento di sorte che fa' anco cantare a' beneplacito loro tre ucelletti al modo d'Herone, e' sonar cinque trombe, tale è dunque l'organo di Tivoli, il quale se ben ho procurato di descriver piu chiaramente che ho possuto non di meno conosco non essermi successo con la felicita che desiderano, onde ragionevolmente Mons.r Barbaro attribuisce la difficolta che è nell' organo di Vitruvio alla materia istessa non a' Vitruvio, il qual secondo il buon costume suo, lo dovette descrivere piu facilmente che fusse possibile, ma la gran distanza de tempi e' la perdita delle cose ci fa' sentir questo con molti altri danni Ma' di tutto sia ringratiato il Signore	a nice and very long madrigal is comparted, the wheel being so wide that it touches 23 keys and the quantity of wind being such that it also lets three small birds sing according to the method of Hero and even five trumpets. This is the organ of Tivoli and, although I tried to describe it in the clearest way I could, I nevertheless know that the result is less successful than what I wished for. In the same way, Monsignor Barbaro explains the obscurity of Vitruvius's description of the organ by appealing to the subject matter and not to Vitruvius's incapacity. According to his habits, Vitruvius had to describe it in the simplest way but the long time that has passed and the loss of things caused this obscurity and many other damages. For everything be God blessed

Hero of Alexandria, and Vannocci Biringucci, Oreste. 1582. Libro degli artifizii spiritali over di fiato d'Herone Alessandrino. In Ms. L. VI. 44. Trans. Matteo Valleriani. Siena: Biblioteca degli Intronati

Bibliography

Sources

Audebert, Nicolas. 1576/77. Le Palais, Jardin, & Fontaines, de Tyuoly. Excerpt from id., Voyage d'Italie, Landsdowne Ms. 720, fols. 334v-337v. London: British Museum. Transcribed in Ronald William Lightbown. 1964. Nicolas Audebert and the Villa d'Este. Journal of the Warburg and Courtauld Institutes 27: 164-190. Nicolas Audebert. 1981. Voyage d'Italie. Ed. Adalberto Olivero, II, 81-84. Rome: Luciano Lucarini.

Della Porta, Giovanni Battista. 1606. I tre libri de' spiritali. Napoli: Giovanni Giacomo Carlino.

Del Re, Antonio. 1611. Dell'antichità tiburtine, Capitolo V. Diviso in due parti. Rome: Appresso Giacomo Mascardi.

Hero of Alexandria, and Vannocci Biringucci, Oreste. 1582. Libro degli artifizii spiritali over di fiato d'Herone Alessandrino. In Ms. L. VI. 44. Siena: Biblioteca degli Intronati.

Ligorio, Pirro. ~1568/9-1573. Trattato di Pirro Ligorio patrizio napolitano cittadino romano di alcune cose appartenente alla nobiltà dell'antiche arti, e massimamente de la pittura, de la scoltura e dell'architettura, e del bene e del male che s'acquistano coloro i quali errano nell'arti e quelli che non sono della professione che parlano troppo per parere dotti di quel che non sanno, e detrattando altrui se istessi deturpano. Ms. a.II.16. Antichità di Roma (Torino, vol. 29). Torino: Archivio di Stato. Transcribed in Anna Schreurs. 2000. Antikenbild und Kunstanschauung des Pirro Ligorio 1513-1583, 402-435. Köln: Walther König.

Montaigne, Michel de. 1983. Journal de Voyage. Ed. Fausta Garavini. Paris: Gallimard.

Schickhardt, Heinrich. 1600. Zweite italienische Reise. In Cod. Hist. 4, ms. nr. 148. Stuttgart: Württembergische Landesbibliothek.

Zappi, Giovanni Maria. 1920. Annali e memorie di Tivoli. Ed. Vincenzo Pacifici. Tivoli: Nella sede della Società in Villa d'Este.

Secondary Literature

Boas, Marie. 1949. Hero's Pneumatica. A Study of its Transmission and Influence. Isis 40: 38-48.

Butzer, Paul Leo, and Dietrich Lohrmann (eds.). 1993. Science in Western and Eastern Civilzation in Carolingian Times. Basel: Birkhäuser.

Coffin, David R. 1960. The Villa d'Este in Tivoli. Princeton: University Press.

Degering, Hermann. 1989. Die Orgel, ihre Erfindung und ihre Geschichte bis zur Karolingerzeit. (Nachdruck) Buren: Fritz Knut.

Dernie, David. 1996. The Villa d'Este at Tivoli. London: Academy Editions.

Franke, Birgit. 1997. Gesellschaftsspiele mit Automaten – "Merveilles" in Hesdin. Marburger Jahrbuch für Kunstwissenschaft 24: 135-158.

Kaiser, Simone M. 2012. Die Verbindung von Antikenforschung und fürstlicher Repräsentation im Villengarten der frühen Neuzeit. In Gipsabgüsse und antike Skulpturen. Aufstellung und Ausstellung seit der Renaissance, ed. Charlotte Schreiter. Berlin: Reimer.

Kaiser, Simone M. 2014. Garden Design as an Artistic Form of Organized Knowledge: The Villa d'Este in Tivoli and its Dragons of 1572. In Art and Knowledge in Rome and the Early Modern Republic of Letters, 1500-1570, (Fragmenta 5, 2011), ed. Thijs Weststeijn, 39-62. Turnhout: Brepols.

Kaiser, Simone M. *Forthcoming*. Villa academica – fabrica della vita. Eine topologische Untersuchung des frühneuzeitlichen Gartenraumbildes am Beispiel der Villa d'Este in Tivoli. PhD, Goethe Universität Frankfurt a.M.

Lamb, Carl. 1966. Die Villa d'Este in Tivoli. Ein Beitrag zur Geschichte der Gartenkunst. München: Prestel-Verlag.

Lightbown, Ronald William. 1964. Nicolas Audebert and the Villa d'Este. Journal of the Warburg and Courtauld Institutes 27: 164-190.
Markovits, Michael. 2003. Die Orgel im Altertum. Leiden: Brill.
Miller, Naomi. 1986. Paradise Regained: Medieval Garden Fountains. In Medieval Gardens (Dumbarton Oaks Colloquium on the History of Landscape Architecture IX), ed. Elisabeth B. MacDougall, 135-151. Washington, DC: Dumbarton Oaks Research Library and Collection.
Occhipinti, Carmelo. 2001. Carteggio d'arte degli ambasciatori estensi in Francia (1546-1553). Pisa: Scuola Normale Superiore.
Valleriani, Matteo. 2007. From Condensation to Compression: How Renaissance Italian Engineers Approached Hero's Pneumatics. In Übersetzung und Transformation, eds. H. Böhme et al. Berlin: DeGruyter.
Valleriani, Matteo. 2009. The Transformation and Reconstruction of Hero of Alexandria's «Pneumatics» in the Garden of Pratolino. In Pratolino, un mito alle porte di Firenze – Pratolino, a Myth at the Gates of Florence, eds. L. Ulivieri and S. Merendoni. Venezia: Marsilio.
Valleriani, Matteo. 2010a. The Garden of Pratolino: Ancient Technology Breaks Through the Barriers of Modern Iconology. In Ludi Naturae – Spiele der Natur. Transformationen eines antiken Topos in Wissenschaft und Kunst, eds. N. Adamowsky et al. München: Fink Verlag.
Valleriani, Matteo. 2010b. Il ruolo della pneumatica antica durante il Rinascimento: l'esempio dell'organo idraulico nel giardino di Pratolino. In La civiltà delle acque dal Medioevo al Rinascimento, eds. A. Calzona and D. Lamberini. Firenze: Olschki.
Van Buren, Anne Hagopian. 1986. Reality and Literary Romance in the Park of Hesdin. In Medieval Gardens (Dumbarton Oaks Colloquium on the History of Landscape Architecture IX), ed. by Elisabeth B. MacDougall, 115-134. Washington, DC: Dumbarton Oaks Research Library and Collection.

Sundials on the Quirinal: Astronomy and the Early Modern Garden

Denis Ribouillault

Abstract

This paper deals with the function and meaning of sundials in Early Modern Rome, more specifically in gardens. It concentrates on two gardens, both on the Quirinal hill and directly facing each other: the papal gardens of Monte Cavallo and the Jesuit garden of the Noviciate of Sant'Andrea del Quirinale. Set on each side of the magnificent Via Pia, these gardens represented two intersecting yet contrasting worlds, a rude juxtaposition of one cosmos clashing against another: that of a Jesuit community and that of the Papal court. Each had developed a specific language to articulate their main concerns and proclaim their truths to garden visitors. By drawing a contrasting picture of the S. Andrea garden and the Papal gardens, in which sundials were given very different meanings, the intent of this paper is to probe the awkward, contradiction-ridden, *spinoso* relationship between religion, science and curiosity in Early Modern Rome.

In 1685, the Dutch civil engineer Cornelis Meyer devised a method to re-erect the obelisk of Augustus, which had formed part of a famous ancient sundial known as the *Horologium Augusti*. Published in a book on hydraulic engineering, Meyer's method was supplemented by an entire section in which he explained how one could transform Rome's *piazze*, with their obelisks and columns, into colossal sundials. He suggested that the obelisk be erected on Piazza Monte Cavallo on the Quirinal hill and that the monument be given back its ancient gnomonic function (Fig. 1). Meyer's elaborate gnomon would determine the hours of the night through the use of the same type of pierced gnomon that would be used a century later, in 1792, when the obelisk was finally erected on Piazza

D. Ribouillault (✉)
Département d'histoire de l'art et d'études cinématographiques, Université de Montréal, Pavillon Lionel-Groulx, 3150, rue Jean-Brillant, Montréal, QC, Canada H3T 1N8
e-mail: denis.ribouillault@umontreal.ca

H. Fischer et al. (eds.), *Gardens, Knowledge and the Sciences in the Early Modern Period*, Trends in the History of Science, DOI 10.1007/978-3-319-26342-7_6

Fig. 1 Gaspar van Wittel (?), Project for the Piazza Montecavallo, in Cornelis Meyer, *Arte di restituire a Roma la tralasciata navigazione del Tevere*, Rome, 1685, pl. XIV

Montecitorio.[1] The obelisk could also be completed, Meyer suggested (1685: pl. XIV), with enormous revolving terrestrial and celestial globes:

> One would represent all Kingdoms and Provinces of the Earth, and the other the constellations of the Heavens. And were these globes to be made of bronze, one could build at their foot a clock, which would make those same globes wheel completely around during the space of twenty-four hours, and similar clocks can be made, without the necessity of using weights, such as one we possess which comes from Holland. On those

[1] This article is a development of a portion of the paper I presented at the conference *Gardening and Knowledge. Landscape Design and the Sciences in the Early Modern Period*, entitled "Measuring Time in the Gardens of Papal Rome" (Hannover, 17–19 September 2012), which is currently being developed as a book with the provisional title *Gardens of the Heavens. Astronomy and the Science of Time in the Gardens of Papal Rome*. I owe a lot to Volker R. Remmert for his constant encouragement in the preparation of this article, Joseph Connors for his generous reading and enlightening suggestions and Mandy MacDonald for revising the English and editing my translations.

On the history of this obelisk, see Pliny (1940: XXXVI, 70–73), D'Onofrio (1967: 280–291), and Iversen (1968: 142–160).

same globes there could be noted again various curious observations, either to know from their revolution the hours of the day and of the night, as also, what time it is at a given hour in any part of the world.[2]

Meyer is still little known to art historians today, yet his book is a strong reminder that astronomy and gnonomics—the term 'dialing' is still used by English specialists—were fundamental fields of study in Rome, and on the Quirinal hill especially, and had been for more than a century.[3] Although in his chapter on sundials, Meyer leaves aside the complex mathematics of the dials, his imaginary description of Rome, transformed into one immense astronomical observatory, is proof that the science of the heavens had a deep impact on the way architectural and urban space were perceived in seventeenth-century Rome.

This paper deals with the function and meaning of sundials in Early Modern Rome, more specifically in gardens. One might imagine that the subject had already been explored, given the long tradition of studies dedicated to Roman gardens in the Renaissance and Baroque periods. Compared to the numerous studies devoted to fountains and statues, however, the presence of clocks, sundials and other such astronomical or mathematical instruments has been surprisingly overlooked by scholars, perhaps because evidence for such ornaments in Rome is scarce compared to that for Northern European gardens, where they are plentiful. David Coffin (1991: 184), in his otherwise comprehensive book on the gardens of Papal Rome, gives only two examples of garden sundials.[4] Since I started researching this topic, I have come to realize that this does not reflect historical reality. Several Roman gardens from the sixteenth to the eighteenth centuries featured sundials prominently. Their presence is significant to an understanding of their function, their cultural meaning and what has traditionally been called their "iconography". Moreover, they strongly suggest that astronomical debates had an important impact

[2] Meyer (1685: pl. XIV) and corresponding chapter: "Si legge appresso Plinio, che Augusto haveva trovato un modo di dimostrare mediante l'ombra di essa Guglia l'hore della notte, *Divus Augustus* (dic'egli) *addidit mirabilem usum ad deprehendendas dierum, noctiumque magnitudines, digna cognita res:* onde mi venne desio d'insinuare, che volendo far servire la medesima Guglia al primevo suo uso, si potrebbe lasciare nella sommità sotto la Croce un apertura à Foggia d'un piccolo Cerchio per il quale traguardando la stella detta Tramontana potrebbesi formare sopra il piano opposto ad essa Guglia un horologgio da sapere l'hore notturne... [...] et in oltre per maggior ornamento della medesima piazza potrebbonsi aggiungere due altre basi à quelle, che vi sono di presente (come si dimostra nel sudetto disegno) per collocare sopra di questi due globi, ò sfere, una che rappresentasse tutti li Regni, e Provincie della terra, e l'altra tutte le Costellationi del Cielo. E quando questi globi fussero fatti di bronzo si potrebbe à piedi di essi fabricare un horologio, che facesse alli medesimi globi fare una rivolutione nello spatio d'hore ventiquattro: e simil horologgio può anche farsi, che vada un anno senza haver bisogno d'alzar li pesi, come Noi ne tenemo uno fatto venire d'Olanda. Sopra li medesimi globi potrebbonsi notare ancora diverse altre osservationi curiose, sì per saper della di loro rivolutione l'hore del giorno, e della notte, come anche, che hora è appunto in quell'istesso tempo in qualsivoglia parte del Mondo".

[3] Further on Meyer, see Van Berkel (2002), Fagiolo (2014) and Connors (2015).

[4] Conversely, Coffin devotes an entire chapter to the subject in his book on English gardens. See Coffin (1994: Chap. I). This is not the place to give an extensive bibliography on the links between astronomy and the Church, but a classic work to seek out is Heilbron (1999).

on the world of the garden in early modern Rome. How astronomical knowledge was appropriated and transformed by this specific context and what this neglected dimension tells us about the world of gardens are among the questions that will be raised. To do this, I will concentrate on only two gardens, both on the Quirinal hill and directly facing each other: the papal gardens of Monte Cavallo and the Jesuit gardens of the Noviciate of Sant'Andrea del Quirinale. Set on each side of the magnificent Via Pia, these gardens represented two intersecting yet contrasting worlds, a rude juxtaposition of one cosmos clashing against another: that of a Jesuit community and that of the Papal court. Each had developed a specific language to articulate their main concerns and proclaim their truths to garden visitors. By drawing a contrasting picture of the S. Andrea garden and the Papal gardens, in which sundials were given very different meanings, the intent of this paper is to probe the awkward, contradiction-ridden, *spinoso* relationship between religion, science and curiosity in Early Modern Rome.

Astronomy and the Quirinal Gardens from Gregory XIII to Urban VIII

The garden at Monte Cavallo had already acquired a strong astronomical dimension during the pontificate of Gregory XIII at the end of the sixteenth century, possibly linked with the presence on the hill of an ancient temple of the Sun.[5] In 1584, a Tower of the Winds was built there to echo the more famous one at the Vatican, linked with the calendar reform and fitted with an elaborate *meridiana* and an anemoscope by the mathematician Egnazio Danti.[6] To my knowledge, no known document demonstrates explicitly that this tower possessed a gnomonic function like that of its twin sister at the Vatican. Yet the fact that it was given the same name as its Vatican counterpart reveals Gregory's will to establish a link between the two monuments, reinforcing his claim to power over the Roman landscape and the cosmos at large.[7] Evidence that such a meaning was well established is provided by a poem by Principio Fabricii, published in 1588 in an emblem book in which astronomical themes are central. The accompanying emblem shows the villa of Monte Cavallo and its tower as the new centre of the pope's power over the entire world (Fig. 2).

> Con ampii giri la superba Mole
> Ch'l Regno importa ben fondato, e retto
> Cinge Drago immortal, Signor perfetto,
> Che i Soggetti ugualmente intender vuole.
> E quale in mezo à i celesti Orbi il Sole

---

[5] On the Quirinal palace and its gardens, see Coffin (1991), Fagiolo (1998), Frommel (1999/2002) and Zampa (2007).

[6] On the Quirinal Tower of the Winds, see Tesi (1991: 90) and Courtright (2003: 177–181).

[7] For instance, Giovanni Baglione, in his *Life* of Martino Longhi, speaks of the tower as the "Torre dei venti": see Baglione (1642: 68). For a discussion of the Vatican–Quirinal axis at the time of Gregory XIII, see Meadows-Rogers (1996: 205–221).

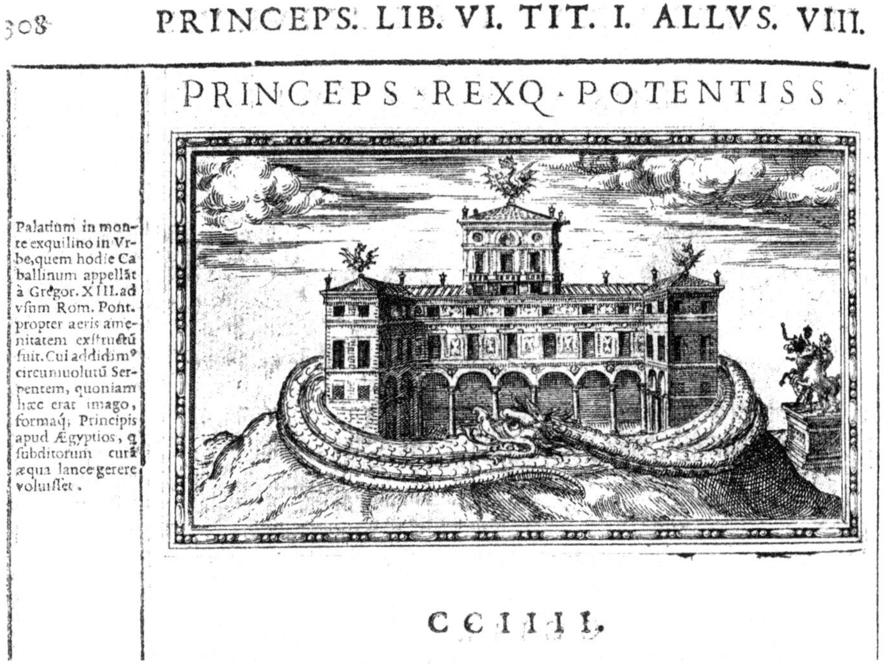

Fig. 2 Villa Montecavallo encircled by the Boncompagni Ouroboros dragon in Principio Fabricii, Delle allusioni, imprese, et emblemi..., Rome, 1588, 308, Bibliotheca Hertziana, Rome

Alluma, e scalda con sereno aspetto
Come à Saturno, à la Sorella il petto,
Tal QVESTI à ogn' Alma, che l'honora, e cole.
Del Mondo Italia è un piccol Mondo, e Roma
E d'ambi il Capo, ove riluce, e splende
Con nova Sfera il Pastor santo, e giusto.
Tal che non men può venerar sua chioma,
Che del foco divin gli Animi accende,
Il Gelato German, che l'Indo adusto. (Fabricii 1588: 308)[8]

In the engraving, the Quirinal tower is topped by an impressively large Boncompagni dragon with open wings, while another large snake-like dragon biting its own tail as in the hieroglyphic *ouroboros*, encircles the whole building, a symbol of cyclicality and eternal return. This kind of symbolism was definitely

[8] "In wide circles the immortal dragon, symbolizing equally the perfect master and his subjects, winds itself around the proud and huge edifice, which is of concern to the well—and rightly—established kingdom. And just as the sun shines in the midst of the celestial bodies and warms the breast with [its] serene countenance, as from Saturn to the Sister, so to each one of these, who honors her, is she the nourishing hill. Italy is a little world within the world, and Rome is the head of both, where the holy and just Pastor shines and sparkles with a new orb. So that no less can frigid German than the parched Indian venerate his Head, who kindles souls with divine fire." Meadows-Rogers (1996: 360–361). On the work of Fabricii, see Ruffini (2005) and Aksamija (2011).

Fig. 3 Medal of 1582 celebrating the calendar reform. Recto: portrait of Gregory XIII; verso: Aries (standing for the vernal equinox) with the Boncompagni dragon shaped like the Ouroboros with the inscription ANNO RESTITVTO MDLXXXII, © Wikimedia Commons. Image is in public domain

associated with the Gregorian calendar reform in a medal of 1582 (Fig. 3). The mechanical clock on the tower is generally considered to have been added later at the time of Urban VIII, but some evidence points to the presence of such a clock already being there in Gregory's time. On the church of Sant'Atanasio dei Greci, on today's Via del Babuino, there is a beautiful clock whose pointer is very interestingly shaped like the *ouroboros* dragon (Fig. 4).[9] In a passage from the *Memorie de Capitoli del Collegio dei Greci* for the year 1677, one can read that the clock's pointer came—"according to hearsay, not documents"—from the palace of Monte Cavallo.[10] Although no other document exists to prove this, it shows none the less that, even a century later, the Quirinal was still associated in people's minds with Gregory's reform and his poetics of time and astronomy.

The villa's connection with astronomy was made even stronger during the pontificate of Pope Urban VIII, when a monumental sundial or *meridiana tetracycla* was installed in the garden in conjunction with two armillary spheres visible in the engraving of the garden made by Giovan Battista Falda: one on the top of a small pavilion, or *padiglione di verzura*, and another very large one set on a balustrade immediately above the Fontana dell'Organo (Fountain of the Organ) (Fig. 5a, b, c). Regarding the localisation of the sundial within the garden, Filippo Camerota and Marcello Fagiolo relied, in their respective articles on the subject, on a document dated November 2, 1628 specifying that the sundial was set "nel mezzo del Giardino novo di Monte Cavallo". Fagiolo suggested a precise spot for this placement, at the

[9] Although the dragon-shaped pointer of this clock refers to the Pope's coat of arms, it is worth recalling the use of the dragon on contemporary astronomical clocks to indicate lunar nodes and determine solar and lunar eclipses.

[10] The two *campanili* of the church are described thus: "Uno per uso delle campane che sono due e di buon suono e l'altro per l'orologio il quale, come corre tradizione ma nelle scritture non si truova notato, era quello che stava nel palazzo di Monte Cavallo che la S.^{tà} N. Di Gregorio XIII comprò da Signori di Casa d'Este". See Angelucci (1991, 1993).

Fig. 4 (**a–b**) Mechanical clock of Sant'Atanasio dei Greci with the pointers shaped like the Boncompagni Ouroboros dragon, Rome. (**a**) © Wikimedia Commons. Image is in public domain. (**b**) Photo author

crossing of the two main axes of the garden. Although this location is consonant with other projects conducted by Carlo Maderno in this area of the garden, the sundial was either never placed there or was rapidly removed, since Giovan Battista Falda depicted it elsewhere in his second engraving of the garden, dated circa 1680: in that work, it is clearly visible next to the aviary (*uccelliera*, corresponding to the ex-Casino dei Grimani) marked 20 on the plan (Fig. 6a, b). This location, close to a view from which one could admire the stupendous panorama of Rome, is consistent with the ways in which sundials were placed in contemporary Roman gardens. The sundial drawn by Falda is topped by the Pope's tiara and keys ("Un Regno con doi chiavi [. . .] che fa finimento sopra per adornamento di detto Horologio") and set on a large square pedestal and not the column with ancient capitals described in the 1628 document mentioned above.[11] In his *Horarium universale perpetuum*, published in 1637, the mathematician Teodosio Rossi explains that he himself suggested to the pope the idea of building a sundial for his garden in 1624, during the ceremonial benediction of the three bells of the new *campanile* that marked the hours on the Tower of the Winds, or *torre dell'orologio*. The sundial would serve to correct the cumulative error of the mechanical clock, which up to the middle of the

[11] On the Quirinal sundial, see Camerota (2000). The document of 2 November 1628, "Misura e stima de lavori diversi fatti per servitio di N.S. in far l'Horologio à Sole Posto nel mezzo del Giardino novo di Monte Cavallo, e fatti da m.ri Agostino Radi e Francesco Castelli, compagni scarpellini," is in Archivio di Stato di Roma, Camerale I, Giustificazioni di Tesoreria, b. 62, fasc. 7, ins. 1. For an hypothesis regarding the original placement of the sundial, see Fagiolo (2000). Here I repeat only the essential information given in these two important studies, to which I refer for further information and bibliography.

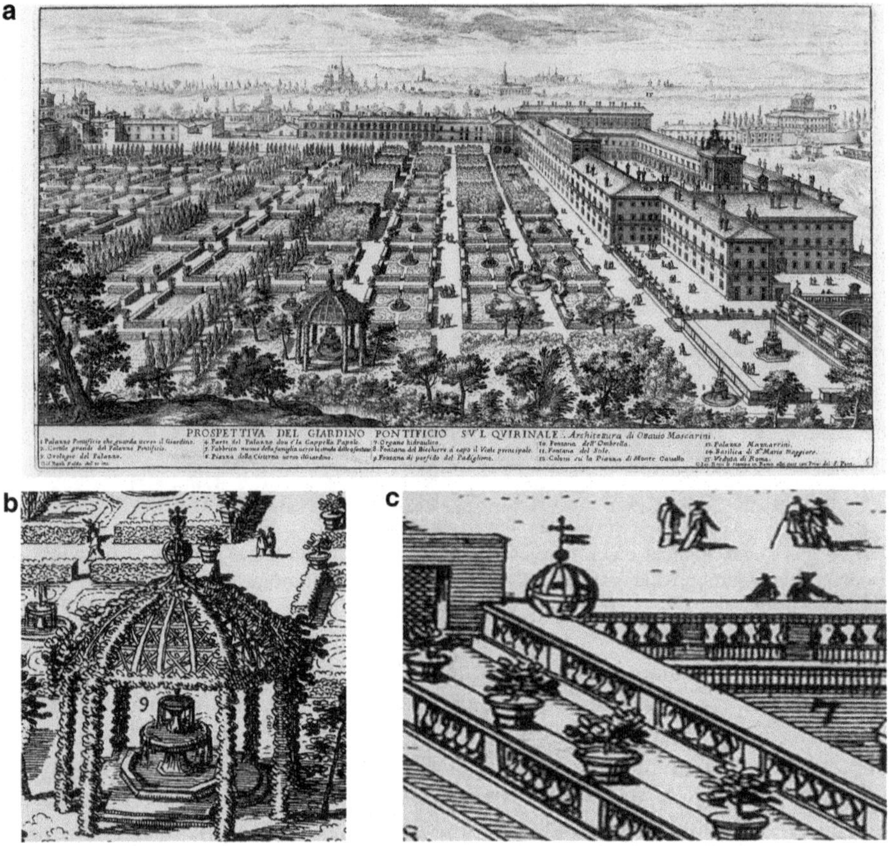

Fig. 5 (**a**) Prospettiva del giardino pontificio sul Quirinale, in Giovan Battista Falda, Li giardini di Roma. . ., Roma, s.a., Canadian Center for Architecture, Montréal. (**b**) Detail of the armillary sphere on top of the Padiglione di verzura. (**c**) Detail of the armillary sphere above the Fontana dell'Organo

seventeenth century was of the order of 15 min per day. The sundial, one of the first Roman works of the famous architect Francesco Borromini, was constructed in 1628 in collaboration with Agostino Radi. Rossi, a student of Christoph Clavius, was considered to be one of the finest experts on gnomonics in Rome. His gnomonic tables, compiled and published in 1593 to guide the construction of sundials for the latitude of Rome, were put to use for the design of the Quirinal sundial.

The sundial is described and illustrated in Athanasius Kircher's treatise on sundials, *Ars magna lucis et umbrae*, published in 1646 (Kircher 1646: 359–360). Originally, it had four gilded bronze gnomons indicating the hours, wrought in the form of the Barberini family's emblems, three bees and a sun. At present, however, the only original elements of the sundial that survive are the four-sided marble drum and the foot supporting it. The imagery was completed with two inscriptions. One, along the upper border of the sundial's foot, suggests an emblematic image of the pope guided on earth by heavenly light, SUPERNI LUMINIS DUCTU. The other, at the base of the foot, is taken from Virgil's *Georgics* and describes bees as "custodians

a

b

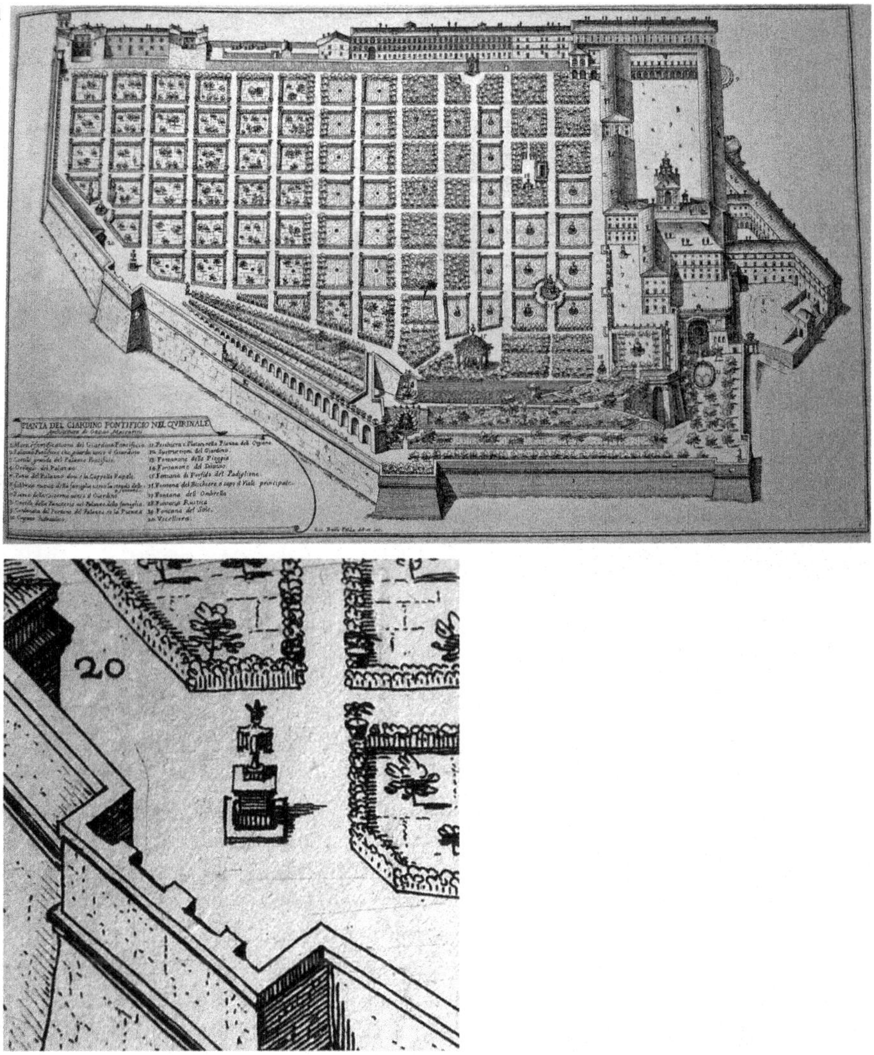

Fig. 6 (**a**) Pianta del giardino pontificio nel Quirinale, in Giovan Battista Falda, Li giardini di Roma. . ., Roma, s.a., Canadian Center for Architecture, Montréal. (**b**) Detail of the Tetracycle sundial

of the doors and observers of the heavens" (Virgil, *Georgics*: IV, 164–165: "Sunt quibus ad portas cecidit custodia sorti. . .") (Figs. 7 and 8).

The link between the sundial and the pope's heraldic emblems is logical and very appropriate. The association between *gnomon*—from an etymological and philosophical point of view, the sundial or *gnomon* designates that which understands, decides, judges or distinguishes (Serres 1986)—and *nomen*, meaning "name", originally *gnomen* and thus deriving from the same root (*gno*) meaning "know", is semantically highly charged, and implies a meaningful confusion

Fig. 7 (**a–b**) Francesco Borromini and Agostino Radi, 'Tetracycle' sundial, marble and bronze, 1628, (formerly in the Gardens of the Quirinal Palace, Rome), modern reconstruction, Museo Galileo, Florence

between identity and cosmology. The rising sun was one of the most widely diffused emblems of the new Pope. It is visible, for instance, in Giovan Ferro's *Teatro d'imprese*, published in 1623 and dedicated to Cardinal Francesco Barberini, in which many sundials are also used as emblems. Claude Lorrain's beautiful landscapes with a rising sun made for Urban VIII may have been the highest artistic expression of this imagery (Beaven 2007).

Bees were also directly related to astronomy. It was believed in antiquity that honey came not from flowers but from the heavens. This view was still held in the Renaissance, when it was thought that honey descended from the sky in the morning dew and was dispersed over the earth, drying in the warmth of the sun to a sweet white powder, which the bees collected and made into honey. This commonplace was repeated, for instance, by Torquato Tasso in a dialogue on *Imprese* of 1594, in which he states: "Bees make honeycombs from flowers, the wax from the tears of trees and the honey from the dew of the air, mostly in the birth of the stars and the vault of heaven".[12] So the sundial's imagery was especially appropriate in the context of a garden.

[12] ". . . l'api facciano i favi de' fiori, la cera de la lacrima de gli alberi, il mele de la rugiada de l'aria il più de le volte nel nascimento de le stelle e de l'arco celeste"; cited in Quiviger (2003: 317–321). The most detailed description of the celestial origin of honey appears in Pliny (1940: XI, xii, 30–32).

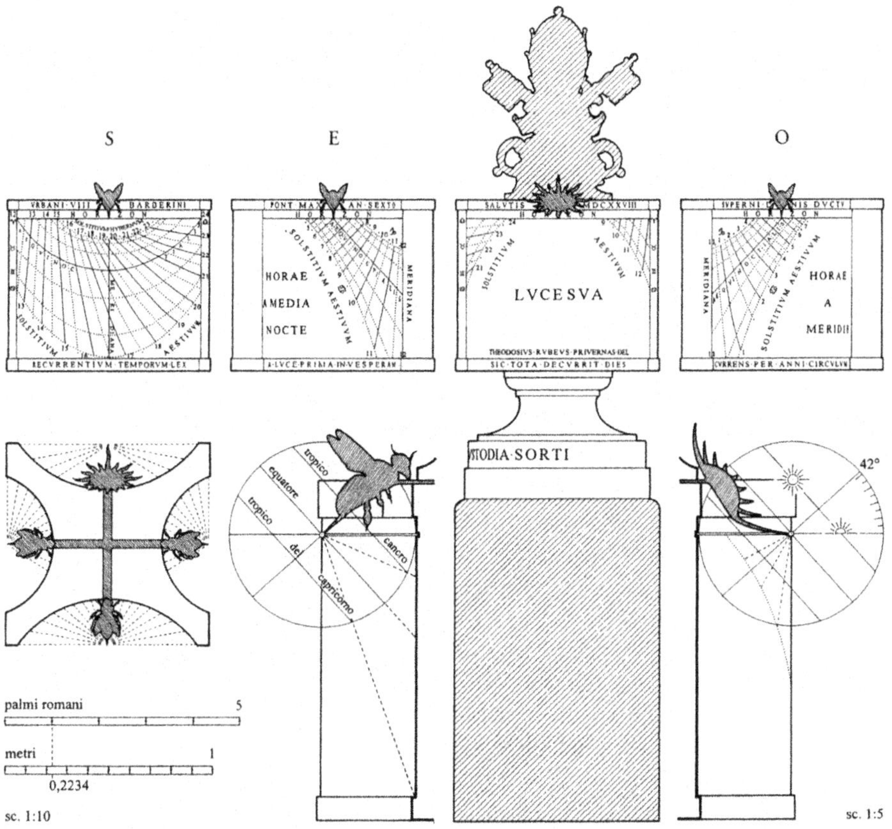

Fig. 8 Reconstruction of the Tetracycle sundial by F. Camerota (2000)

Galileo in the Gardens of Baroque Rome

Urban VIII's decision to strengthen the astronomical dimension of the Quirinal gardens is undoubtedly related to his recognition of the original link between the Quirinal tower and Gregory XIII's reform of the calendar. The fact that he had both the gallery of maps and the Vatican tower restored proves that he fully endorsed Gregory's attention towards astronomy and cosmology.[13] Yet the link between astronomy and the Quirinal gardens had never ceased to be strong. Indeed, Galileo himself presented his latest discoveries on sunspots in the Quirinal garden, as he recalls in a letter to Antonio Antonini dated February 20, 1638: "nel tempo che io mi trovavo in Roma, dove più volte le feci vedere a molti prelati grandi negli Orti Quirinali". The information is confirmed by Angelo De Filiis in his *Istoria e dimostrazione intorno alle macchie Solari e loro accidenti di Galileo* (1613)

[13] For the restoration of the Vatican tower by Urban VIII, see Mancinelli and Casanovas (1980: 12).

and again by Vincenzo Viviani, who, in his biography of the scientist, gives the date of the demonstration as April 1611.[14] Another of Galileo's demonstrations took place in the garden of Cardinal Ottaviano Bandini, very close to the Quirinal papal gardens.[15] These observations were made with the aid of an instrument that would be later termed a helioscope, whereby the image of the sun was projected through a telescope onto a sheet of white paper on which it was then easy to record the different spots observed. The instrument is described by Galileo's great rival at the time, the Jesuit Christoph Scheiner (1630: 137–138).

If the Quirinal gardens were a stage for Galileo's success, they were also to be the site of his downfall: from 1615 to 1633, most meetings of the Congregazione del Santo Uffizio, which conducted his trial, took place at the Quirinal palace (Galluzzi 2009). Certainly many other astronomical experiments took place in gardens, which, together with high towers, were the logical place for them to be conducted. The armillary sphere set in the Quirinal gardens, visible in the engravings by Giovan Battista Falda, was also probably used for such a purpose, as this instrument was particularly useful for teaching the concepts of astronomy. Its location at a point offering a magnificent and unimpeded view over the entire city of Rome is proof that the instrument must have been used for practical demonstrations (Fig. 5c). However, situated below the Tower of the Winds and above the marvellous hydraulic organ of the Fontana dell'Organo, later restored by Athanasius Kircher (Barbieri 1981),[16] it almost certainly also played an important symbolic role, linking the space of the gardens with the heavenly spheres. Here, astronomy, hydraulics and music, sciences which were all derived from mathematics, performed together for the edification and pleasure of the visitor. It is well known that, during his visit to Rome in 1611, Galileo displayed his instruments in various places, often in the setting of an evening meal and to the accompaniment of music. This was most probably the case with the famous outdoor banquet organized by Federico Cesi on April 14 of that year, in the gardens of Monsignor Malvasia on the Janiculum hill, where Galileo showed his new telescope to a group of eminent scientists (Shea and Artigas 2003: 33). In a letter to his friend Filippo Salviati, Galileo enthusiastically recalls his visits to Roman gardens: "I have been received and fêted by many illustrious cardinals, prelates and princes who wanted to see the things I have observed and were much pleased, as I was too, on my part, in viewing the marvels of their statuary, paintings, frescoed rooms, palaces, gardens, and so on" (Shea and Artigas 2003: 39).

The Sundials of the Jesuit Garden of Sant'Andrea del Quirinale

Before further investigating the possible significance of the Barberini sundial, we may pursue this brief overview of the Quirinal gardens' astronomical dimension by mentioning the presence of two sundials in the nearby Jesuit garden of Sant'Andrea

[14] Galileo (1890–1909: XVII, 297; V, 81–82; and XIX, 612). See Galluzzi (2009).

[15] Galileo (1890–1909: XII, 175).

[16] The Organ was made in 1596 by Luca Blasi. The hydraulic organ is described and illustrated in Kircher (1650: t.2, 343, Iconismus XXI).

del Quirinale. They are described quite precisely by the Jesuit father Louis Richeome in *La peinture spirituelle*, published in Lyons in 1611, and can be seen on a print showing the garden attributed to Matthias Greuter inserted into the book (Fig. 9). The description is worth citing at length:

> I have finished what I had to say about the pictures of this first garden. As you are leaving, I advise you to look at two noble pieces by the great Christopher Clavius, the Euclid and Archimedes of our century, and great ornament of the celestial sciences in our Company. One is the horizontal clock drawn on slate, planted on the pillar in the corner of the second square [indicated by the letter E in the view of the garden], and the other [is the] vertical [sundial] on the façade of the house [indicated by the letter C]; both of the old invention, not of Anaximenes of Miletus, who was a disciple of Thales, one of the seven sages of Greece, and lived in the year 3560, but of the Egyptians and Hebrews, who used them long ago, namely in the year 3356, more than two hundred years [earlier], as [appears] from the history of the Kings, where it is said that the prophet Isaiah, who lived in that time, gave a sign of the truth of his prophecy by making the shadow of the royal sundial move back by ten lines, which was a great miracle, even greater than that of Joshua who only stopped the [course of the] sun [...]. Now, these two dials are the same picture and serve the same purpose with their shadows; [they] mark the course of the sun, the days and hours in every possible way, the astronomical, Babylonian, Italic, French, ancient and modern hours. They warn you in all these ways that time passes unceasingly and rolls at extreme speed without returning, although the sun keeps returning every day and every year. They tell you when twenty-four hours are completed, when the day is over and is no more. These [periods of] twenty-four hours end at different times, depending on the variety of clocks and the limitations of the natural day consisting of its twenty-four members and parts. In Babylon, Judaea and [in] Eastern countries, where they count the beginning of the day at sunrise, they end at daybreak. On the contrary, in Italy, now as formerly in Athens, they end at sunset, because their day is from one sunset to the next. Depending on the astronomers' dials, they end at noon, for their day is done at noon. On the contrary, in ancient Rome and for the ancient Egyptians, they end at midnight, [...], a practice still used by the Roman Church and in all Christendom and for the feast days and mysterious days, because they start and end at midnight (Richeome 1611: 499–502).[17]

[17] I have transcribed the original text in modern French in the passages from Richeome quoted here, changing the punctuation to facilitate reading.

TABLEAU TROISIÈME DES IARDINS. LES HORLOGES, OU QUADRANS. J'ai achevé ce que j'avais à dire des tableaux de ce premier jardin. Je vous avertis en sortant de regarder deux nobles pièces du grand Christophe Clavius, l'Euclide et l'Archimède de nos siècles, et grand ornement des sciences célestes en cette compagnie. L'une est l'horloge horizontale, tracée sur cette ardoise, plantée sur le pilier en l'angle de ce second carré [indicated by the letter D in the view of the garden]; Et cet autre vertical, aligné en la muraille du logis [indicated by the letter C], tous deux de l'invention ancienne, non d'Anaximène Milésien, qui fut disciple de Thalès, l'un des sept sages de Grèce, et vécut l'an du monde 3560, mais des Égyptiens et Hébreux, qui, longtemps auparavant, à savoir l'an 3356, plus de deux cents ans [auparavant], en avaient l'usage, comme il [ressort de] l'histoire des Rois, où il est dit, que le Prophète Isaïe, qui vivait en ce temps-là, donna un signe de la vérité de sa Prophétie, [en] faisant rétrograder l'ombre du cadran Royal de dix lignes, [ce] qui fut un grand miracle: voire plus [grand encore] que celui de Josué qui arrêta seulement le Soleil [...]. Or, ces deux cadrans sont un même tableau et un même office par leurs ombres, et [...] marquent le cours du Soleil, les jours et les heures en toutes façons; à l'astronomique, à la Babylonienne, à l'Italique, à la Française, à l'antique, et à la moderne. Ils vous avertissent en toutes ces manières, que sans pause le temps coule, et roule d'une extrême vitesse sans revenir, encore que le Soleil revienne, et

Fig. 9 Attributed to Matthias Greuter, The Garden of the Jesuit Noviciate of Sant'Andrea al Quirinale, Rome, in Louis Richeome, *Peinture spirituelle...*, Richeome (1611: 64), © Bibliothèque Municipale de Lyon

Richeome's description of the sundials is typical of his book, which is both a meticulous description of the garden, the house, the church and their decoration and an allegorical reading of the world, a sort of exegetical promenade.[18] The sundial maker, the Jesuit Christoph Clavius, was by far the most renowned astronomer of his time and his *Gnomonices libri octo*, published in Rome in 1581, was considered the ultimate work on the subject of gnomonics. It was written during the very years when Clavius was working on the calendar reform, which was established in 1582.[19] The garden sundials were extremely elaborate, especially as they indicated the time in many different places, in keeping with the Jesuits' presence all around the world. This geographical dimension echoed that given to the scenes of martyrdom painted by Tarquinio Ligustri in 1599 in the nave of the church of San Vitale, which was an integral part of a visit to the Noviciate complex.[20] The choice and arrangement of the plants and trees in the garden of Sant'Andrea were also deliberately symbolic, reflecting martyrological subjects.[21] Pilgrims and students would have immediately made the connections between these different parts of the garden or were likely to be reminded of them by their Jesuit guides. An illustration of the interrelation between the Jesuits' geographical and astronomical pursuits would later be given by the indefatigable Jesuit scientist Athanasius Kircher, who depicts a universal sundial in his *Ars magna lucis et umbrae* of 1646 (Fig. 10).[22] It is in the shape of an olive tree rooted in Rome, whose branches extend to locations

reprenne ses tours, et retours journaliers, et annuels. Ils vous disent, que quand vingt quatre heures sont accomplies, le jour est fini, et n'est plus. Ces vingt quatre heures finissent en divers temps, selon la diversité des horloges, et limites du jour naturel composé de ces vingt quatre membres et parcelles. En Babylone, en Judée, et [dans les] pays orientaux, où l'on compte le commencement du jour au lever du Soleil, elles finissent à la pointe du Jour. Au contraire, en Italie, maintenant, comme jadis en Athènes, [elles finissent] au Soleil couchant, parce que leur jour est d'un couchant à l'autre. Selon les cadrans des Astronomes, elles finissent à midi, car leur jour se fait en midi. Au contraire, selon l'ancienne Rome et les vieux Égyptiens, elles se terminent à la minuit, [...], ils marquaient les points de leurs jours, [une] façon qui se garde encore en l'Église romaine, et par toute la Chrétienté et [pour les] fêtes, et [les] jours mystérieux, car on les commence et les finit à la minuit.

[18] For a brief but useful description of the Quirinal garden and Richeome's description of it in English, see Witte (2011: 102–108). See also Davidson (2006). On Richeome, see Fabre (1995), Guiderdoni-Bruslé (2001), Van Assche (1999), Gijsbers (1999), Bailey (2003a: 48–52), Bailey (2003b: 129–137), Höltgen (2004), Dekoninck (2005: 68–75, 78–81, and 119–121), and Loach (2012).

[19] Clavius'most important works regarding sundials and the calendar reform are *Gnomonices libri octo*, Rome, 1581; *Fabrica et usus instrumenti ad horologiorum descriptionem peropportuni*, Rome 1586; *Astrolabium*, Rome, 1593; *Horologium nova descriptio*, Rome, 1599 and *Romani calendarii a Gregorio XIII restituti explicatio*, Rome, 1603. Further on Clavius, see Lattis (1994: 67–91).

[20] On Ligustri's martyr cycle at San Vitale and its link with Jesuit geography, see Bailey (2003a: 166–186); on the Noviciate of Sant'Andrea, (ibid.: Chap. 2: 38–106). On Jesuit geography, see Harris (1999).

[21] Macioce (1990: 122–123), Macioce (1995), Van Assche (1999), and Bailey (2003a: 167).

[22] On this print, see Godwin (1979: 78–79) and Harris (1999: 219).

Fig. 10 Universal dial, in Athanasius Kircher, *Ars magna lucis et umbrae*, Rome, 1646, fol. 553 (Iconismus XX), © from Remmert (2011, 75)

around the world; each "olive" on the branch contains a miniature sundial set for the specific latitude. Surrounding the tree, the eight winds, corresponding to the different directions and parts of the world, are blowing not air but words in the respective language for each region. At the four corners, Jesuit saints are seen praying below four vertical sundials related to the cardinal directions and comparable to the quadrant visible on the garden façade of the Noviciate in Greuter's image

(Fig. 9). Richeome also underscores the universality of the Christian message at the end of the book in his enthusiastic description of the Jesuits' pastoral activities, which recalls the Pentecostal moment (Acts 2:4–11) when all those on whom the Holy Spirit has descended understand each other, though speaking in many languages: "You will see Italians, Germans, Poles, Bohemians, Hungarians, Flemings, Walloons, Irish, Scots, Englishmen, Greeks, Spaniards, Frenchmen, and other nations of Christianity, and all are taught in their mother tongue, and each of you has your separate group and your school in the same place, and without confusion." Let the message of San Vitale, he concludes, reach "not only you, but also everyone who dresses like you, in other similar houses placed on the five parts of the habitable world" (Richeome 1611; Bailey 2003a: 186).

As in Rome, sundials were given a very prominent place within Jesuit houses all around the world. Another extraordinary example of the use of sundials for the purpose of demonstrating the universality of the Jesuit mission is the elaborate universal sundial of the Jesuit Collège de la Trinité of Villeroy near Lyons in France, where astronomy was taught to novices as early as 1604. Dated 1662, it gave the time in the 37 colleges of the Company, and was part of an elaborate pictorial programme celebrating the college as a Temple of Wisdom.[23] Also worth mentioning within the framework of this study is the famous pyramidical sundial built in 1669 by Jesuit Father Francis Line, also known as Francis Hall, for Charles II of England's garden at Whitehall (Line 1673; Coffin 1994). Line's sundial boasted the traditional royal symbols, as well as portraits of the king, queen and other members of the nobility. It is an interesting example to compare with Borromini's Quirinal sundial, not only for its use of heraldic symbols reinforcing social and political power, but also for its links with knowledge, mathematics and curiosity. Because this elaborate sundial was especially useful for demonstrating astronomical theories, Line was even requested by a friend of Galileo to have a similar dial sent to Rome to help defend the theory of heliocentrism. Before that, Line had devised several other sundials which caught the attention of European *literati*. Some of those served as ornaments in the garden of the *Collegii Anglici societatis Jesu Leodii*, the English Jesuit College in Liège, with Father Line using them to teach the principles of astronomy and mathematics. They were described in his *Tractatus de horologiis* (University of Liège, Ms. 377; Sauvenier-Goffin 1958).[24]

But let us return to Sant'Andrea. Some of Richeome's remarks about the placement of the sundials in the garden are particularly interesting. For him, the two sundials are to be considered together ("ces deux cadrans sont un même tableau, et un même office"), although they are not placed near each other. Likewise, in the papal garden we have just left, the various astronomical instruments created a web of interconnected objects, which could be linked thematically by the moving visitors. For instance, the astronomical ornaments entered into

[23] For a full description, see Menestrier (1663: 101ff) (*Description et explication des montres solaires tracées dans les trois faces de la grande cour du Collège de la Trinité*). For an analysis of the cultural context, see Van Damme (2005).

[24] On Father Line, see further Reilly (1962, 1969).

dialogue with some of the fountains in the garden, such as the Fountain of Apollo and the Fountain of the Sun (numbered 11 on Falda's view), both with explicit solar imagery (Fig. 6a). In the Jesuit garden, the two sundials not only resonated with San Vitale's martyr cycle, as mentioned above, but were also echoed by the presence of an obelisk located on the lower square of the garden (Fig. 9). Supplied with *imprese* and inscriptions directly pertaining to astronomy, the sun and the winds, this obelisk possessed a complex moral and exegetic significance, which Richeome explains in great detail in his book (Richeome 1611: 667–671): for instance, the eastern face shows the figure of Christ and the sun and the words, "His name is Orient" (Zechariah 6:12–13), thus recalling the very symbol of the Jesuit order that is also depicted on the vertical sundial.

Sundials and Astronomical Debates in Baroque Rome

As might be expected, the significance given to the sundials in the papal and Jesuit gardens varies considerably between them, but in each case reflects perfectly the context for which the sundial was produced. As Marcello Fagiolo (2000) has argued, the tetracycle sundial in the papal gardens may be understood in the framework of the interest of both Urban VIII (Maffeo Barberini) and the Barberini family in scientific culture and astronomical theories. Specifically, Fagiolo's interpretation of the Quirinal gardens as an image of the "City of the Sun", administered under the enlightened and absolutist government of Pope Barberini, insists on the affinities existing between certain elements of the gardens and Tommaso Campanella's famous eponymous work written in 1602 and published for the first time in 1623.[25] In this sense, and for obvious reasons of propaganda, the sun is given a centrality within the gardens' imagery which is fully consonant with a sympathetic approach to heliocentrism. As such, it can be compared to similar works of art with astronomical subjects made for the pope, such as Andrea Sacchi's *Allegory of Divine Wisdom* in the Palazzo Barberini, painted between 1629 and 1633, probably according to a written programme devised by Campanella, in which, although the Earth looms large in the foreground, the Sun takes centre stage (Fig. 11).[26] The chronology would fit with such an interpretation, since the placement of astronomical instruments in the papal garden around 1624–1628 corresponds to the very period during which Maffeo Barberini's accession to the papacy gave Galileo cause to hope for papal support.[27] If we are to follow this interpretation, however, it must be emphasized that the heliocentrism discussed here referred to Galileo's theories only on a secondary level and that it was the propaganda value of the pope's heliocentric heraldry that was really decisive.

[25] On Campanella and Urban VIII, see further Bolzoni (2001).

[26] See Scott (1991: 38–94), for this work and the interest in astronomy and astrology among the Barberini circle, as well as the earlier work of Lechner (1976).

[27] (Fagiolo 2000: 227–229). On the relationship between Galileo and Urban VIII, see further Scott (2007).

Fig. 11 Andrea Sacchi, Allegory of Divine Wisdom, Rome, Palazzo Barberini, 1629–1630, fresco, vault, © Wikimedia Commons. Image is in public domain

While still a cardinal, Maffeo Barberini, in a poem entitled *Adulatio perniciosa*, had praised the "lens" of the "learned Galileo" and his "skill" in using it to make new astronomical discoveries, but his position was no different from that of other prelates and members of the Church who posited the subordination of science to Holy Scripture. The fact that Tommaso Campanella stressed this passage of the poem in his *Commentaria* of 1628 provoked the ire of the pope, who was afraid of being associated with supporters of the Copernican theory (Bolzoni 2001: 265).[28] Urban VIII wanted Galileo's prestige without accepting his most cherished beliefs. To be sure, the armillary spheres set in the garden, topped respectively by a cross and the pope's coat of arms, as can be seen in Giovan Battista Falda's engraving, were almost certainly geocentric Ptolemaic models, with the earth in the centre, not Copernican (Fig. 5b, c).[29] In short, if the garden was indeed Galileo's panegyric, it was first and foremost the pope's universe.

[28] On the great difficulty of proclaiming heliocentrism in seventeenth century Rome, see further Beretta (2009).

[29] Most armillary spheres are constructed according to the Ptolemaic (geocentric) system and have a small terrestrial globe in the centre. Examples survive from medieval times to c. 1700, when a few Copernican (heliocentric) armillary spheres were made. Furthermore, the armillary spheres in

To clarify this vexed issue of how astronomical sources were translated into encomiastic monuments, we may turn to an interesting—and I believe unpublished—Latin poem. In 1623, Ignazio Bracci, a Jesuit, protonotary apostolic and a friend of Federico Cesi and the members of the Accademia dei Lincei, published a book on the etymology of the pope's name, anticipating in a way the extraordinary publication in 1626 of Federico Cesi's *Apiarium*. The book, entitled *La occulta corrispondenza tra l'arme, e 'l cognome de' Barberini*, focuses on the bees and, like the Quirinal sundial with its rich heraldic decoration, is a celebration of the pope's good government. One poem is dedicated to Galileo; it is quoted here in English translation (Bracci 1623: 26):

> ABOUT THESE SAME THINGS, TO GALILEO, THE FIRST AMONG MATHEMATICIANS
> Galileo, to whom no part of the sky is unknown
> and who can call each star by its name,
> you open the constellations to ignorant mortals,
> while the earth marvels that you have only counted its lights.
> Say, what star shines in the sky with three bees,
> a star which, when it rises, you know, a great expanse of ground is covered with flowers?
> Do you think that these bees have their origin in the rotting flesh of the bull of Europa?
> Do you think the high table of Jupiter, abundantly supplied with ancient banquets, calls for a new nectar and a new ambrosia?[30]

The mention of Europa's bull is a reference to the ancient belief that bees were born from the inside of the bull (*bugonia*), as explained in Virgil's *Georgics* (IV, 299–314 and 548–558). The association of this legend with the bull of Europa, however, is a modern invention and must be understood in light of Galileo's astronomical discoveries, in 1609, of Jupiter's satellites in the constellation of Taurus (in antiquity, the constellation represented Jupiter disguised as a bull after his rape of Europa). Semantically, the poem is thus constructed in a manner very similar to the Quirinal sundial. It celebrates Galileo's scientific discoveries by using ancient sources, notably Virgil, but is essentially aimed at celebrating the pope's power and the dawn of a new golden age, a new era of abundance and prosperity made explicitly visible in the magnificent papal gardens. In both examples, the heraldic symbols of the pope are the pivotal elements in the construction of meaning.

Now let us return to Richeome's exegesis of Christoph Clavius' sundials in the Jesuit garden. Richeome uses two important biblical sources relating astronomical miracles to explain to the novices and to garden visitors how those sundials should be

---

Falda's engraving have a tilted cage and a large horizontal band representing the horizon, which is typical of Earth-centred armillary spheres and is often missing when the Sun occupies the centre.

[30] "DE EISDEM, AD MATHEMATICORVM / Principem Galilaeum. Pars caeli quem nulla latet, Galilaee, potesque / Nomine qui, stellas dicere quamque suo, / Sideraque ignaris aperis mortalibus, unum / Mirante orbe suas te numerasse faces. / Dic age, quodnam apibus caelo tribus emicat astrum, /Quo exoriente, vide'n, plurima floret humus? / Anne Europaei dissecto viscere tauri / Ortum credis apes has habuisse suum? / An Iovis alta, putas, dapibus saturata vetustis / Nectar mensa novum poscit, & ambrosiam?" I thank Lise Ottis for her help in translating and interpreting this passage.

understood: the story of the Sun standing still in Joshua 10:12, and that of the Sun reversing its course in 2 Kings 20: 8–11, also known as *Horologium Ahas*. What is especially interesting here is that these were the main sources used by Jesuit theologians to reject the Copernican theory of terrestrial motion, as Volker R. Remmert (2006, 2011) has shown, and that they were used for the same reason by Clavius himself, first in his Sacrobosco commentary of 1570 and again as illustrations on the title-page of his *Opera mathematica* of 1612 (Fig. 12). It is likely that Richeome knew this from Clavius' works, or even directly from Clavius himself, and that the novices propounded those same arguments to the students and pilgrims who visited the garden.[31] As in the papal gardens, it thus appears that the debate about Copernican theories had a bearing on the way those sundials were used and understood. Nonetheless, I would argue that the scientific debate does not seem to have been the main point here, only an implied one. Richeome—like Clavius on his title-page—does not fully decode the implicit meaning of these biblical references. In the same way that the sundial becomes the object of heraldic praise in the papal gardens, relegating the question of its link with heliocentrism to mere conjecture, Richeome is primarily interested in the moral dimension of the sundials, which takes up the whole of the second section he dedicates to sundials in his book. The sundial is presented as a reminder of the inevitability of death, at which point the precise measure of time becomes totally irrelevant. Thus, for Richeome (1611: 501–502), one must above all follow "the Divine Sun Jesus Christ", which "does not set for those who want to be wakeful and fight like brave Joshua, who stopped the course of the Sun". Time, quadrants and sundials will be useless, he concludes, for those who have entered victoriously into the "city of light", for eternity.[32]

[31] The garden and its ornaments were used by the Novices for teaching the Christian doctrine. In the engraving by Greuter, we see children and visitors being guided and taught in the different parts of the garden (Fig. 9). On this educational dimension of the Jesuit garden, see Witte (2011: 102–106).

[32] "Donc, selon les divers points de ces horloges diverses, les vingt-quatre heures finissent, au levant, au couchant, au midi, et à la minuit; Et selon le cours des horloges de la vie humaine, elles peuvent finir à chaque heure du jour, sans avoir égard aux cadrans, et points artificiels du jour: Car quand la mort viendra soit à dix heures, soit à douze, soit à quatorze ou à vingt, soit à midi, ou à la minuit; à la Diane, ou à la serée [*sic*], voilà nos vingt-quatre heures passées, et notre Soleil couché, et notre jour mortel, et naturel fini; et partant, mes bien aimés, tenons-nous prêts à chaque point du jour, à chaque heure, voire à chaque minute; Car à tous points nous pouvons mourir; Et comme fleurs Héliotropes de sens, et de raison, suivons le Soleil divin Jésus-Christ, qui ne couche point pour ceux, qui veulent veiller, et combattre à guise de vaillants Josué, qui arrêta le Soleil pour avoir son jour à grande mesure, et donner à souhait contre les ennemis de Dieu, et demeurant maîtres du champ, emporter la victoire entière; suivons ses conseils et avertissements, afin qu'en la lumière de sa grâce, nous voyions le chemin de salut, voyions nos ombres, nos heures, et nos inconstances, nos vanités, nos imperfections grandes, et petites, et ayant la victoire de nous-même, marchant toujours avec notre Soleil, et faisant toujours progrès à la vertu envers Dieu, et gardons-nous d'être surpris de la nuit; Afin que quand cette nuit viendra, et qu'il faudra quitter la vie de ce corps, et sortir de l'inconstance du temps coulant, nous entrions victorieux en la cité de lumière, au pourpris [*sic*] de l'éternité permanente, et recevions la couronne de gloire, qui vole par dessus les cadrans, et les temps, et les siècles des siècles". Interrestingly here, the use of sundials runs parallel to the moralizing interpretations that David Coffin sees in English sundials and in the English emblem literature. Coffin (1994: Chap. I).

Fig. 12 Frontispiece to Christoph Clavius, *Opera mathematica*, 1612, © from Remmert (2011, 75)

The Image of the Sundial as an Architectural Source?

Before leaving the Quirinal hill, one should note that the importance of astronomical measurements—especially sundials—for the Jesuits may have had an impact on the architecture of the main church of the Noviciate adjacent to the garden. The church of Sant'Andrea del Quirinale, conceived by Gianlorenzo Bernini and Pope Alexander VII Chigi, was built much later, between 1658 and 1672, on the site of the former chapel of the Noviciate. Concerning Bernini's famous oval plan for the church, Geraldine Dunphy Wind (2013) has observed a "remarkable congruence of the radial unity of Bernini's Church and geometric properties found in Vitruvius's instructions for making a sundial, described in his *De architectura*" (IX, 7 and 2–7).[33] According to her reading, the entire plan is strictly based on the image of the Vitruvian sundial or *Alemma*, a hypothesis which would fit with the solar appearance of the vault itself, with its ribs and the illuminated central lantern (Fig. 13). The image of the sundial would have had deep symbolic meaning. It celebrated Pope Chigi, who also used solar imagery, and functioned as part of the Jesuit poetics of light, entering into dialogue not only with the various sundials, obelisks, flowers and plants in the adjacent garden, but above all with the other main iconographic elements of the church, namely the 'heliotropic' Saint Andrew ascending into heaven.[34] The staging or 'phenomenology of light' (Fagiolo 2013: 113) at Sant'Andrea indeed recalls, Dunphy Winds (2013) argues, the text from the liturgy for Saint Andrew's feast day, which is central to the entire program of the church: "An exceeding Bright light (*luce perfusus*) surrounds him for nearly a half an hour and then in this light (*cum luce*), and in this joy, he mounts to the realms of light (*lucis atria*)" (Dunphy Wind 2013: 22).

Although the idea is appealing, there is no hard evidence to prove that Bernini designed the church with Vitruvius's sundial in mind. What Dunphy Wind's argument demonstrates, at best—and this is what is most interesting to us—is that *it is possible* that Jesuit novices and other people familiar with sundials made that connection, whether it was intended or not. They knew the sundials that had been in the garden for at least half a century and had been exposed to the long Jesuit tradition of astronomical studies, of which they had almost certainly learned at least the first principles in the course of their apprenticeship.

The Jesuit novices at Sant'Andrea del Quirinale would have been taught astronomy and mathematical sciences at the nearby Collegio Romano, the most important centre of astronomical research in Rome together with the monastery of the Minim friars at Trinità dei Monti. Such was the case, for example, of Orazio Grassi, who entered the novitiate of Sant'Andrea in 1600 and then studied at the Collegio Romano between 1603 and 1610 with Clavius, Christoph Grienberger and Odo van Maelcote. Grassi was not only well versed in astronomy (*Tractatus tres de*

[33] On the planning of the church, see Connors (1982).

[34] For a thorough reading of the iconographic programme of Sant'Andrea, see Careri (1990: 149–183).

Fig. 13 Gianlorenzo Bernini, Antonio Raggi, Vault of the church of Sant'Andrea al Quirinale, Rome, 1661–1666, © Wikimedia Commons. Image is in public domain

sphera, de horologis ac de optica, 1617) and a fierce opponent of Copernican theories and Galileo, but was also expert in architecture—as *consiliarius aedificiorum* he supervised the construction of every building for the Company of Jesus and especially that of Sant'Ignazio. He published a treatise on Vitruvius' chapter on sundials *In primum librum de architectura M. Vitruvii et in nonum eiusdem de horologiorum solarium descriptione duo brevissimi tractati*, in 1624 (Milan, Biblioteca nazionale Braidense, AF.IX.33). This chapter is the source, according to Dunphy Wind's proposal, of the oval plan of the church of Sant'Andrea. Grassi is also known to have carried out thorough research into the physics of light. Since he died in 1654, he could not have been responsible for this idea, but, again, his work demonstrates the *plausibility* of establishing links between astronomy and architecture in early modern Rome.[35]

Piazza San Pietro is another case in point. Since the time of Sixtus V, when the obelisk was re-erected in the center of the Piazza, plans had been formed to turn the huge monument into the needle of a large sundial, based on the prestigious model of the *Horologium Augusti* (*Avviso* di Roma of December 24, 1588; D'Onofrio 1967:

[35] Further on Orazio Grassi, see Presti and Ercolino (2002) and Bösel (2004).

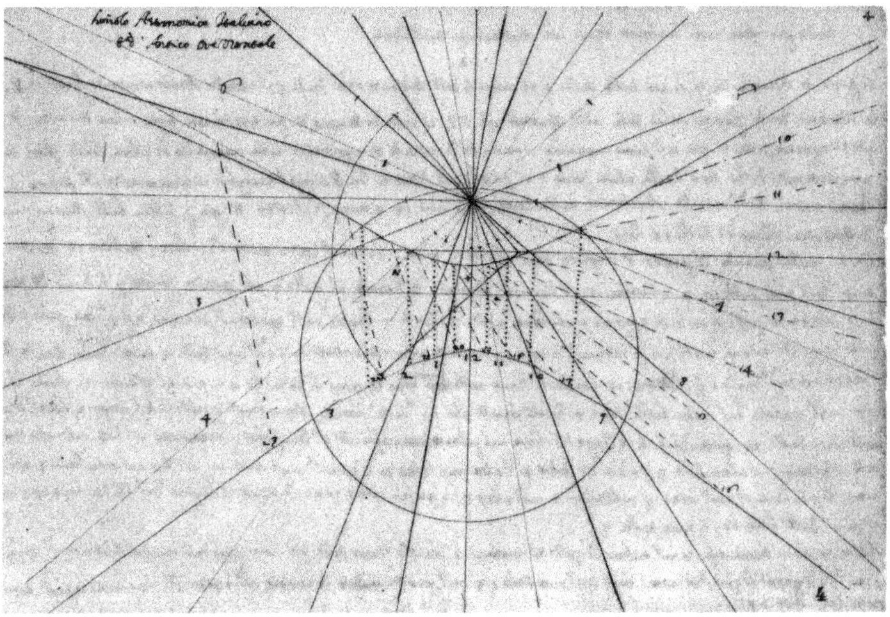

Fig. 14 Fabio Chigi (Alexander VII), Drawing of a sundial, Biblioteca Apostolica Vaticana, Chig. A I 18, f. 4 (from Morello 1992: Fig. 2)

90).[36] Again, in a letter dated from 1667 to Cardinal Giacomo Filippo Nini, Athanasius Kircher suggested creating 'l'horologio più grande del mondo' (Bartola 1989: 99) on Piazza San Pietro and this grandiose idea was given graphic form in Cornelis Meyer's book of 1685 (Pl. XV) (Fig. 15). We know that Fabio Chigi (Alexander VII) himself had a deep interest in gnomonics. For instance, he left us a drawing of a sundial probably based on Augustus's sundial in the Campomarzio (BAV Chig.a.I.18, f. 4; Morello 1992: Fig. 2) (Fig. 14). This has led some art historians to suggest that the pope, with the genial help of Bernini, conceived the great oval of Piazza San Pietro as a magnificent sundial, or 'Theater of the Sun', "for the measure of time and space, with its obelisk conceived as both a gnomon and *axis mundi*" (Fagiolo 2013: 115–121). In truth, only in November 1817 would a meridian line be built and the obelisk used as a gnomon.[37] As for Sant'Andrea del Quirinale, Alexander VII's interests in the sun and sundials cannot explain in itself the entire conception of Piazza San Pietro, but it points to the possibility that he and some of his contemporaries could *imagine* it functioning as a sundial. This leap

[36] See also s.a. *Piazza San Pietro* (2007).

[37] "A cura di mons. Gilii, astronomo vaticano, venne disegnata una meridiana per movimenti mensili e giornalieri del sole all'ora del meriggio, alla quale questo obelisco serve di gnomone". See D'Onofrio (1967: 90).

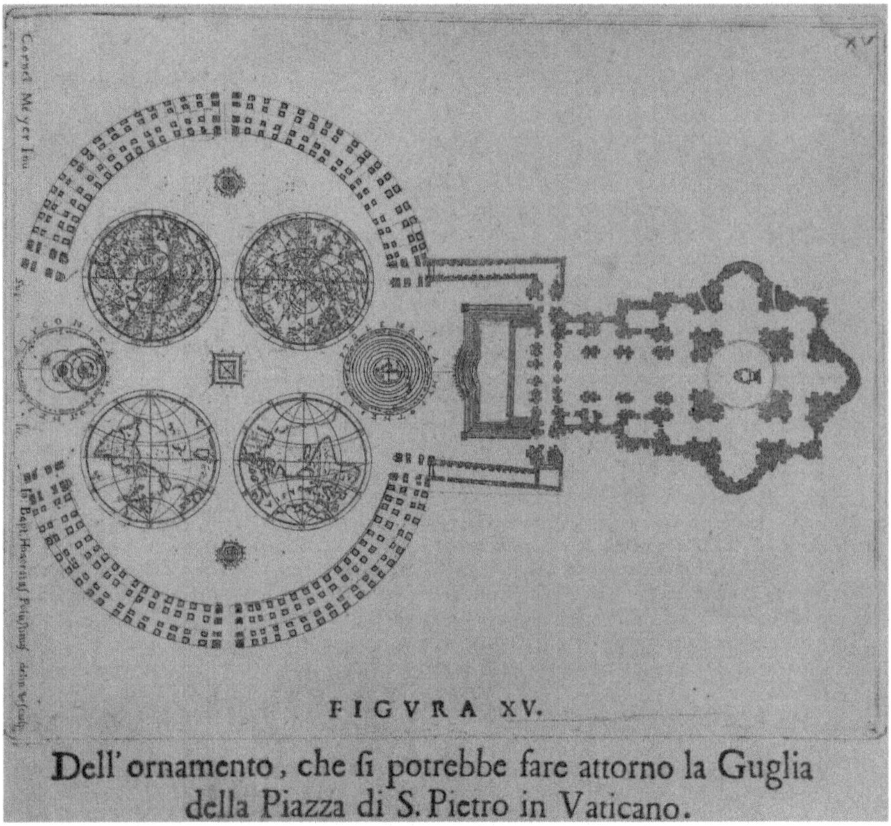

Fig. 15 Gaspar Van Wittel (?), Project for Piazza San Pietro, in Cornelis Meyer, *Arte di restituire a Roma la tralasciata navigazione del Tevere, Rome*, 1685, pl. XV, © Photo author

from the world of concrete appearances to that of the scientific and curious imagination is exactly what Cornelis Meyer's wonderful book of 1685 makes tangible.

Conclusion

The study of mathematical instruments used as garden ornaments offers an interesting occasion to question the interaction between the history of material culture and the history of theories and ideas. Sundials are, primarily, useful instruments to measure time but, in the early modern period, they were also 'useful' in many other ways, as objects of meditative practices, as teaching tools or as encomiastic monuments. To borrow Koen Vermeir's conclusion about Athanasius Kircher's clocks and sundials and their metaphysical meaning, they were "at once measuring instruments and representations of political and metaphysical meanings [...]

meanings related with entertainment, spectacles and courtly culture" (Vermeir 2007: 394). Michel Serres makes an even stronger statement: "[T]he gnomon or sundial is less useful as an instrument to tell the time of day, one which nobody has cared about between Antiquity and our grandparents' time, than it is as a geometrical model of the Universe: at the same time an observatory and a cosmographic schema of the world".[38] In the case of the Quirinal sundials we have discussed, their central presence in large and famous gardens is obviously linked with the scientific debates that took place during this period, and more specifically with important astronomers such as Galileo or Christoph Clavius, who had direct associations with those places. Yet, by becoming a part of the gardens, these sundials took on meanings which added a new complexity to their connection with the world of professional scientists. The scientific aspect of the instruments somehow became entangled with—and perhaps even secondary to—their essentially symbolic, propagandistic or moral function. In this sense, gardens, with their astronomical imagery, can be compared in many ways to the title-pages of early modern mathematical and astronomical books. They shaped the way users or viewers understood Nature and the Universe, reminding them that the main goal of science, even in its most refined applications, must be the search for God and the building of a new Paradise.

Besides provoking wonder and delight, like an automaton, a marvellous fountain or a precious statue, these mathematical instruments materialized a fundamental connection between Heaven and Earth, between Nature as God's infinite cosmic creation and nature in the garden.[39] Being derived from mathematical knowledge, they echoed the geometrical features of the garden and emphasized the human capacity not only to rule over nature but also to explore the boundless sky. They boosted the owners of these gardens as patrons of the sciences and, most importantly, I believe, provided elite non-scientists—for instance, Urban VIII or Alexander VII—with an opportunity to "talk about science", to convey what was truly important about their engagement with science, a *discourse-about* which was of primary importance in the early modern period.[40] In Stefano Guazzo's *La civil conversazione*, one of the key manuals on social behaviour of the time, "knowledge begins in conversation and ends in conversation" (Guazzo 1993: I, 30). In the same way that *discourse-about* painting and *discourse-about* music became central as a badge of identity for the social elite, one may argue that *discourse-about* science, and astronomy in particular, was also of great importance for the creation

[38] Serres (1986): "Le gnomon ou cadran solaire sert moins à dire l'heure dont tout le monde se moque depuis l'Antiquité jusqu'à nos grands-parents, qu'à construire un modèle géométrique del'Univers: observatoire à la fois et schéma cosmographique du monde". Further on this issue, see Mosley (2006).

[39] On the role of entertainment and public enjoyment in the development of early modern technology, see Findlen (1990) and Bredekamp (1995: 69–80).

[40] For a recent and thorough discussion of science in early modern Rome, see Romano (2009), with ample bibliography.

of meaning in the early seventeenth century. Speaking of music, Andrew Dell'Antonio has remarked (2011: 6):

> Effective *discourse-about* (a marker of the civilized conversationalist) enables a change in the status of the object—indeed, in the case of music, encourages the idea that music (through careful assessment of the listener's experience) can somehow be "objectified" as an item for discussion. More importantly, *discourse-about* also brings attention to discourse as crucial to the creation of status and meaning (though the discourse itself may ostensibly be a transparent account of the item under discussion). Thus the sophisticated or elevated nature of the discourse can be seen as more important than its "objectivity" or "accuracy" (in technical terms) concerning the item around which discourse is taking place.

An interpretative framework of this kind could prove especially useful for discussing the iconography *of* and *on* (that is, a form of *discourse-about* on) scientific instruments and their link with larger decorative and/or architectural ensembles, such as the magnificent gardens of Rome that Galileo had appreciated so much in 1611.

Bibliography

Sources

Baglione, Giovanni. 1642. Le Vite de' pittori, scultori et architetti. Dal pontificato di Gregorio XIII nel 1572 in fino a' tempi di Papa Urbano Ottavo nel 1642. Rome.

Bracci, Ignatio. 1623. La occulta corrispondenza tra l'arme e il cognome dei Barberini. Roma: Per Guglielmo Facciotti.

Fabricii, Principio. 1588. Delle allusioni, imprese, et emblemi del Sig. Principio Fabricii da Teramo sopra la vita, opere, et attioni di Gregorio XIII. Pontefice massimo, Libri VI. Nei quali sotto l'allegoria del Drago arme del detto Pontefice si descrive anco la vera forma d'un Principe Christiano; et altre cose, la somma della quali si legge dopo la dedicatione dell'opera all'Ill(ustrissi)mo et Ecc(ellentissi)mo S(igno)r Duca di Sora. Rome: Bartolomeo Grassi.

Galileo, Galilei. 1890-1909. Opere di Galileo Galilei. Ed. Antonio Favaro. Firenze: G. Barbèra.

Kircher, Athanasius. 1646. Ars magna lucis et umbrae. Rome.

Kircher, Athanasius. 1650. Musurgia universalis, sive ars magna consoni et dissoni. Rome.

Line, Francis. 1673. An explanation of the Diall sett up in the King's garden in 1669 / Explicatio Horologii in Horto Regio Londini in Anglia an. 1669. erecti, in quo plurima horologiorum sciatericorum genera continentur : quibus praeter omnis generis horas diversimode expressas, multa etiam ad Geographiam Astrologiam et Astronomiam spectantia, per Solis umbram oculis cernenda subjiciuntur. Inter quae, plurima, et potissime magis curiosa, noviter inventa, et a nemine hactenus tradita reperiuntur. Quae omnia breviter et dilucide publicae utilitati exponit Reverendus Pater Franciscus HALLUS, alias Linus Societatis Jesu, Matheseos professor. Liège: Streel [Coll. Univ. Liège, I. 105.2].

Menestrier, Claude François. 1663. Le Temple de la sagesse ouvert à tous les peuples. Lyon: Chez Antoine Molin.

Meyer, Cornelius. 1685. Arte di restituire a Roma la tralasciata navigazione del Tevere. Rome: s.n.

Richeome, Louis. 1611. La peinture spirituelle ou l'Art d'admirer aimer et louer Dieu en toutes ses oeuvres, et tirer de toutes profit salutere. ... Lyon: Pierre Rigaud.

Scheiner, Christoph. 1630. Rosa Ursina. Braccianum: s.n.

Secondary Literature

s.a., 2007. Piazza San Pietro: una meridiana e una rosa dei venti per la Basilica Vaticana. La Basilica di S. Pietro 19: 1-2.

Aksamija, Nadja. 2011. The Emblem Book as the World Reformed: Symbolic Landscapes in Principio Fabricii's Delle allusioni, imprese, et emblemi. In Sacred Landscape. Landscape as Exegesis in Early Modern Europe, eds. Denis Ribouillault and Michel Weemans, 133-159. Florence: Olschki.

Angelucci, Sergio. 1991. La lancetta dell'orologio: restauro e significati. Bolletino d'arte 66: 101-103.

Angelucci, Sergio. 1993. Il tempo del drago. L'orologio di Sant'Atanasio dei Greci a Roma. Art & Dossier 80: 38-41.

Bailey, Gauvin Alexander. 2003a. Between Renaissance and Baroque: Jesuit Art in Rome, 1565-1610. Toronto et al.: University of Toronto Press.

Bailey, Gauvin Alexander. 2003b. La contribution des jésuites à la peinture italienne et son influence en Europe 1540-1773. In L'art des Jésuites, ed. G. Sale, 123-168. Paris: Mengès.

Barbieri, Patrizio. 1981. L'organo idraulico del Quirinale. L'Organo 19: 7-61.

Bartola, Alberto. 1989. Alessandro VII e Athanasius Kircher S.J. Ricerche ed appunti sulla loro corrispondenza erudita e sulla storia di alcuni codici chigiani. Miscellanea Bibliothecae Apostolicae Vaticanae 3: 7-105.

Beaven, Lisa. 2007. Claude Lorrain's Harbour Scenes: Sun, Science and the Theatre in the Barberini Years. In Art, Site and Spectacle: Studies in Early Modern Visual Culture, ed. D.R. Marshall, 146-161. Melbourne: Fine Arts Network.

Beretta, Francesco. 2009. L'héliocentrisme à Rome, à la fin du XVIIe siècle: une affaire d'étrangers?. In Rome et la science moderne entre Renaissance et lumières, ed. A. Romano, 529-554. Rome: École française de Rome.

Bolzoni, Lina. 2001. Urbano VIII, Campanella e la censura dei 'Commentaria'. In Il piacere del testo. Saggi e studi per Albano Biondi, ed. Adriano Prosperi, 265-284. Rome: Bulzoni editore.

Bösel, Richard. 2004. Orazio Grassi: architetto e matematico gesuita; un album conservato nell'archivio della Pontificia Università Gregoriana a Roma. Rome: Argos.

Bredekamp, Horst. 1995. The Lure of Antiquity and the Cult of the Machine. The Kunstkammer and the Evolution of Nature, Art and Technology. Princeton: Markus Wiener.

Camerota, Filippo. 2000. La meridiana tetracycla del Quirinale. In Francesco Borromini, eds. Christoph L. Frommel and Elisabeth Sladek, 233-241. Milan: Electa.

Careri, Giovanni. 1990. Envols d'amour: le Bernin; montage des arts et dévotion baroque. Trad. Michelle Coquet. Florence: Usher.

Coffin, David. 1991. Gardens and Gardening in Papal Rome. Princeton: Princeton University Press.

Coffin, David. 1994. The English Garden. Meditation and Memorial. Princeton: Princeton University Press.

Connors, Joseph. 1982. Bernini's S. Andrea al Quirinale: Payments and Planning. The Journal of the Society of Architectural Historians 41: 15-37.

Connors, Joseph. 2015. The One-room Apartment of Cornelis Meijer. In Artistic Practices and Cultural Transfer in Early Modern Italy: Essays Dedicated to Deborah Howard, eds. Nebahat Avcıoğlu and Allison Sherman, 45-64. Burlington VT: Ashgate.

Courtright, Nicola. 2003. The Papacy and the Art of Reform in Sixteenth Century Rome. Gregory XIII's Tower of the Winds in the Vatican. Cambridge: Cambridge University Press.

Davidson, Peter. 2006. The Jesuit Garden. In The Jesuits II. Cultures, Sciences, and the Arts, 1540-1773, ed. John O'Malley, 86-107. Toronto et al.: University of Toronto Press.

Dekoninck, Ralph. 2005. 'Ad imaginem': Statuts, fonctions et usages de l'image dans la littérature spirituelle jésuite du XVIIe siècle (Travaux du Grand Siècle, 26). Genève: Droz.

Dell'Antonio, Andrew. 2011. Listening as Spiritual Practice in Early Modern Italy. Berkeley, Los Angeles, London: University of California Press.

D'Onofrio, Cesare. 1967 (1965). Gli obelischi di Roma. Roma: Bulzoni editore.
Fabre, Pierre-Antoine. 1995. Lieu de mémoire et paysage spirituel. In Le jardin, art et lieu de mémoire, eds. M. Mosser and P. Nys, 135-148. Besançon: Éditions de l'Imprimeur.
Fagiolo, Marcello. 1998. I giardini papali del Vaticano e del Quirinale. In Giardini Regali, Fascino e immagini del verde nelle grandi dinastie: dai Medici agli Asburgo, ed. Monica Amari, 69-80. Milan: Electa.
Fagiolo, Marcello. 2000. Il segno del giovane Borromini nella 'Città del Sole' di Urbano VIII (1624-1631). In Francesco Borromini, eds. Christoph L. Frommel and Elisabeth Sladek, 215-232. Milan: Electa.
Fagiolo, Marcello. 2013. Bernini e il 'gran teatro' della Roma barocca. In Roma barocca. I protagonisti, gli spazi urbani, i grandi temi, ed. Marcello Fagiolo, 97-125. Roma: De Luca.
Fagiolo, Marcello. 2014. Roma di Innocenzo XI: la città, la scienza, le feste e l'incontro con Le Nôtre. In Innocenzo XI Odescalchi. Papa, politico, committente, eds. Richard Bösel et al., 275-288. Rome: Viella.
Findlen, Paula. 1990. 'Quanto scherzevole la natura'. La scienza che gioca dal Rinascimento all'Illuminismo. Intersezioni. Rivista di Storia delle Idee 3: 413-436.
Frommel, Christoph Luitpold. 1999/2002. Il palazzo del Quirinale tra il XV e il XVII secolo. Quaderni dell'Istituto di Storia dell'Architettura, 34/39: 275-284.
Galluzzi, Paolo. 2009. Galileo e gli 'spettacoli del cielo' sul colle del Quirinale. Il Quirinale 5: 81-90.
Gijsbers, P.-M. 1999. Claudio Aquaviva, Louis Richeome, and Durante Alberti's Altarpiece for Sant'Andrea al Quirinale. In Docere, delectare, movere: affetti, devozione e retorica nel linguaggio artistico del primo baroco romano, 27-40. Rome: De Luca.
Godwin, Joscelyn. 1979. Athanasius Kircher: A Renaissance Man and the Quest for Lost Knowledge. London: Thames and Hudson.
Guazzo, Stefano. 1993. La civil conversazione (1574). Ed. Amedeo Quondam, 2 vols. Ferrara: Franco Cosimo Panini.
Guiderdoni-Bruslé, Agnès. 2001. Les formes emblématiques de 'l'humanisme dévot': une lecture du 'Catéchisme royal' (1607) de Louis Richeome, S.J. In An Interregnum of the Sign: The Emblematic Age in France. Essays in honour of Daniel S. Russell (Glasgow Emblem Studies, 6), ed. D. Graham, 227-235. Glasgow: University of Glasgow.
Harris, Steven J. 1999. Mapping Jesuit Science: The Role of Travel in the Geography of Knowledge. In The Jesuits: Cultures, Sciences, and the Arts, 1540-1773, ed. John W. O'Malley, vol. 1, 212-240. Toronto et al.: University of Toronto Press.
Heilbron, John L. 1999. The Sun in the Church. Cathedrals as Solar Observatories. Cambridge: Harvard University Press.
Höltgen, Karl Joseph. 2004. The Illustrations of Louis Richeome's 'La peinture spirituelle' (1611) and Jesuit Iconography. In Florilegio de Estudios de Emblemàtica, ed. S. López Poza, 447-458. El Ferrol: Sociedad de Cultura Valle Inclàn.
Iversen, Erik. 1968. The Obelisks of Rome. In Obelisks in Exile, vol. 1. Copenhagen: G.E.C. Gad Publishers.
Lattis, James M. 1994. Between Copernicus and Galileo: Christoph Clavius and the Collapse of Ptolemaic Cosmology. Chicago: University of Chicago Press.
Lechner, George S. 1976. Tommaso Campanella and Andrea Sacchi's Fresco of Divina Sapienza in the Palazzo Barberini. The Art Bulletin 58: 96-108.
Loach, Judi. 2012. An apprenticeship in 'spiritual painting'. Richeome's La peinture spirituelle. In Ut pictura meditatio. The Meditative Image in Northern Art, 1500-1700, eds. Walter S. Melion, Ralph Dekoninck and Agnes Guiderdoni-Bruslé, 337-399. Turnhout: Brepols.
Macioce, Stefania. 1990. Undique splendent: Aspetti della pittura sacra nella Roma di Clemente VIII Aldobrandini. Rome: De Luca edizioni d'arte.
Macioce, Stefania. 1995. Aspetti simbolici nel giardino del noviziato di S. Andrea al Quirinale in Roma. In Memor fui dierum antiquorum: Studi in memoria di Luigi di Biasio, ed. P.C. Ioly Zorattini, 381-395. Udine: Camponotto.

Mancinelli, Fabrizio, and Juan Casanovas. 1980. La Torre dei venti in Vaticano. Vatican city.
Meadows-Rogers, Robert D. 1996. The Vatican Logge and their Culminating Decorations under Pius IV and Gregory XIII: Decorative Innovation and Urban Planning before Sixtus V. Ph. D. University of North Carolina.
Morello, Giovanni. 1992. I rapporti tra Alessandro VII e Gian Lorenzo Bernini negli autografi del Papa (con disegni inediti). In Documentary Culture. Florence and Rome from Grand-Duke Ferdinand I to Pope Alexander VII, eds. E. Cropper and G. Perini, 185-207. Bologna: Nuova Alfa Ed.
Mosley, Adam. 2006. Objects of Knowledge : Mathematics and Models in Sixteenth-Century Cosmology and Astronomy. In Transmitting Knowledge: Words, Images, and Instruments in Early Modern Europe, eds. S. Kusukawa and I. Maclean, 193-216. Oxford: Oxford University Press.
Pliny. 1940. Natural History. Trans. H. Rackham. Cambridge: Harvard University Press.
Presti Cesare, Ercolino Maria Grazia. 2002. Orazio Grassi. Dizionario biografico degli italiani, 58. http://www.treccani.it/enciclopedia/orazio-grassi_%28Dizionario-Biografico%29/. Accessed 15 September 2013.
Quiviger, François. 2003. Honey from Heaven: Aspects of the Topos of the Bees in Renaissance Artistic Literature. In Visuelle Topoi. Erfindung und tradiertes Wissen in den Künsten der italienischen Renaissance, eds. Ulrich Pfisterer and Max Seidel, 317-321. München-Berlin: Deutscher Kunstverlag.
Reilly, Connor. 1962. Francis Line, Peripatetic (1595-1675). Osiris 14: 222-253.
Reilly, Connor. 1969. Francis Line SJ: An Exiled English Scientist, 1595-1675. Rome: Institutum historicum S.I.
Remmert, Volker R. 2006. Picturing Jesuit Anti-Copernican Consensus. Astronomy and Biblical Exegesis in the Engraved Title-page of Clavius's Opera mathematica (1612). In The Jesuits II. Cultures, Sciences, and the Arts, 1540-1773, ed. John O' Malley, 291-313. Toronto et al.: University of Toronto Press.
Remmert, Volker R. 2011. Catholic Biblical Exegesis and the Origins of the Galileo Affair: The Engraved Title Page of Clavius's Opera mathematica (1612). In Picturing the Scientific Revolution: Title Engravings in Early Modern Scientific Publications, 17-45. Philadelphia: Saint Joseph's University Press.
Romano, Antonella (ed.). 2009. Rome et la science moderne entre Renaissance et Lumières. Rome: École française de Rome.
Ruffini, Marco. 2005. Le imprese del drago: Politica, emblematica e scienze naturali alla corte di Gregorio XIII (1572-1585). Rome: Bulzoni editore.
Sauvenier-Goffin, Elisabeth. 1958. Une page de l'enseignement des sciences exactes de l'ancien pays de Liège: le tractatus de horologiis du Père Linus. Bulletin de la Société Royale des Sciences de Liège 27^{e}année, 11-12: 280-284.
Scott, John Beldon. 1991. Images of Nepotism: the Painted Ceilings of Palazzo Barberini. Princeton: Princeton University Press.
Scott, John Beldon. 2007. Galileo and Urban VIII. Science and Allegory at Palazzo Barberini. In I Barberini e la cultura europea del Seicento, eds. L. Mochi Onori, S. Schütze and F. Solinas, 127-136. Roma: De Luca.
Serres, Michel. 1986. Gnomon. Les débuts de la géométrie en Grèce. In Éléments d'histoire des sciences, ed. Michel Serres, 95-153. Paris: Bordas.
Shea, William R., and Mariano Artigas. 2003. Galileo in Rome. The Rise and Fall of a Troublesome Genius. Oxford: Oxford University Press.
Tesi, Valerio. 1991. Il Palazzo del Quirinale nel Cinquecento. In Il palazzo del Quirinale, ed. Franco Borsi, 77-114. Roma: Editoriale Lavoro.
Van Assche, Kristof. 1999. Louis Richeome, Ignatius, and Philostrates in the Novice's Garden: Or, the signification of Everyday Environment. In The Jesuits and the Emblem Tradition (Imago Figurata Studies 1a), eds. J. Manning and M. Van Vaeck, 3-10. Turnhout: Brepols.

Van Berkel, Klaas. 2002. 'Cornelius Meijer inventor et fecit'. On the Representation of Science in Late Seventeenth Century Rome. In Merchants and Marvels. Commerce, Science and Art in Early Modern Europe, eds. P. H. Smith and P. Findlen, 277-294. New York: Routledge.
Van Damme, Stéphane. 2005. Le temple de la sagesse. Savoirs, écritures et sociabilité urbaine (Lyon, XVIIe-XVIIIe siècles). Paris: Éditions de l'EHESS.
Vermeir, Koen. 2007. Athanasius Kircher's Magical Instruments: An Essay on 'Science', 'Religion', and Applied Metaphysics. Studies in History and Philosophy of Science 38: 363-400.
Wind, Geraldine Dunphy. 2013. Sant'Andrea al Quirinale and the sundial. Source. Notes in the History of Art 32: 19-26.
Witte, Arnold. 2011. The Power of Repetition. Christian Doctrine and the Visual Exegesis of Nature in Sixteenth- and Seventeenth-century Painting. In Sacred Landscape. Landscape as Exegesis in Early Modern Europe, eds. Denis Ribouillault and Michel Weemans, 91-112. Florence: Olschki.
Zampa, Paola. 2007. 'Per comodità, diletto e sicurezza': il complesso del Quirinale nei progetti e nelle sistemazioni di Urbano VIII. In I Barberini e la cultura europea del Seicento, Atti del Convegno Internazionale, Roma, palazzo Barberini, 7-11 dicembre 2004, 501-512. Rome: De Luca.

Jacques Lemercier's *Scenografia* of Montjeu: Architectural Prints, Cartography, and Landscape in 1620

Anthony Gerbino

Abstract

The Archives Nationales holds an extraordinary drawing from the early seventeenth century: a bird's-eye view of the château and gardens of Montjeu by the architect Jacques Lemercier. The drawing is remarkable partly because of its minute precision and rigorous use of perspective. The complex topography of the site—particularly its shifting elevation with respect to the château—is presented with remarkable economy. Lemercier's view is also extraordinary for another reason: it is one of the earliest working drawings we have of any French garden. It even stands out among contemporary prints of aristocratic landscapes, very few of which depict the château and its surroundings at such a large scale and as an integrated ensemble. The image sheds light not only on the formal evolution of the French garden, but also on the way that novel drawing conventions may have driven that evolution. As a project drawing, the Montjeu *scenografia* provides valuable information about how such sites were designed, as well as the technical and mathematical capabilities they required.

Introduction

The Archives Nationales holds an extraordinary drawing from the early seventeenth century: a bird's-eye view of the château and gardens of Montjeu in Burgundy (Fig. 1).[1] The image is small but detailed. Set within a larger field of blank paper, it measures only about 60 × 15 cm; yet, it is dense with information. The complex

[1] N III Saône-et-Loire 32. The inscription reads "Scenografia, Du Chasteau de Montjeu".

A. Gerbino (✉)
Art History and Visual Studies, University of Manchester, Mansfield Cooper Building, Manchester M13 9PL, UK
e-mail: anthony.gerbino@manchester.ac.uk

H. Fischer et al. (eds.), *Gardens, Knowledge and the Sciences in the Early Modern Period*, Trends in the History of Science, DOI 10.1007/978-3-319-26342-7_7

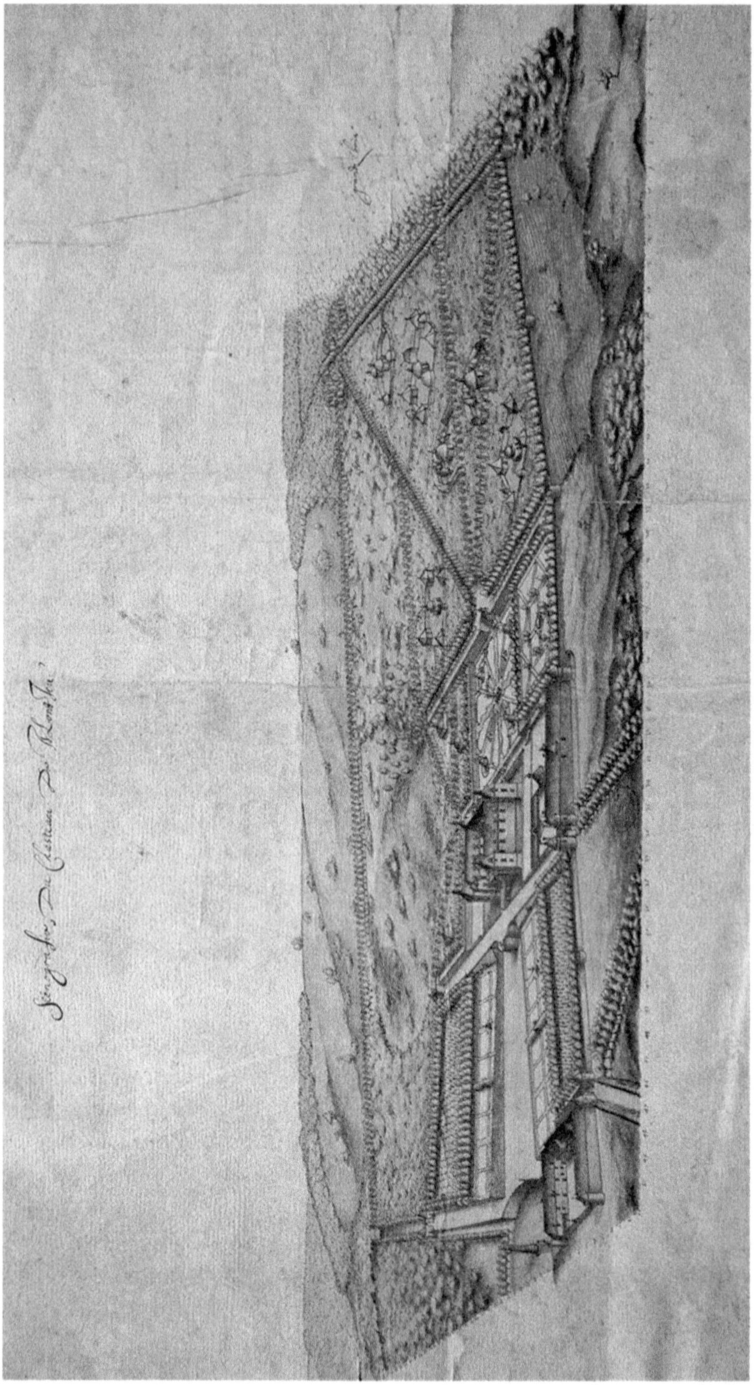

Fig. 1 Jacques Lemercier, *Scenografia du Chasteau de Montjeu*, 1619–1620 (Archives Nationales, Paris)

topography of the site, both natural and man-made, is presented with remarkable economy. The viewer looks across a low-lying valley, the far side rising gently toward the horizon. The house and gardens are set in a sequence of terraces along the foot of the slope. The most striking feature of the drawing is the rigorous use of perspective, which serves to clarify both the shifts in elevation and the way the château sits in the wider landscape. The gardens rise not only along the central axis but also on either side in a way that mirrors and exploits the sloping walls of the valley. The drawing's large scale is crucial to the effect. The house and its surroundings have been subordinated to a common measure, permitting both elements to be seen in reciprocal relationship.

The drawing's critical history is recent. It was first published in 1986 by Kenneth Woodbridge, who identified the château as the property of Pierre Jeannin, *conseiller d'état* and ambassador to Holland for Henri IV and subsequently *surintendant des finances* during the regency of Maria de'Medici (Woodbridge 1986: 161–162). Woodbridge also correctly dated the drawing to the period immediately after 1619. Jeannin had sold his post to Henri de Schomberg in September of that year and retired to his estates near Autun, where the minister, according to Tallement de Réaux, "amused himself by building". Woodbridge placed Montjeu among the early seventeenth-century houses of the "new financial elite", a group of royal ministers and *intendants* close to Richelieu, who commissioned the gardens at Wideville (1630–1636), Tanlay (1642), and the early country houses of François Mansart. For Woodbridge, this group of buildings—Mansart's in particular—proved to be a decisive influence on the work of André Le Nôtre in the second half of the century. It was the theatrical character of châteaux such as Balleroy and Maisons—sited in carefully controlled, but immense landscapes—that presaged the creation of Vaux-le-Vicomte.

In 2005, Alexandre Gady considerably advanced our understanding of the Montjeu drawing, when he attributed it—convincingly—to Jacques Lemercier. Gady recognized Lemercier's handwriting in the inscription and buttressed this observation with an important discovery: a contract for the service buildings of the lower court (distinguished in the drawing by their red roofs). Dated March 13, 1620, the contract calls for a local mason to undertake the work according to Lemercier's designs.[2] The architect had already provided plans for Jeannin's nearby château at Dracy-Saint-Loup; the "scenografia" suggests that Lemercier also had, or was searching for, a greater role at Montjeu. The unusual use of perspective and the lively coloring suggest that it is a type of presentation drawing, presumably intended for Jeannin to envision the completed project before construction. Gady's attribution serves to link two seemingly unrelated aspects of Lemercier's early career: the architectural engravings produced in Rome during his sojourn in 1607–1608 and his designs for Cardinal Richelieu's eponymous château in Poitou, executed in the 1630s.

[2] Minutier Central, Étude VII, liasse 9, Archives Nationales, Paris. The service court described in the contract and ultimately built differs from the one shown in the drawing, which must therefore represent an earlier scheme. Gady has dated the drawing on this basis between late 1619 to early 1620. See Gady (2005: 31 and 224–225).

In this paper, I want to draw attention to an unremarked aspect of this drawing, particularly regarding its rarity. Our knowledge of sixteenth and early seventeenth-century gardens comes almost exclusively from engravings, usually intended to record or publicize existing sites. The famous suite of prints by Jacques Androuet du Cerceau, *Les plus excellents bastimens de France* (1576–1579), contains many images of just this sort, as do the garden treatises of the early seventeenth century. These engravings contain essential information about the composition, dimensions, and planting of several famous gardens, but they necessarily stand at a remove from the circumstances under which those sites were created. In contrast, actual project drawings—those produced in view of a specific commission—are almost unknown until the second half of the century. Lemercier's perspective view of Montjeu is one of the earliest working drawings we have of any French garden. It even stands out among earlier, engraved representations, very few of which depict the château and its surroundings at such a large scale and as an integrated ensemble. The image sheds light not only on the evolution of the French garden in terms of its formal elements, but also on the way that novel drawing conventions may have driven that evolution. As a project drawing, it provides valuable information about how such sites were designed, as well as the technical and mathematical capabilities they required.

The Site

Jeannin purchased the *seigneurie* of Montjeu in 1586, 5 years after becoming président à mortier de la Parlement de Bourgogne. It was early in his career, but his ambitions for the property were grand. In 1596, he acquired the nearby *seigneurie* of Montjeu-en-Autun, and in 1614 joined the two properties to a third, the châtellenie de la Thoison. He continued to enlarge the estate with many smaller purchases until his death in 1623. Further extended by his successors, the entire domain would ultimately cover some 6000 ha (Ganay 1925).[3]

Although the house still stands largely as it was built, we know only the bare outlines of its history. Begun in 1606, it was still unfinished in May 1611, when the Jesuit architect Étienne Martellange came to the site during a visit to Autun.[4] He made two pen-and-wash views that provide valuable evidence for the state of construction (Figs. 2 and 3). Set within a moat, the building was to be composed of three ranges in the form of a U, the whole bracketed by salient, square towers at the corners. The aisles were still unfinished, but Martellange seems to have been more struck by the still-untamed nature of the site. The view from the rear emphasizes the steep slope of the hill and the dense surrounding forest. Indeed,

[3] This article—largely unreferenced—abstracts the work of local historian Charles Boëll, whose research drew heavily on the archives of the château. See Boëll (1912, 1935, 1940).

[4] Boëll gives 1606 for the start of construction, but without support. The chapel was decorated in 1623, which may provide an end date. See Boëll (1912: 254, 1935: 245).

Fig. 2 Étienne Martellange, *Veüe du Château de Monjeu près d'Autun*, dated 6 may 1611 (Bibliothèque nationale de France)

the moat seems barely to fit on the space cleared for it. The modest garden forecourt in the second view was old-fashioned even for the time: square in plan, closed with walls and hedges, and planted with rectangular compartments. Its only nod to ornament was a small fountain and a covered trellis. Set below the house, the forecourt too appears as something of an island in the largely unkempt landscape, still dotted with huts of building workers and tenant farmers. When Lemercier arrived in 1619 or 1620, the château itself was close to completion, but its surroundings must have looked largely as they had to Martellange. In response, Lemercier proposed an ambitious scheme, in which his commission for the service court formed the basis of an entirely *ex novo* setting for the château.

In the absence of further documentation, we cannot say for certain whether Lemercier's plan was adopted, but the current configuration of the site—which dates from the second half of the seventeenth century—does strongly suggest it. As Gady has remarked, a late seventeenth-century map published by de Ganay bears enough resemblance to Lemercier's proposal to suggest that it had formed the basis for later improvements. Several common elements are still visible in online satellite imagery: the large rear parterre with its curved hemicycle, the entrance and elevated terrace on the cross-axis of the house, a round belvedere and stairs, the splayed

Fig. 3 Étienne Martellange, *Maison de Mr Le President Janin Proche d'Autun Dicte Monjeu*, dated 7 May 1611 (BNF)

forecourt and flanking quincunxes, and the great basin overlooking the valley (Ganay 1925: 198).[5]

The Design

In many ways, the design of Montjeu was traditional. Even a brief perusal through Androuet du Cerceau's *Les plus excellents bastiments de France* makes clear that Lemercier was drawing on a formal repertoire established in the sixteenth century (Guillame 1999).[6] The forecourt and the rear parterre, for example, are larger than the footprint of the château. Nor are these elements separate from the house,

[5] De Ganay did not provide the source or the whereabouts of the plan. According to Boëll, the stairs flanking the existing *demi-lune* were built by an architect named Bussin in 1623. Later improvements were undertaken by Jeannin's grandson, Nicolas Jeannin de Castille, after the estate was elevated to a marquisate in 1655. These include the park wall completed in 1660 and the *communs* on the south side of the château, constructed from 1669 to 1670. See Boëlle (1935: 246, 1940: 314–321).

[6] On Du Cerceau's publication project, see Du Cerceau (1988) and, now Boudon and Mignot (2010: 11–33).

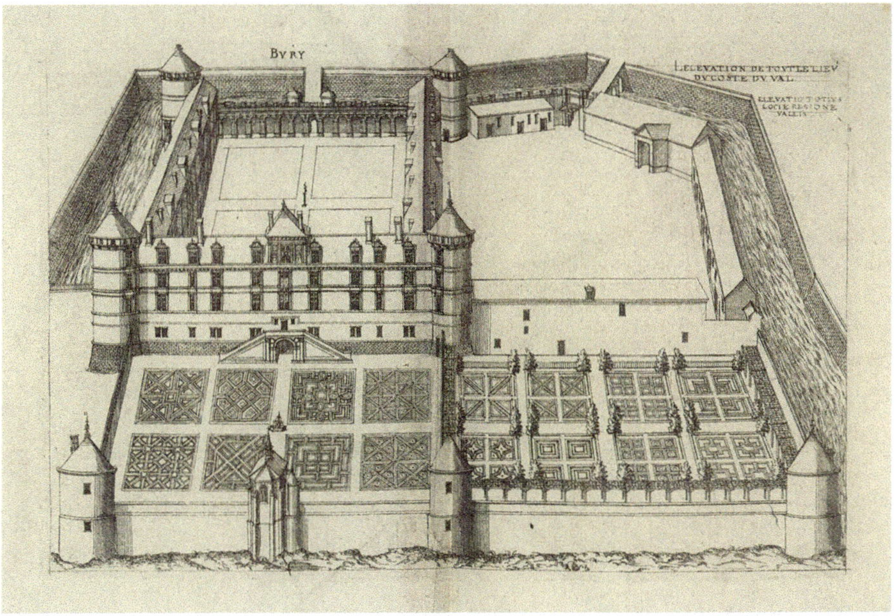

Fig. 4 Jacques Androuet Du Cerceau, Bury, from *Les plus excellents bastiments de France*, 1576–1579 (University of Manchester)

enclosed and distant; rather, the house is set within the gardens, which radiate from it on three sides. Most important, the composition of the site as a whole is regulated by axes established by the château itself. In this respect, Montjeu follows in a line begun by Bury in 1512–1513, the earliest example of a French château in which the house and garden were designed in reciprocal, symmetrical relationship (Fig. 4). At Montjeu, the axes serve another traditional function: they extend into the park and beyond in the form of tree-lined *allées*. The cross-axis serves as the main approach road, while the long, central axis plunges through the park, as the middle prong of a trident. This latter feature, though rare, finds a precedent at François I's royal château at Villers-Cotterêts around 1540 (Fig. 5).

Montjeu therefore draws on well-established tradition, elements of which date to the early sixteenth century, but other precedents are more recent. Indeed, Lemercier was also channelling some of the most influential gardens of the previous generation. The steeply sloping site, for example, would have brought to mind at least two possibilities: the complex garden designs for Verneuil-sur-Oise or Henri IV's additions to the Château-Neuf at Saint-Germain-en-Laye from the late 1590s. Although more grandiose than Montjeu, both of these sites incorporated a dramatic sequence of descending platforms. Of the two, Verneuil is the more apt comparison, both for the smaller size and for the gentler slope. The gardens were never completed, but Lemercier would have known of them via Du Cerceau, who

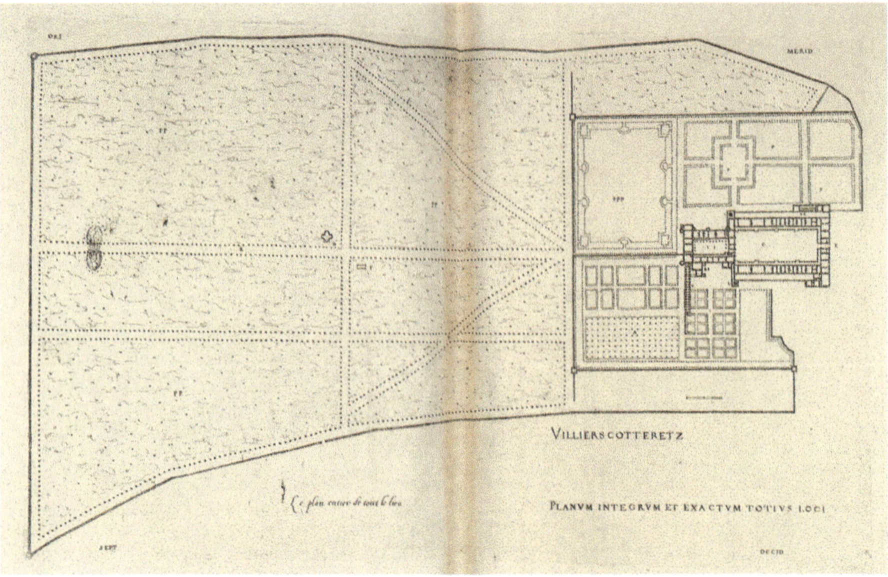

Fig. 5 Jacques Androuet Du Cerceau, Villers-Cotterêts, from *Les plus excellents bastiments de France*, 1576–1579 (University of Manchester)

represented the play of terraces both along and—as at Montjeu—on either side of the central axis (Fig. 6).[7] At both Verneuil and Saint-Germain-en-Laye, the transitions between levels are marked by monumental staircases, an imperial motif that goes back to Bramante's design for the Belvedere court. Montjeu, too, gave Lemercier room to experiment with such set-pieces: the visitor would have descended to the rear parterre via a flight set within a curved exhedra, then to the château along the arms of a grand horse-shoe, and finally to the forecourt via a tri-lobed belvedere.

Perhaps the most recent developments in garden design are reflected in the rear, where the four compartments cohere to produce a single, unified parterre. Individuated square compartments, although still used in the early seventeenth century, were being gradually supplanted by larger, integrated compositions. The parterre at the Luxembourg palace, installed by Jacques Boyceau sometime after 1625, is usually identified as the turning point in this development.[8] Set in a vast

[7] Verneuil, begun in the 1560s, was still not finished in 1576, nor after a second campaign in 1600–1616. See Du Cerceau (1988: 122–123) and Boudon (1999: 155–156).

[8] Sara Galletti has recently redated Tommaso Francini's plan of the Luxembourg gardens (BNF, Estampes et photographie, Va 419 FT 4) to 1625 on the basis of the contract for the surrounding terrace. This plan does not show Boyceau's parterre, which must have been laid out sometime after that year. See Galletti (2012: 127–132). Boyceau published the plan of the parterre in 1638, but the actual design may have come from Claude Mollet. See Boyceau (1638) and Karling (1974).

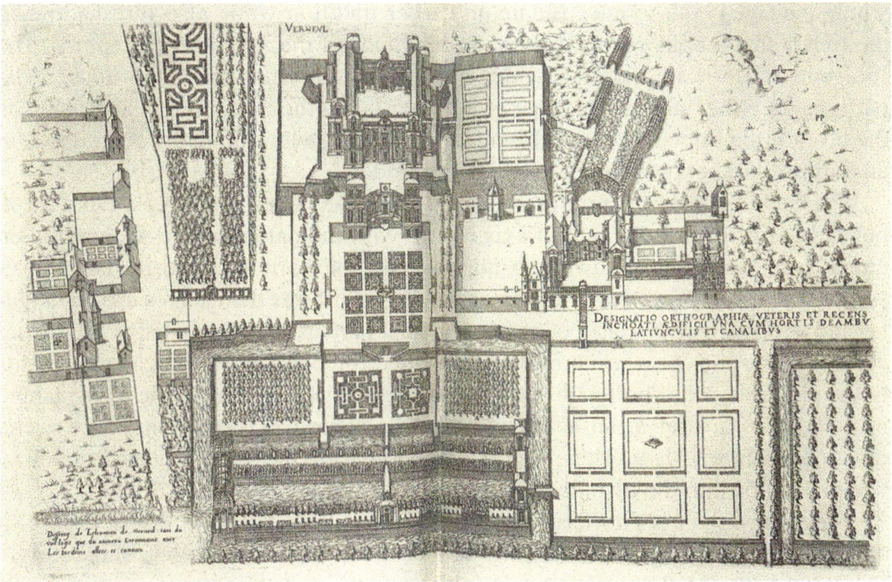

Fig. 6 Jacques Androuet Du Cerceau, Verneuil, from *Les plus excellents bastiments de France*, 1576–1579 (University of Manchester)

square terminated by a wide hemicycle, the whole ensemble measured about 125 × 166 m. The compartments were organized geometrically on the axis of the château and planted with flowering boxwood trained into curvilinear, arabesque patterns borrowed from textiles. The parterre was meant to be seen and comprehended at a glance from a fixed point, typically the center of the rear upper story of the château. As Jean Guillaume has noted, the *parterre en broderie* represented a new principle of composition, whereby large ensembles of elements were arranged around a central view, allowing them to be apprehended as a tableau. Lemercier appears to have been fully aware of this trend. In fact, his design for the rear parterre at Montjeu anticipates that of the Luxembourg by at least five years. Although not as large, it proceeds from the same principle. The whole forms a sunburst motif, divided into four quarters by sinuous paths, with each quarter further trisected into wedge-shaped segments. A fountain jet set in a central lobed basin and four others at the corners serve to further unify the design. The compartments are empty, but the serpentine paths—to my knowledge unprecedented—suggest that the parterre was not intended to be planted *en broderie* and may therefore represent an alternative and independent solution to that found at the Luxembourg.[9]

The principle of the central view has also been used to structure the gardens of the forecourt. Indeed, this area is the most characteristic of the way the design as a

[9] On this point, see Guillame (1999: 122).

whole manages both to encapsulate an earlier tradition while also foreshadowing the trends that would coalesce in the second half of the seventeenth century. The forecourt itself is trapezoidal, widening with the distance from the house. The parterres to either side are laid out at a higher level, their retaining walls rising as the level of the forecourt drops. The result frames the view along the central axis in an almost literal expression of a visual pyramid.

The flanking elements reinforce this axiality. The lateral parterres are not level, but rather tilt toward the central axis, rising with the natural slope of the hills on either side (Lemercier shows two alternatives, the parterre opposite the viewer sloping more rapidly toward the forecourt). The parterres themselves are rectangular and decorated with a grid of compartments, perhaps to acknowledge the earlier garden on the site. As a result of the fan-shaped forecourt, the orthogonal elements of the parterres—the short paths of the grid, the trellised *berceaux*, the trees planted in quincunx—are set at an acute angle to the central axis, pointing, as it were, toward the distant view. The effect is emphasized by the approach road opposite, which extends into the distance along the far end of the parterre.

The vast *mirroir d'eau* terminates the composition. Set on a lower terrace, it would have been partly hidden by the parapet of the forecourt. In view of the shift in elevation, the tall helicoidal fountain in the center serves two purposes: like a rifle sight, it projects the central axis out into the landscape beyond, while also signalling the presence of the concealed basin. The pool—wider than the house itself—would have acted as an artificial horizon, underlying the view of the valley in the distance. Admittedly, the design of the forecourt differs from the style of Le Nôtre and the later seventeenth century. As a comparison with the present-day site makes clear, Lemercier's proposal is more complex and differentiated. The transitions between the levels have been smoothed out and the angled elements for the most part rectified. Yet, Lemercier's aim underlies both designs: the desire to organise the landscape as an expression of the seignieurial residence, to extend the architecture of the house into its surroundings on a scale approaching that of vision itself.

The Drawing

We can therefore define Lemercier's project with reference to a clearly definable tradition, but the drawing also helps to explain how this tradition evolved in the way that it did. The clues are to be found not only in the formal qualities of the design, but also in the way the design has been represented. In this respect, the scenography is more novel and innovative than it might at first appear. In the first place, it shows the house and the surrounding landscape—the park included—as an integrated ensemble. Axiality and symmetry, tree-lined *allées*, the play of elevated terraces, central views projected into the distance: many of the sites in *Les plus excellents bastiments* include one or more of these elements, but none depict all of them as the combined product of a coherent overriding design. In Lemercier's drawing, the house and landscape respond symmetrically along a central axis, which is projected not only through the front and rear gardens but also, crucially, through the park. The park is not merely an afterthought; the drawing's purpose is to situate the park itself in the topography of the site as a whole. That decision underlies the scale chosen for the drawing, which serves very consciously to frame the scene at the limits of the park, particularly at the upper and right edges. Indeed, the house and gardens themselves—ostensibly the most important elements—are virtually restricted to the lower left quadrant.[10]

The scenography is also novel in that it is not an engraving, but a project drawing. It is, in fact, one of the earliest project drawings we have for any French garden.[11] Given the rarity of such documents, it is likely that the novel unity of the design follows from the drawing itself, in particular, its scale. The scenography, in other words, appears to be a new type of project drawing, one tailored expressly to conceive and depict an entire seigneurial landscape. It is not difficult to see where Lemercier would have gotten the idea. Du Cerceau illustrated plans for at least five châteaux on the scale of the park, including Vallery, Gaillon, Villers-Cotterêts, Chantilly, and Dampierre (Du Cerceau 1988: 109–111, 148, 202–203, 246–247, and 284–285). These engravings constitute one likely source of inspiration for Lemercier, but it is important to underline their essential difference from the Montjeu drawing. None of the five depict an *ex novo* project. Rather they are the product of survey drawings—*relevées* made in the mid-1560s or early 1570s—of sites where improvements and additions had accreted piecemeal, often from the château's medieval origins as a fortified castle or keep. Although we know that Du Cerceau used models and preparatory drawings as the basis for some of his engravings—at Verneuil, in particular—it is unlikely that master plans like Lemercier's were generally made at a scale large enough to represent the park.[12] What was lacking was not the technical skill in surveying, but rather the general aim of compositional unity at such long distances. In the absence of such a desire, even large elements—*murs de clôture*, *grands allées*—could be planned and laid out on site, with no need to coordinate them on paper. Of Du Cerceau's five extended plans, Villers-Cotterêts is the only one to exhibit a planned composition on a central

[10] The emphasis on the park may have been intended to draw attention to the six clearings scattered within it, very unusual features for the time. Although recent scholarship has emphasized the economic role of the park in the life of such country estates—it was typically the principal source of game and timber—Lemercier's clearings do not have an obvious practical or productive purpose. Their domed pavilions and exhedrae in formal, hierarchical layout suggest instead a social, courtly function, perhaps as early versions of the *bosquet*. On the role of the park, see Farhat (2006).

[11] There are at least four surviving project drawings for gardens that predate Lemercier's scenography. The first two—a plan and a perspective view—relate to a villa project by Étienne Dupérac for Catherine de' Medici in 1586. See Grodecki (2005) and Deswarte-Rosa (2005). The third and fourth are projects for the expansion of the Jardin neuf at the Tuilleries under Henri IV, dated around 1595 and 1608 (BNF, Estampes, Rés. Ve 53i ft. 5). See Fonkenell (2010: 52). The earliest surviving project drawing to postdate Lemercier's is Tommaso Francini's plan (c. 1625) for the Luxembourg gardens. See note 8.

[12] On the chronology of Du Cerceau's publication project and on his use of models, see Boudon and Mignot (2010: 17).

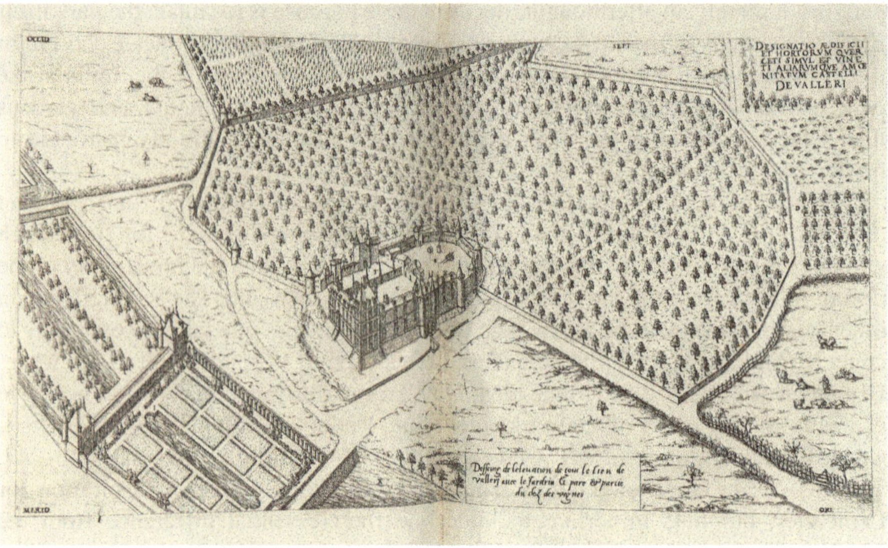

Fig. 7 Jacques Androuet Du Cerceau, Vallery, from *Les plus excellents bastiments de France*, 1576–1579 (University of Manchester)

axis, which terminates as an *allée* through the park. It is, however, an approximate arrangement. The main elements—house, garden, *allée*—are not precisely aligned. Nor do the gardens as a whole respond to the axis, reflecting the fact that the three elements were planned and laid out at different times.[13]

The other inspiration for the scenography also came from the world of architectural prints: the bird's-eye view. Du Cerceau was a master of this genre, too. The *Plus excellents bastiments* contains many examples, but the more likely influences are the ones that zoom out to an almost cartographic scale. Of the five sites mentioned above, the views of Vallery (Fig. 7) and Dampierre are represented in this way. Lemercier had been fully attuned to this print genre, at least since his Roman sojourn, when he had tried his own hand as an architectural printmaker. His 1608 view of the Farnese palace at Caprarola (Fig. 8)—also entitled *scenografia*—is a masterpiece of architectural representation, copied repeatedly throughout the seventeenth century and all the more astonishing for the fact that Lemercier was only 23 years old at the time. As Francisco Martinez Mindeguía has shown, its real interest lies in the way it absorbs and reinterprets a wide array of graphic sources, in which Étienne Dupérac's *oeuvre* appears to have been particularly influential (Mindeguía 2009).[14] Among other innovations, Dupérac had perfected the bird's-

[13] The château—built on the site of a medieval *maison forte*—was planned sometime after 1528 and built from 1532–1535. The gardens appear to have been designed separately in late 1538 or early 1539. Work was largely complete by 1546. See Riboulleau (1991: 37–39 and 100).

[14] On this print and the others Lemercier produced during his stay in Rome, see Gady (2005: 19–26).

Fig. 8 Jacques Lemercier, *Scenografia generale del palazzo di Caprarola dell'illustrissimo signor cardinal Farnesio*, 1608 (BNF)

eye view. Examples turn up throughout his work, but Lemercier seems to have zeroed in on his famous prints of contemporary Italian gardens—especially the Belvedere Court and the Este gardens at Tivoli. The high angle of view, the dramatic use of perspective, and the inclusion of the building's immediate surroundings in the landscape underlie both the Caprarola engraving and the Montjeu drawing.

Dupérac's career and influence saw something of a rebirth in the early seventeenth century, following his return to France as a court architect for Henri IV in the mid-1590s. His work on the Château-Neuf at Saint-Germain-en-Laye would engender a third group of engravings that may have had some impact on Lemercier. This project was—not coincidentally—replete with imperial and Italianate motifs of the kind that Dupérac often emphasized in his prints: grand scale, axiality, symmetry, hierarchy. The large engraving by Michel Lasne, based on a drawing by Alessandro Francini, clearly imitates Dupérac's visual style (Fig. 9).[15] Lasne's view is the best known representation of the site and would have a long afterlife in later views of the Château-Neuf. However, it is a nearly contemporary view by

[15] Dupérac's project for Catherine de'Medici's villa at Chaillot (see note 11) was conceived in a similar spirit. Framed by a descending series of monumental terraces, it also privileged the view from the opposite bank of the Seine. Nor does it seem coincidental that one of his drawings for the project took the form of a bird's-eye view.

Fig. 9 Michel Lasne (after Alessandro Francini), *Château royal et parc de Saint-Germain-en-Laye*, 1614 (BNF)

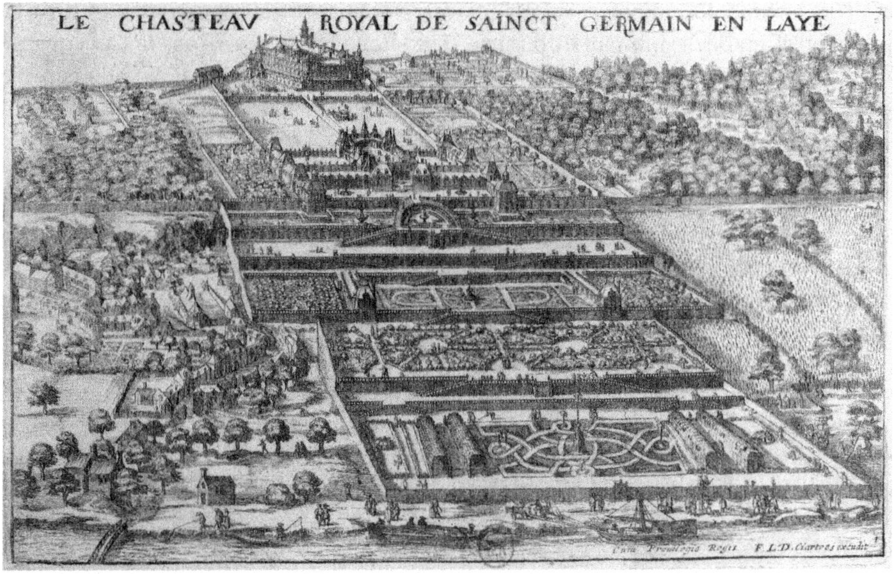

Fig. 10 Claude de Chastillon, *Le Chasteau Royal de Sainct Germain en Laye*, c. 1615 (BNF)

Claude de Chastillon that comes closest to the particular effect of the Montjeu drawing (Fig. 10). In Chastillon's print, the angle of vision has been lowered; we see the scene from an imagined, elevated point on the opposite bank of the Seine. The view has also been turned from the central axis, which makes the slope of the hillside and the stepped effect of the terraces more palpable. In addition, the gardens are given a wide berth within the landscape; the houses and vineyards are set into the bank opposite, which rises to the walled park in the upper register.[16]

As these sources make clear, the inspiration for Montjeu came not from an existing garden or series of gardens, but from the world of architectural prints. These engravings furnished Lemercier less with formal elements of the design, than with a conception of the site as a complete entity. More important than the sources, however, is the new purpose to which Lemercier has put them. The scenography, while borrowing the visual language of the bird's-eye view, transforms its purpose; the image does not serve to celebrate an existing site, but rather to envision a new one. The drawing thus exploits the viewer's familiarity with its precedents: the project is presented with the visual conventions normally used for already-famous palaces and châteaux. In some sense, that is what any presentation drawing would do, but the difference here lies in the scale. The bird's-eye view encompasses the entire landscape and allows the composition to be considered as a totality at that scale. The innovative unity of Lemercier's proposal flows from this decision.

[16] See Lurin (2010: 46 and 142–143), which assembles the site's rich visual documentation.

The technical challenges involved were the same for preparing a bird's-eye view, but with one important difference. The site to be surveyed was not an existing complex, but a raw, forested hillside. Lemercier would have had to integrate the proposal with the existing built features—namely the château—and with the topographical data from the site itself. The drawing shows how he did this. The numbers along the bottom and left edges indicate that the perspective was made from a measured ground plan overlaid with a grid. The units are large and appear to correspond to two *perches* (13 m or 2 × 6.5 m), a standard unit of measurement for surveyors.[17] Although unusual for architectural drawings, it was apt for Lemercier's purpose. At 20 *pieds*, the *perche* could conveniently accommodate the representation of both larger buildings as well as the surrounding landscape; the site as a whole measures roughly 660 × 585 m. In addition to the measurements in plan, Lemercier would also have had to account for the changing topography with a separate elevation survey, requiring specialized instruments. The terrain drops 30 m from the end of the park to the rear parterre and another 12 m to the level of the basin.[18] The drawing thus implies the use of a preparatory section, which would have allowed Lemercier to plot the descent for each of his staircases (and from the forecourt to the basin) on a notional vertical axis.

From a cartographic point-of-view, the skills involved were specialized, particularly as the size of the project fell between professional camps. Architectural drawings typically required a much larger scale—that is, they covered a smaller area—and did not usually address complex changes in terrain unless they affected the built work. (Indeed, Du Cerceau himself was often guilty of this lapse. His engravings do not usually show shifts in elevation between buildings, unless they are linked by terraces.) Maps by professional cartographers and military engineers, on the other hand, were usually prepared at a smaller scale, that of the city or region. Estate and forest surveyors, in contrast, generally had very rudimentary graphic ability. Indeed, they often did not make maps to scale at all.[19] Nor were gardeners typically responsible for designing sites as large as Montjeu. The statutes of the corporation of gardeners, ratified in 1599, describe them not as artists or designers, but primarily as suppliers of trees, flowers, and fruit plants. This picture is complemented by contemporary contracts, which typically take the form of lease

[17] The numbered units appear to represent a length of about 13 m. The estimate was made using the measuring tools on Google Earth and IGN's Géoportail, comparing results from the sides of the chateau and the moat with Lemercier's numbered divisions. The estimate must, of course, remain tentative until it can be corroborated with measurements taken on site. It is worth noting, however, that the closest regular multiples of the *toise* that come closest are hardly convenient (6 × 1.95 m = 11.7 and 7 × 1.95 m = 13.65).

[18] The IGN's relief plan of the site, also available on Géoportail, provides altitude markings at 5-m intervals.

[19] On military maps, see Buisseret (2007). On field and forest surveying, see [w.n.] (1987). This exhibition catalogue, which draws heavily from the cartographic holdings of the Archives Nationales, reproduces Lemercier's drawing along with other examples from the Series N. The direct comparison underlines Lemercier's very advanced drawing abilities.

agreements for the maintenance of specific sites. On the rare occasions that drawings are mentioned in these sources, they usually refer only to the design of flowerbeds and parterres.[20]

Lemercier's scenography thus represented an advance in architectural representation that mixed in equal parts surveying, printmaking, and cartography. It heralded a practice that would become normalized by the second half of the century: the creation of large-scale project drawings that functioned, at the same time, as a *plans d'arpentage*, measured surveys approaching the scale of an entire *seigneurie*. This breakthrough allowed the scale of the house to be applied consistently throughout the estate—permitting both architecture and landscape to be seen in reciprocal relationship. It is difficult to know how widespread this practice was. Was Lemercier an outlier? Unfortunately, we do not know of any comparable project drawings for the gardens of the first half of the century. If Mansart made any such large-scale plans for his châteaux, none survive. Nor do Lemercier's own drawings for the environs of Richelieu or Rueil. Yet, by the time Le Nôtre and his workshop produced the master plan for Vaux-le-Vicomte in the 1650s (MS 1040, Bibliothèque de l'Institut de France), the practice may have been well established. In that plan—one of the first surviving project drawings after Lemercier's—the park appears as the logical and natural extension of the garden.

Contemporaries seem to have been attentive to this change in scale and to the designed character of the domain as a whole. When John Evelyn visited Maisons in 1649, he described the gardens in his diary as "very magnificent, with extraordinary long Walkes set with Elmes, & a noble Prospect on the Sienne [sic] toward Paris." Another visit a year later drove him to further raptures. It was a confection on the scale of an entire estate: "[T]he prospect on the river, & towards the forest is incomparable... Taking it altogether," he wrote, "the Meadows, Walks, River, forest, wheatfields & Vineyards, I hardly saw anything in Italy exceed it." [21]

Bibliography

Sources

Boyceau, Jacques. 1638. Traité du iardinage selon les raisons de la nature et de l'art, Michel Vanlochom. Paris.

Du Cerceau, Jacques Androuet. 1576-1579, repr. 1988. Les plus excellents bastiments de France, ed. David Thomson. Paris: Sand & Conti.

Evelyn, John. 1955. The Diary of John Evelyn. Ed. E. S. de Beer, 6 vols. Oxford: Clarendon.

Mollet, André. 1651. Le jardin de Plaisir. Stockholm: H. Keyser.

[20] See Lespinasse (1886–1897: 703–711) and Grodecki (1985: 329–344). André Mollet's *Le jardin de plaisir*, H. Keyser, Stockholm, 1651 was the first treatise to depict the area beyond the parterre, as a unified conception of the garden based on grand axes. See Rostaing (2006).

[21] Evelyn (1955: Vol. 2, 563, Vol. 3, 18) (Sept 10, 1649, Sept. 5, 1650).

Secondary Literature

[w.n.]. 1987. Espace français: Vision et aménagement, XVIe-XIXe siècle. Paris: Archives nationales.

Boëll, Charles. 1912, 1935, 1940. Montjeu et ses seigneurs, deuxième partie: Le president Jeannin et ses descendants. Mémoires de la Société Eduenne 40: 227-300; 47: 217-264; 48: 287-329.

Boudon, Françoise. 1999. Jardins d'eau et jardins de pente dans la France de la Renaissance. In Architecture, Jardin, Paysage: l'environnement du château et de la villa aux XVe et XVIe siècles, ed. Jean Guillame, 137-83. Paris: Picard.

Boudon, Françoise and Claude Mignot. 2010. Jacques Androuet du Cerceau: Les dessins des Plus excellents bâtiments de France. Paris: Picard, Cité de l'architecture et du patrimoine, Le Passage.

Buisseret, David. 2007. French Cartography: The *ingénieurs du roi*, 1500-1650. In History of Cartography, Cartography in the European Renaissance, ed. David Woodward, 2 vol., 1504-1521. Chicago: University of Chicago Press.

Deswarte-Rosa, Sylvie. 2005. Le dernier caprice architectural de Catherine de Médicis: une villa à hippodrome sur la coline de Chaillot par Etienne Dupérac. Revue de l'art 150: 32-46.

Farhat, Georges. 2006. Au-delà du 'terroir sterile': Le parc dans l'organisation du domaine seigneurial (1550-1700). In André Le Nôtre, fragments d'un paysage culturel: institutions, arts, sciences & techniques, ed. Georges Farhat, 88-113. Sceaux: Musée de l'Île-de-France.

Fonkenell, Guillaume. 2010. Le Palais des Tuileries. Arles: H. Clair.

Gady, Alexandre. 2005. Jacques Lemercier: Architecte et ingénieur du Roi, Éditions de la Maison des sciences de l'homme. Paris.

Galletti, Sara. 2012. Le Palais du Luxembourg de Marie de Médicis 1611-1631.Trans. Julien Noblet. Paris: Picard.

Ganay, Ernest de. 1925. Le château et les jardins de Montjeu. La Revue de l'art ancien et moderne 47: 197-203.

Grodecki, Catherine. 1985. Documents du Minutier central des notaires de Paris: Histoire de l'art au XVIe siècle (1540-1600). Vol. I, Architecture, vitrerie, menuiserie tapisserie, jardins. Paris: Archives nationales.

Grodecki, Catherine. 2005. De l'Ermitage de la Reine à la villa à l'antique. Revue de l'art 150: 21-31.

Guillame, Jean. 1999. Le jardin mis en ordre: jardin et château en France du XVe au XVIIe siècle. In Architecture, Jardin, Paysage: l'environnement du château et de la villa aux XVe et XVIe siècles, ed. Jean Guillame, 102-136. Paris: Picard.

Karling, Sten. 1974. The Importance of André Mollet and His Family for the Development of the French Formal Garden. In The French Formal Garden, eds. Elisabeth B. MacDougall and F. Hamilton Hazlehurst, 1-25. Washington, D.C: Dumbarton Oaks.

Lespinasse, René de (ed.). 1886-1897. Les métiers et corporations de Paris. 3 vols. Paris: Imprimérie nationale.

Lurin, Emmanuel (ed.). 2010. Le Château-Neuf à Saint-Germain-en-Laye. Paris: Les Presses Franciliennes.

Mindeguía, Francisco Martinez. 2009. Anatomia de un dibujo: el Palacio de Caprarola, de Lemercier. Annali di architettura 21: 115-25.

Riboulleau, Christiane. 1991. Villers-Cotterêts: un château royal en forêt de Retz. Amiens: L'Inventaire générale.

Rostaing, Aurélia. 2006. La bêche ou le compas? Le métier de jardinier dans la première moitié du XVIIe siècle. In André Le Nôtre, fragments d'un paysage culturel: institutions, arts, sciences & techniques, ed. Georges Farhat, 74-87. Sceaux: Musée de l'Île-de-France.

Woodbridge, Kenneth. 1986. Princely Gardens: The Origins and Development of the French Formal Style. London: Thames & Hudson.

Utopia, Science and Garden Art in the Early Modern Era

Hubertus Fischer

Abstract
Utopia and the theory of architecture in Renaissance humanism have certain concepts of towns and urban life in common. Against this backdrop, this paper intends to explore the question of how gardens were integrated into urban architecture and what role science played in this. Emphasis will be placed on the little-known gardens in Filarete's treatise on architecture of the 1460s; the radially-arranged city discussed in that work returns in the first Italian literary utopia one hundred years later. In the Protestant and Puritan utopias, *Christianopolis* and *Nova Solyma*, from the first half of the seventeenth century, science then forms a close alliance with religion. On the one hand, this bestows additional significance on the gardens; on the other hand, it puts them into perspective by comparing them with the "divine arts of nature". As both utopias show, progress in understanding nature can indeed be religiously motivated. Finally, the further need for research will be sketched [cf. (Fischer 2012), in which the gardens in More's *Utopia* and Bacon 's *New Atlantis* are examined in depth; hence the present article places emphasis on other aspects; a first outline of the topic in Fischer et al. 2011.]

Urban Utopia and the Theory of Architecture

The utopias of the Early Modern Era, from Thomas More's *Utopia* (1516) to Johann Valentin Andreae's *Christianopolis* (1619), present themselves in the shape of urban utopias. Nonetheless, concepts and plans for the urban environment

My thanks are due to Ms. Kurth, department librarian, Bau- und Stadtbaugeschichte, TU Berlin, for her kind support, and to Catherine Atkinson for the excellent translation of this paper.

H. Fischer (✉)
Centre of Garden Art and Landscape Architecture, Leibniz University Hannover, Herrenhäuser Straße 8, 30419 Hannover, Germany

H. Fischer et al. (eds.), *Gardens, Knowledge and the Sciences in the Early Modern Period*, Trends in the History of Science, DOI 10.1007/978-3-319-26342-7_8

only take on a more concrete and diversified form in the course of a fairly long process. This begins with More's *Utopia* (Fig. 1), which gave the genre its name, and continues through Francesco Doni's *Il mondo savio e pazzo* (1552) and Caspar Stiblin's *Commentariolus de Eudaemonensium Republica* (1555) (Fig. 2) to Tommaso Campanella's *Civitas Solis* (1602–1603/1623) and Andreae's *Christianopolis*. Seen in this light, it is logical that the utopias of Campanella and Andreae are the first to refer explicitly, in their titles, to the town, namely the City of the Sun and Christianopolis, as the place of perfect community life.

The fact that the utopian polity is constituted as a town is due not least to the models provided by Antiquity, especially Plato's *Politeia* and Aristoteles' *Politics*, with the polis as the best form of living together (Plato 1925: 2 and 92–394; Aristoteles 1965: 226–267). The citing of works of Antiquity—a sign constitutive for this genre since More—even includes the legendary Atlantis mentioned in Plato's dialogues, *Timaeus* and *Critias* (Plato 1925: 405 and 494), to which Francis Bacon alludes with his scientific utopia, *New Atlantis* (1624/1627). This, together with the optimism relating to science and education, are clear signs of the humanist impulses that were the source of utopian thought in the Early Modern age. Science and education do indeed visibly inform the town's design, its planned layout, the forms and purposes of its buildings, and its architecture's programme of imagery and educational function.

This poses the question as to the relationship that existed between the architectural treatises and designs for planned towns that had been produced since the second half of the fifteenth century on the one hand and utopian townscapes on the other. "Likewise in the Renaissance, in the course of humanist enquiry into Antiquity, a vision of the town emerges in the manner of a topos, but one that is also influenced by European traditions and is reflected in the theories of town planning in architectural treatises [...]" (Rahmsdorf 1999: 23). From this, one can deduce that there must be an inner link to literary utopias—one that was upheld for more than a century. "Since the authors of literary utopias were moulded by the educational canon of humanism and late humanism until well into the seventeenth century, these urban concepts were fostered here a long time after they had first been expressed in the Renaissance humanism of the fifteenth century and were only gradually replaced by the social visions of later periods, which were more aligned towards the territorial state" (ibid.).

So, if there is one characteristic aspect that the theory of architecture and humanist urban utopias have in common, expressed in pithy examples, then it is the idea of *integrating gardens into architecture*, if possible both in form and dimension, in order to lend visible expression to harmony and proportion. This can lead to solutions that point well beyond their own times and into the future, as in the *Trattato di architettura* (between 1461 and 1464) by the Florentine sculptor and architect Antonio Averlino (ca. 1400–ca. 1469), who was called Filarete ("friend of virtue"[1]). He dedicated this treatise to Francesco Sforza (1401–1466), Duke of Milan, and another copy to the Medici, to whose state he returned after his stay in Milan. He also shaped it in such a way that it differed from the usual treatise form in

[1] Cf. Hubert (2006) and Tiller (2006).

Fig. 1 Thomas Morus, *De optimo reipublicae statu deque nova insula Utopia*, Basel 1518, p. 12, Herzog August Bibliothek Wolfenbüttel, 31. Quod

Fig. 2 Caspar Stiblin, *Commentariolus de Eudaemonensium Republica*, Basel 1555, p. 120f, Herzog August Bibliothek Wolfenbüttel, 1198.9 Theol. 8

two respects: it was the first theoretical work of this type written in the vernacular and it was the first attempt at giving the treatise a new shape—one that stimulated the imagination through its narrative structure and numerous inserted drawings.

Both aspects indicate that such dialogues on building (cf. Tönnesmann 2004), which were presumably inspired by Plato, and with the courtly episodes as interludes, wished to address a public among the governing class and at court and to interest them in architecture, rather than attempting to win over a learned public. "It is Filarete's intention to push through the use of architectural forms that are oriented towards Antiquity—and were already known in his home-town [Florence,

HF]—in his place of work, too, in Milan—thereby replacing the *moderni* that still prevailed there, i.e. the gothic architecture" (Rahmsdorf 1999: 35). Thus, an imaginary town emerges, backed up by descriptions of the building work's progress and an account of the surrounding Arcadian landscape—a town with buildings, streets and parks, developed from a strictly symmetrical plan, the first of its kind in Europe, laid out as a regular octagon with roads radiating out and a square in the centre with a palace and a cathedral (Figs. 3 and 4).

It is remarkable that a reality fiction is being created as early as this and in the fashion that became constitutive for the literary utopia genre from More's times onwards. *Sforzinda*, as the town was called, the cornerstone ceremony for which was supposed to have taken place on April 15, 1460, combines social order with the mathematical shaping of space in such a way—again comparable with the

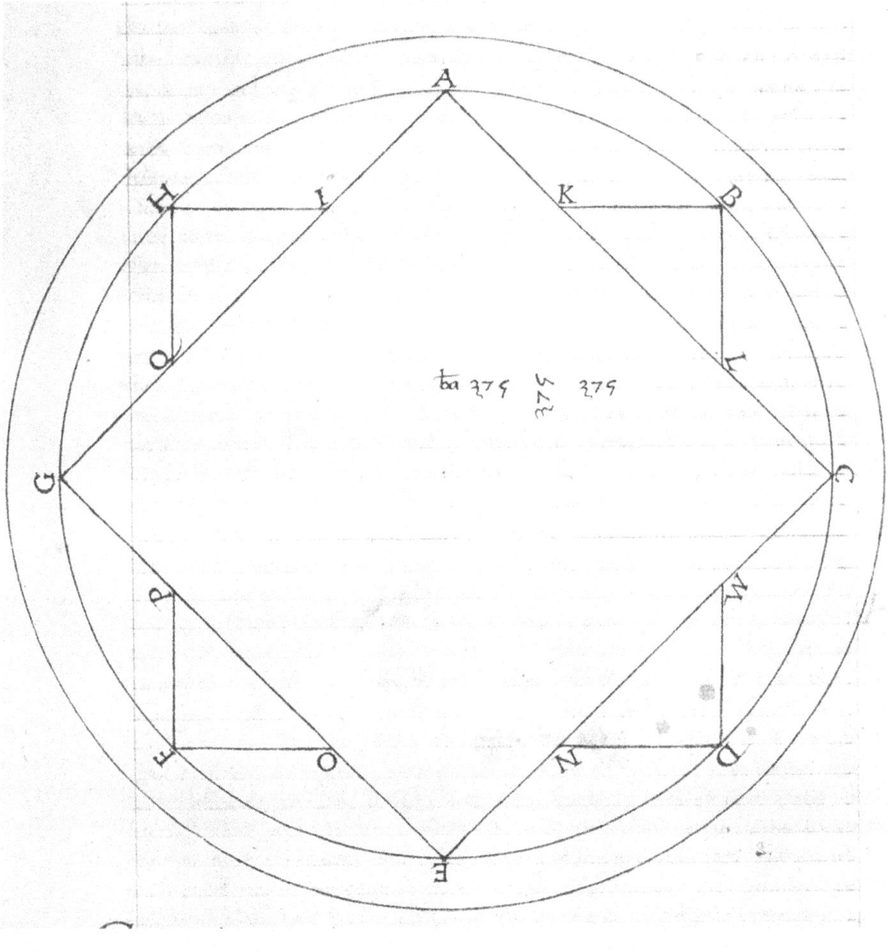

Fig. 3 Filarete, Construction drawing for Sforzinda, facsimile, ed. Spencer, fol. 13v, TU Berlin, Bibliothek des Fachgebiets für Bau- und Stadtbaugeschichte

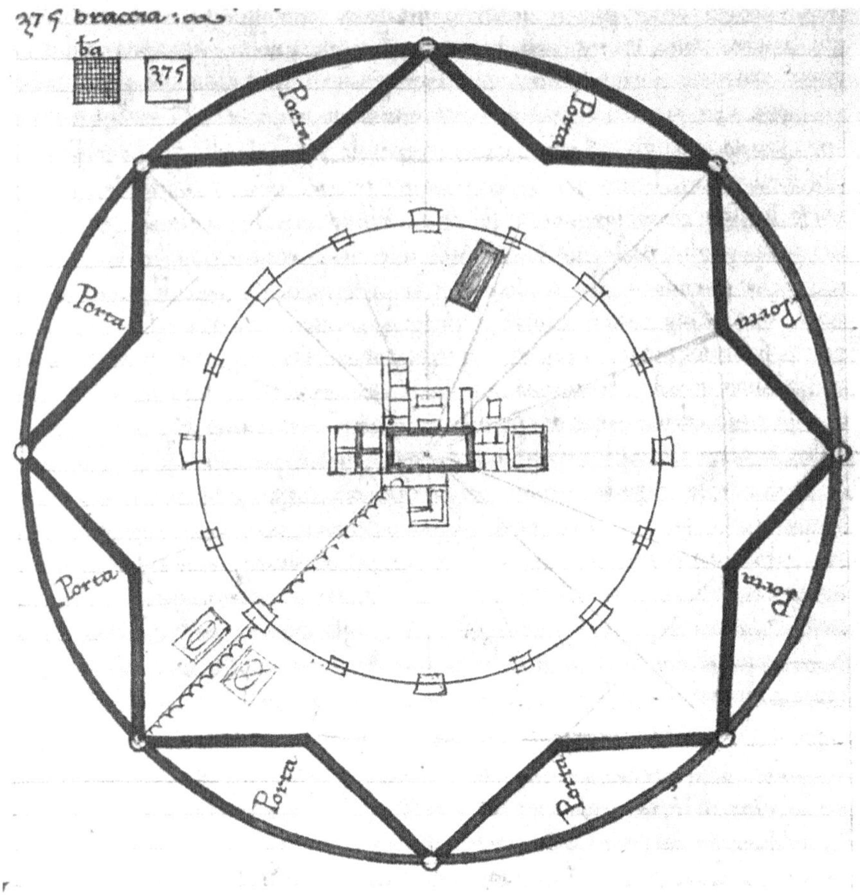

Fig. 4 Filarete, Ground plan of Sforzinda, facsimile, ed. Spencer, fol. 43r, TU Berlin, Bibliothek des Fachgebiets für Bau- und Stadtbaugeschichte

utopias—that the town's layout and its buildings represent the spatial expression of a model polity. Without going into detail here, it should be pointed out that in the *archicodomus*, the monastery-like educational institute for boys with its cloisters, courtyard and loggias,[2] the aim was to create a synthesis of practical and scientific training, of the *artes mechanicae* and the *artes liberales*, teaching in the arts and crafts and fine arts (Filarete 1965: 230–241). This serves as an example for a general concern, that of: "[. . .] providing the artists with fixed rules based on the disciplines of the quadrivium for their work—this is the goal of the new quattrocento literature on art theory" (Tigler 1963: 34).

[2] Built on the basic plan of the inner square, the quadrangle, like the universities in Cracow, Salamanca, Oxford and Cambridge; cf. Jellicoe and Jellicoe (1988: 152f).

Architecture as a Science: The Forms and Dimensions of the Gardens in Sforzinda

In the shape of Filarete's *Trattato*, we have a work that links the reality fiction of an ideal town with an exemplary social order and ascribes an important role to education and science—and all of this half a century before More's *Utopia*. Filarete regards architecture itself as a science (Filarete 1965: 188) and the architect as the comprehensively trained shaper of the polity (ibid.: 198). This also extends to the gardens as an architectural subtask, for which—as for buildings, too—representing and explaining the forms and dimensions (*modi i misure*) are of particular importance. Thus, there is hardly a building that Filarete does not equip with a garden—an exactly surveyed garden of strict proportionality and harmonious measurements, usually provided with a fish pond and often surrounded by a colonnade or a wall, which, as in the case of the Prince of Sforzinda's palace garden, can also be furnished with a garden house and a fountain. In order to understand the measurements stated in the following quotation, one needs to know that the *braccio* (plural *braccia*), an old Italian yard-stick, measures between 58 and 69 cm.

> Perhaps his lordship your father will want the gardens [to be] as they appear here in this line drawing [...]. They will be 60 braccia wide and 120 braccia in length with a portico all around six braccia deep. At the foot of the garden there will be a place 12 braccia square where one can make a room or a dwelling. In the middle of the garden we will make a fish pond 40 braccia long and 30 wide with a fountain in the center 12 braccia in circumference. As I have said, around each of these two gardens there will be a vaulted portico six braccia deep and ordered in such a way that one can go to this passage. The middle portico, the one that divides one garden from the other, will be 12 braccia wide, and the two on the sides will be six. They will be arranged so that the water [from the fountain] will not fall on them but will drain back into the pool (ibid.: 100).

The monastery of the Friars Minor on the town's main street leading to the prince's palace is developed by Filarete from a basic square, likewise with a garden. This takes up the third strip of the whole area (which is divided into three equal strips) and is interrupted perpendicularly in the middle by a building for the gardeners' flat, the barns and stables (ibid.: 132f). Equally, the boys' school has an extensive garden, for which there is mention of not only the measurements but also of the access, the enclosure, the terrain's layout and the plants.

> As you have understood, the garden will be 240 braccia wide and 270 [braccia] long. It will not be possible to enter this garden except through the dwellings. The garden will be built in this form [...]. First of all there will be a fishpond in the middle of it 30 braccia wide and 60 long. The garden will also serve for other [kinds of] training if needed. A space of 105 braccia on every side will be left over in the garden. The wall that separates it, on the sides, from the shops of the artisans will be only eight braccia, as I have said. The sides will be terraced all around in such a way that it can all be planted with useful plants (ibid.: 232).

The terraced garden, planted with herbs, is not only intended for this useful function alone, for the curriculum and daily plan are explicitly stated: "After lunch all the children have permission to amuse themselves for an hour [. . .] The master's assistant [. . .] will take them into the garden. Since it is the custom of children to run and play, it will be permitted" (ibid.: 237). A surprisingly modern pedagogical concept, which one later finds in Andreae. In contrast, in the girls' school garden, a number of precautions have to be taken:

> "The garden with its fish pond will be in this form, as you can see here [Fig. 5]. Between the wall of the dwellings and the garden there will be a passage where carts and horses can go to supply the house with wood, wine, and things necessary for their sustenance. As you can understand, the passage to the garden will be above by open stairs which will give on to the garden. They cannot enter the garden by any other way. The walls of the garden will be not less than 20 braccia high. The wall will be smooth all around, both on the interior and the exterior. The water, which will clean many things, will go here. The house of the gardener will be on the interior at one end." "Yes, but I want the gardener to be a married man. He can never bring any one into it without the permission of the headmistress" (ibid.: 243).

In this reform project for integrating young women into society, the garden, as a *hortus conclusus*, serves to preserve morals, decency and virtue, while at the front, where it faces the road, the convent-like building is equipped with great windows with iron fittings to be able to "exhibit" the female pupils at lessons and thus lure suitable, eligible young men. So, it is not in vain that the prince calls this institution *domus honestatis*; after all, with the assistance of this building and its sheltered garden, respectable marriages are to be brought about.

The most striking gardens are to be found in the ancient town of King *Zogalia*, *Plusiapolis* ("rich town"). The layout and fortification of the harbour town is to be executed according to the description handed down in the *libro d'oro*. There is the palace garden, the size and furnishings of which immediately attract attention: "[There was] a garden in the middle 320 braccia on one side and 150 on the other with a portico all around on columns like other cloisters on the interior side. It was all painted with noble ancient and modern histories. They were done by most worthy masters" (ibid.: 185). It is the other Antiquity, that of old Egypt and the Orient, that is expressed in this corridor: the origins and beginnings of the pharaohs, the life story of Cyrus, Semiramis's memorable deeds and buildings, the story of Cambyses and Oropastos, the cunning of Darius, the Egyptian pyramids and the city of Thebes with its hundred gates (ibid.: 185ff). The painted frescos conjure up the fascinating phenomenon of ancient times submerged in an enigmatic semi-darkness; one feels oneself reminded of Ranke's foreword to his *Weltgeschichte* (*Universal History*): "In our day the ruins of buried cities have been disinterred, and buildings have been discovered, on the walls of which the mightiest of monarchs of their day caused their deeds to be inscribed" (Ranke 1884: ix). Such a background was very fitting for any ruler who had only recently assumed power. The condottiere, Francesco Sforza, gained the duchy in 1450 by marrying the heiress, Bianca Maria Visconti. To this knowledge of ancient times with its shadowy figures, one can add—as water technology advances—the knowledge of hydraulics.

Fig. 5 Filarete, Ground plan of the girls' school with garden, facsimile, ed. Spencer, fol. 140v, TU Berlin, Bibliothek des Fachgebiets für Bau- und Stadtbaugeschichte

> In the courtyard there was a fountain arranged in such a way that the water shot up as high as the court, because the water came from [a place] as high as the house and even higher as its source. It went up to the top of the house in a conduit and then separated to go to different places. One part was sent to the fountain and then to all the other fountains. It ran from these fountains into the fish pond in the garden (ibid.: 187f).

The integrated system of water pipes also plays an important role in other parks. In addition, the palace garden radiates an aura of unmistakable opulence: "Our garden was in the center of the building. It was as large as I have stated before, with a loggia, or portico, all around, and contained many different sorts of fruit, apples, oranges, and very noble citrons." One can say the same for the fish stocks: "In the middle of it there was a beautiful large fish pond 40 braccia square. It was all vaulted around the sides where the fish could go. It was arranged in such a way that, when necessary, all the water could be emptied out of the pond. There were many different kinds of fish in great quantity in it" (ibid.: 188).

Let us take stock before continuing. The garden, integrated into the architecture, corresponds with it in dimension and shape, and generally has in the center a *peschiera*, a rectangular water basin in which fish are held, with a fountain. The water supply is guaranteed by rainwater, which has been collected and feeds both the basin and the fountain in the courtyards. Harmony is expressed by the fact that the jet of water of each of these fountains reaches the height of the courtyard. Moreover, one can say that the garden (or double garden) is a closed, subdivided rectangle, the smaller part of which is generally half the size of the larger one; it is bordered by a (vaulted) colonnade, a peristyle or by a wall or part of a building. Above the columns, this passage can be formed into a gallery or walkway. In a palace garden, one might also find a hall or other building at the end of the garden and a decorated peristyle; moreover, a flight of steps can lead down from another building into the garden. Architectural considerations largely overshadow the question of how the garden is planted. Only in one case does the corresponding description cause a truly vivid effect—as if one were looking through into a garden in a contemporary painting.

> We entered into the garden behind the house, which was more than 300 braccia wide on every side. Around it were cypresses and laurels that cast a shadow over the fish pond in some places. There were many different fruit trees as well, as for example oranges, lemons, and grape arbors. At the head of the garden some pines were scattered which offered shade with their branches. Under them grew a grass that seemed like green velvet. Some roebucks and other small animals went about grazing on it, which was a pleasure to see (ibid.: 225).

Constructed Divertimenti

Filarete holds an extremely unusual garden in store in the building complex of King *Zogalia*, which, as already mentioned, was to be reconstructed in the harbour city. If Filarete has hitherto received little attention as a garden theoretician, this is not true of this design, but despite the spectacular site, interest usually dwindles after a few sentences.[3] For all of that, it seems to be a kind of link in garden architecture between Babylon of the sixth century BC and the present, between the 'hanging gardens of Semiramis' and the *bosco verticale*, which currently reaches to the sky in the middle of Milan.[4] Before you reach this impressive, terraced architecture-cum-nature construction, you first have to walk across various outlying bridgeheads, buildings and other structures. They deserve our attention simply for the fact that they are integral elements of the garden and lend it a special appearance.

"We will now continue with a description of the garden. It was on the plain [and] was a mile on each side" (ibid.: 209). The huge site—the Italian mile ("miglio") measured 1486 m—is laid out as a fortified square, surrounded by extensive fortifications, while the dimensions of its buildings are subtly coordinated and carefully composed, right down to the base, shaft and capital of the columns. Two-arched, strictly proportioned bridges with piers in the centre and on land, and with a large gateway, lead over the wide watery ditch and grant access to the garden from four sides (Fig. 6). The bridges, in turn, end in a gate, through which one approaches the multi-storeyed entrance building and the adjacent building at a higher level. Admittedly some elements are adopted from military architecture—which were still familiar to Joseph Furttenbach (1591–1667) when designing the "pleasances" in his *Architectura recreationis*—but by enclosing the garden with a ditch, bridge, gate, house and wall, Filarete develops a well-proportioned divertissement of openings and closures, which leads up into a constructed *curiositas*. This is an impressive and surprising sight (Fig. 7):

Fig. 6 Filarete, Bridge to the garden, facsimile, ed. Spencer, fol. 120v, TU Berlin, Bibliothek des Fachgebiets für Bau- und Stadtbaugeschichte

[3] Cf. e.g. Vercelloni and Vercelloni (2010: 37), Fabiani Giannetto (2008: 94), Golombek (2008: 246) (with the idea of there being similarities to Timurid-Turkmenian sites); Impelluso (2007: 35f), Lauterbach (1996: 130) and Tagliolini (1988: 62f).

[4] http://www.stefanoboeriarchitetti.net/?p = 207

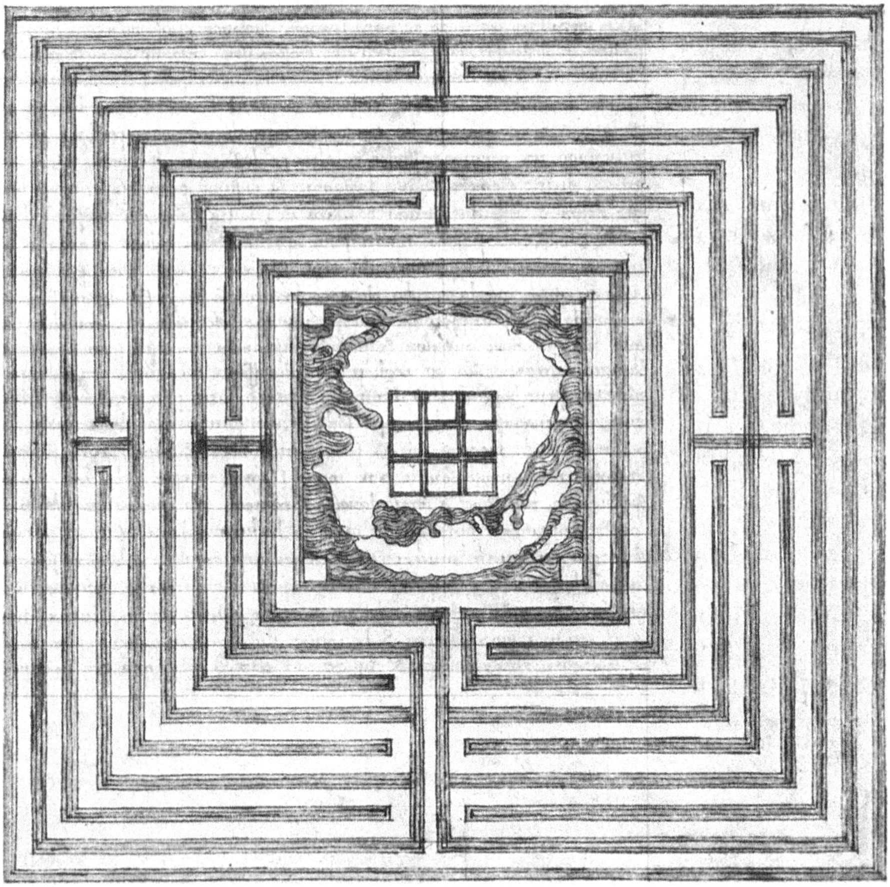

Fig. 7 Filarete, Garden labyrinth with a world map, facsimile, ed. Spencer, fol. 121r, TU Berlin, Bibliothek des Fachgebiets für Bau- und Stadtbaugeschichte

> The garden was laid out as is described here. First of all [it was] a square of 3000 braccia and divided into seven parts, 100 land braccia wide. Each of these streets had a little shelter at its corners 12 braccia square. From one to the other is a passage on columns like the one above. Thus one can go under cover at the lower level and in the open on the upper level from one to the other of these places. They are placed between the two streams of water in such a way that whether one chooses the upper or lower level of these seven ways, he can always see the water. One stream runs on the outside and one on the inside [. . .] (ibid.: 209).

Although there were garden labyrinths as early as the fourteenth century, Filarete was presumably the first theoretician of architecture to discover the labyrinth for the garden as well as for other buildings, certainly a while before Francesco di Giorgio Martini (1439–1501), who reckons "the labyrinths among the 'fantasie' that delight the eye with their great diversity" (Tausch 2003: 83). One can certainly say this of Filarete's ingeniously devised system of corridors; you have to wander through them for a long time before reaching the exit. "In the

Renaissance [. . .] labyrinths are both seen as a curiosity, enriching the garden with a further object to incite curiosity and admiration, and they are discussed theoretically as an object that needs to be designed ingeniously, for which the theory of architecture declares itself responsible—thus they are understood in a precise sense as architecture" (ibid.). It involves creating a geometric riddle, which must appear as an initiation to one who walks through it, accompanied by water-bearing canals. It was not in vain that one was supposed to enter Furttenbach's "Irrgarten" (maze) through the door called *Diligentia*, "finding one's way out, in turn, through *Patienzia*/having learnt much of *Scienzia* by discovering the correct path" (Furttenbach 1988: 77). Besides, he confirms not only that the labyrinth is a task of the architectural theory, but also that architecture, by means of "science, adds beautiful shapes to this work" (ibid.: 78).

At this point, I would like to point to the collection of statues, the context of which, however, is only revealed when one progresses to the inner garden, the garden "in a new manner" (Filarete 1965: 209). "Also on the four corners of the water, that is, at the end of the streets, as you can understand from the drawing, there were four little shelters, you might say as a simile for the winds. They were in the form you see here [. . .] with a large bronze figure at the summit of each one representing that wind" (ibid.: 221). The gods of the winds, which in the Greek myths are mostly portrayed as winged gods, were the sons of Astraios and Eos: Aeolus, as their ruler, then Boreas, Eurus and Zephyrus. The wind god is likewise shown here winged (Fig. 8a).

Fig. 8 Filarete, Wind deity (**a**) and garden in a new fashion (**b**), facsimile, ed. Spencer, fol. 122v, TU Berlin, Bibliothek des Fachgebiets für Bau- und Stadtbaugeschichte

In the Centre of the World: Hanging Gardens

It is obvious that all that has been mentioned so far requires no small degree of mathematical and technical knowledge; but this is also quite in accord with the view of architecture as a science. To be sure, it is a means to an end and not an end in itself, and in this case, the purpose seems to be to lead into a particular world via the said system of openings and closures, as well as via the labyrinthine initiation. 'World' is meant here quite literally, because, as one reads further on: "It says that there was a square of 1,000 braccia on each side which was reduced to a circle. This, in turn, was laid out like a map of the earth [*nel mappamondo*, Filarete 1965: folio 120v]. All the streams flowed in and flowed out from the center of this" (ibid.: 209). Thus, it becomes clear that the "inner garden" is a constructed *mappa mundi*. These were, as a rule, orientated towards the east and in accordance with the

> most widespread scheme of medieval world maps [. . .] [corresponded to] the T inscribed in an O (hence also called T-O maps). The T's beams separate Asia, i.e. the eastern (or upper) half of the inhabited earth from Africa (below right) and Europe (below left). The T's vertical beam represents the Mediterranean, [. . .] while the left part of the crossbeam symbolises the Don (Tanais in the Middle Ages), which divides Asia from Europe, and the right part of the crossbeam the Nile, which separates Asia from Africa (Simek 1992: 154 f.).

So, this is how one has to imagine the division of the inner garden in a circle. The fact that all the water issued forth from here, and also returned to this place, has undoubtedly to do with the wish to imitate 'realistically' the oceans and rivers that separate the continents—perhaps even with *refluxiones* (ocean currents). In addition, one should consider that on the world maps "the ring of water [. . .] shown encircling the earth, which is represented as a globe, on its outer circumference [is] the visible part of the watery sphere, of which the inhabited continents only take up a relatively small part [. . .]" (ibid.: 155). This all presupposes a geographical or at least cartographical knowledge in the sense of the *mathematicae mixtae* that could not be obtained without learned sources. One gradually recognises that this involves more than a garden; it is, namely, the reproduction or recreation of the world with the elements earth, water and air (fire only makes an appearance indirectly in that it has gone into the metal ores used to make the numerous sculptures and bronze pipes). It is well known that the four winds are often shown in the corners of the medieval and early modern world maps; here, in Filarete's plan, they are revolving on the huts in their corresponding positions. And when there is talk of mountains (Filarete 1965: 209), this is likewise equivalent to "the mountain chains", often shown on the *mappae mundi* "as rows of symbolic mountains" (Simek 1992: 156). However, in the center of this world, one finds neither Jerusalem nor—also conceivable—Rome as the *caput mundi*; one finds the triumph of a new garden art (Fig. 8b).

The *disegno* illustrates and gives more definite shape to the extraordinarily detailed description (Filarete 1965: 209–211), whereby one should note that Filarete "underlines the careful graphic design of every construction project as indispensable for the erection of a fine building. We have this to thank for the numerous illustrations, ground plans and elevations" (Rahmsdorf 1999: 34; cf. Hubert 2003; Beltramini 2001; Carpo 2001: 131–139). In the middle, there is a five-storied, centrally planned building on a square ground plan, which reaches a height of roughly 60 m. Four two-storeyed buildings, with a height of approximately 24 m, each stand at an angle to the corners, built on a surface area of identical size. Between these lie courtyards enclosed by colonnades, each of which takes up the same area as the five buildings. In the walls of these buildings, there are cavities for smaller stairs and water pipes, while their vaults are supported by columns of roughly half the height of these. The gardens lying on the five terraces are accessed by spiral staircases, the pillars of which carry inside them the rising water piping. Access is also provided by the outer flights of stairs, which one can even ride up.

If one follows the *disegno*, then the gardens consist, as far as the planting is concerned, mainly of symmetrically arranged trees, which are larger on the high central building than on the four side-buildings. A thorough sense of rhythm is achieved by the alternation of trees with statues, owing to their different heights and shapes. The gilt, bronze equestrian statues at the four corners of the centrally planned building remind one that the princely equestrian image had undergone excellent refinement in such paintings and frescos as the *Battle of San Romano* (ca. 1450) by Paolo Uccello (ca. 1397–1475) and the *Procession of the Magi to Bethlehem*, produced by Benozzo Gozzoli (1420–1497) between 1459 and 1463, and that a few years later, in 1479, the equestrian statue of the condottiere *Bartolommeo Colleoni* by Andrea del Verrocchio (ca. 1435/1436–1488) set standards that would remain for a long time. As the embodiment of sovereign and heroic power and dignity, the equestrian statues crown the garden's centrally planned building in triumphal fashion: the triumph of the new garden art is also the triumph of the prince who has this garden built as a 'recreation of the world'. One can relate the text's double function as a theory of architecture and an homage to a ruler in a very general way to the whole of the text: "In this way the treatise reconfirms its dual nature as theoretical work and courtly tribute" (Giordano 1998: 65).

The fact that princely rule requires the veneration of gods and inventors to thrive and prosper is symbolised by other statues: in Saturn, one finds the god of agriculture, in Bacchus, of winegrowing, in Minerva, the goddess of the wool and cloth trade, in Carmenta, who in Roman mythology was still the prophetess, the inventor of the Latin characters. The twelve deeds of Hercules (Greek *Dodekathlon*), likewise represented in statues, constituted a programme of imagery both in Antiquity and then again beginning in the Renaissance, in which strength, courage and valour were embodied, so that this popular hero served as an example of the victorious helper who freed people through his labours and tenacious courage, but also served the philosophers as a model of someone who earned immortality through his life and his suffering for others. Hardly any of the well-known Italian princely dynasties of the time were willing to do without the *Dodekathlon*: neither in Florence, where the Medici needed them in their palazzo, decorated with frescos

Fig. 9 Isola Bella garden, Photo Hubertus Fischer 2013

by Piero del Pollaiuolo (1443–1496) and his brother Antonio (1432–1498) ca. 1460, nor in Mantua, where the Gonzaga used them in their Palazzo Te, built by Giulio Romano (ca. 1499–1546).

The supporting structure of this elaborate programme of sculpture is the combination of architecture and nature, as magnificent as it is carefully proportioned, and bearing a complex system of water pipes in the walls and pillars. One cannot rule out the possibility that Filarete used the so-called 'hanging gardens of Semiramis' as a model. As far as we know today, these derive from Nebuchadnezar II and were erected in about 600 BC. The description by the Greek historian, Diodorus Siculus, whom Filarete definitely had known,[5] has at any rate significant aspects in common with Filarete's garden, including the terrace-shaped, layered site, supported by rows of halls. It contains, in addition, the explicit statement: "On all this again earth had been piled to a depth sufficient for the roots of the largest trees; and the ground, which was levelled off, was thickly planted with trees of every kind that, by their great size or any other charm, could give pleasure to beholder" (Diodorus 1933: vol. I, 387). There is no doubt that the gardens of *Sforzinda* involved an ambitious building programme: "When they were all built, the gardens were completed not without great expense and difficulty" (Filarete 1965: 222).

Around two hundred years later, between 1632 and 1671, Carlo III Borromeo (1586–1652) and his heirs had the palace and garden of Isola Bella built on an island in Lago Maggiore. Begun originally by the Milan master builder Giovanni Angelo Crivelli (* first half of the seventeenth century), and completed by the architect, sculptor and engineer Carlo Fontana (1638–1714), who came from the Ticino, a garden—bearing a fair amount of similarity to Filarete's garden—rose up on ten successive terraces, furnished with balustrades and sculptures. The similarity starts with the terraced structure and the balustrades (Fig. 9), continues in the statues of ancient deities alternating with tree groups (Fig. 10), and culminates in

[5] Gianfrancesco Poggio Bracciolоni (1380–1459), from 1453 on head of the Florentine chancellery under the Medici, translated the first five books of Diodorus's *Library* at the instigation of Pope Nicholas V (1447–1455), finishing it in 1449; the first printing was undertaken in Bologna in 1472; cf. Walser (1974: 219–234) and Grassi (1987).

Fig. 10 Isola Bella garden, Photo Hubertus Fischer 2013

the sculpture of a horse, a leaping unicorn. But above all, it is the underlying idea of a perfect synthesis of architecture, nature and sculpture that materialises in both gardens as a clearly structured, layered arrangement and that presents the observer with 'hanging gardens' as a total art work. It would be worth examining comprehensively the mathematical, engineering and hydraulic achievements that were necessary to create this.

From Sforzinda to Christianopolis: The Increasing Trend Towards Science and Edification

The reality of the garden Isola Bella goes well beyond what the early modern utopia had ever dreamt up for gardens. What links it with the utopian garden is its island character, for this is the prerequisite for all spatial utopias from More to Bacon. The spatial utopia's concrete expression as an urban utopia in the first Italian utopia from 1552 by Doni (1513–1574) shows an astonishing similarity to Filarete's *Sforzinda*: "[. . .] A large town, which was built in a perfect circle like a star" ([. . .] *vna gran Città, la quale era fabricata in tondo perfettissimo, a guisa d'una stella* [Doni 1552: 94r/v]). One could even think that Doni was actually thinking of Filarete's *Trattato* (Fig. 3). The wood engraving on the title page of the French edition (Fig. 11) shows the system of the circular radial town as a view or elevation, but thereby leaves out the star-shaped incisions.

Admittedly, as far as the integration of the gardens into the planned town's architecture is concerned, this is most impressive in the case of Andreae's *Christianopolis*: in its strictly symmetrical, square ground plan (Fig. 12), which, alongside the *templum* in the centre and *collegium* bordering the square, also outlines the surrounding *hortus physicus* and the *hortuli civium* in a double row. Incidentally, in *Christianopolis*, we are dealing with a three-fold reform: a reform of the church and religion in the sense of a living belief and active Christianity, a reform of the secular order in the sense of a truly Christian society, and a reform of the sciences in the sense of overcoming the barren, abstract learnedness in favour of the *realia* and of nature, with an emphasis on the empirical, on experiment and invention. Since all true science, especially the science of nature, is seen as a reconciliation with Christ, religion and science are closely connected; they also take up by far the most space in *Christianopolis*. This is represented, for example, by the highly mathematical treatment of urban space, even down to the exact measurement of all its buildings.

More than any other author of a utopia, the Lutheran theologist, clergyman and author Andreae (1586–1654) investigated questions of town planning, and most probably found guidance in the plans of the master builder Heinrich Schickhardt (1558–1634), for Freudenstadt (Fig. 13), a new foundation of Duke Friedrich I. It is revealing that Schickhardt's "'populist' attempt at giving the town a good organisation by consulting with the citizens"—an attempt that failed due to Friedrich's veto—"[. . .] is picked up again in Christianopolis and carried out, an example of which is the revival of the citizens' gardens" (Rahmsdorf 1999: 218).

Fig. 11 Anton Francesco Doni, *Les Mondes* [. . .], Lyon 1583: 204, Herzog August Bibliothek Wolfenbüttel, 45.5 Eth

Thus, this utopian "revival" can be seen as an element of the civic improvement of Christian society. Since there is no house ownership in *Christianopolis*, the gardens that belong to the houses can also be subject to changing usage: "The house has one door and the head of the house is responsible for it. [. . .] At the rear of each building is a garden, kept with much care and nicety, inasmuch as the gardens are conducive to health and furnish fragrance" (Andreae 1916: 169).

There are no generally accessible communal gardens as there were in More's *Utopia*. It is not by chance that, in contrast to *Utopia's* easily opened double door mechanism, Andreae employs a 'supervised' door. But these *hortuli civium* are not the only gardens. Immediately next to these was the surrounding belt of the *hortus physicus*, as is hinted at in the view between the blocks of houses (Fig. 14). The fact that this garden is assigned to the *collegium*, the size and dimensions of which are

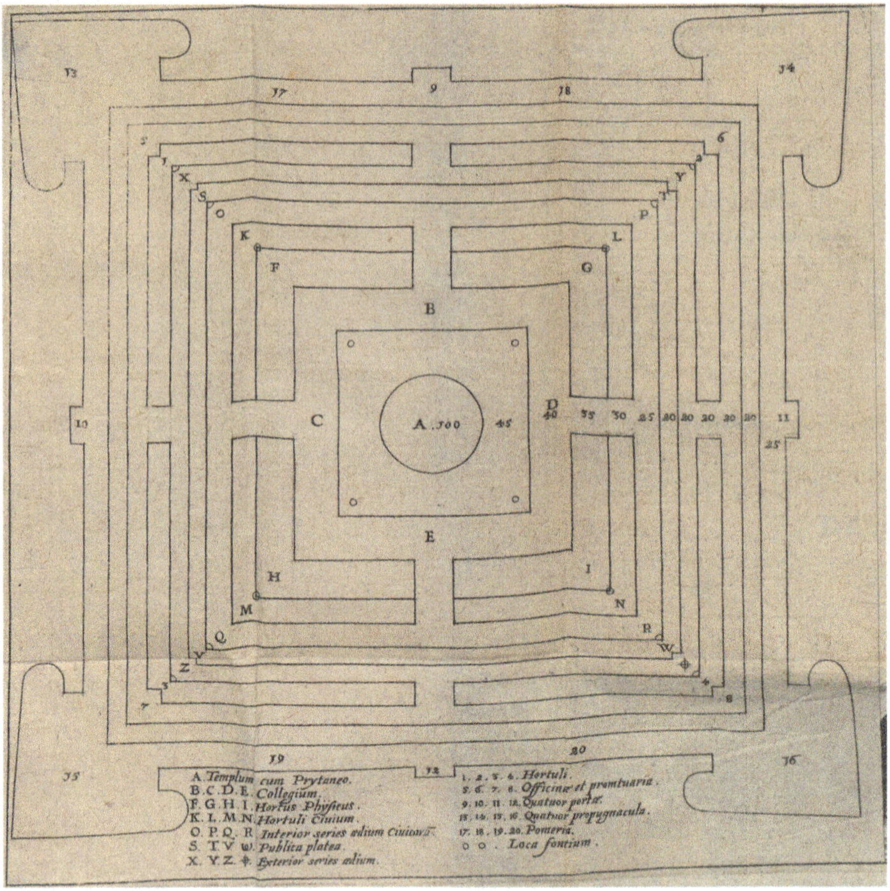

Fig. 12 Johann Valentin Andreae, *Reipublicae christianopolitanae descriptio*, Strasbourg 1619, between p. 24 and 25, Gottfried Wilhelm Leibniz Bibliothek Hannover, T-A 217

described in detail, tallies with the high esteem in which mathematics is held in *Christianopolis*. It houses a cosmos of sciences: a library, an archive, a laboratory, where "the properties of metals, minerals, and vegetables, and even the life of animals are examined, purified, increased, and united, for the use of the human race and in the interests of health" (ibid.: 190); an apothecary, an anatomy institute, an exhibition room for nature (*theatrum physicum*), where there are "even boys at their play recognizing, naming, and investigating according to their characteristic marks and signs thousands of herbs, classifying them with respect to diseases" (ibid.: 201); vaults for mathematical instruments, an exhibition room for mathematics (*theatrum mathematicum*), lecture halls, where "all is open, sunny, and happy, so that with the sight of pictures, even, they attract the children, fashion the minds of the boys and girls, and advise the youths" (ibid.: 206). As in the concept of Filarete's

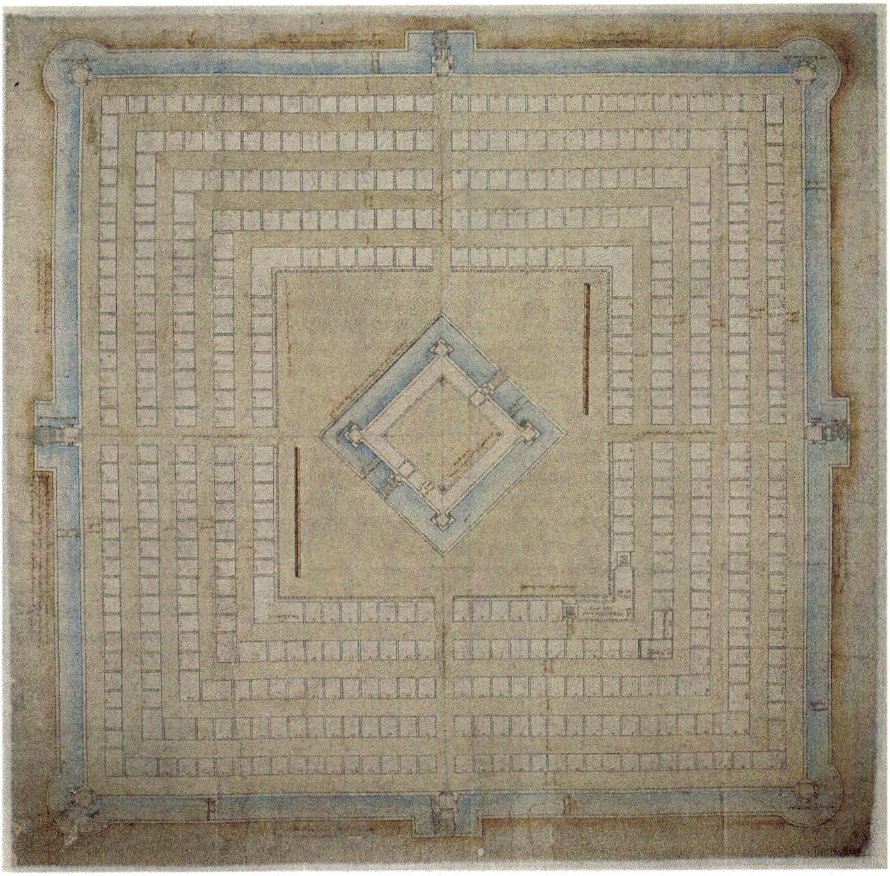

Fig. 13 Heinrich Schickhardt, Fourth design for Freudenstadt, Hauptstaatsarchiv Stuttgart, N 220 B 14

archicodomus, the gardens also serve a recreational purpose: "When they have vacant time, they are permitted to engage in honorable physical exercises either in the open spaces of the town or in the field" (ibid.: 210). There are lecture halls for each of the *septem artes liberales* and further disciplines, so that in the case of *Christianopolis*, one can speak of a far-reaching trend towards education and science in the utopia.

The gardens fulfil several purposes. Firstly they offer a scientifically arranged view of the flora in accordance with the points of the compass, which, to be sure, is, at the same time, as pleasurable as a painting[6]: "Around the college is a double row

[6] On its effect as a painting cf. also Gregory Grämiger in "Reconstructing Order: The Spatial Arrangement of Plants in the Hortus Botanicus of Leiden University in Its First Years".

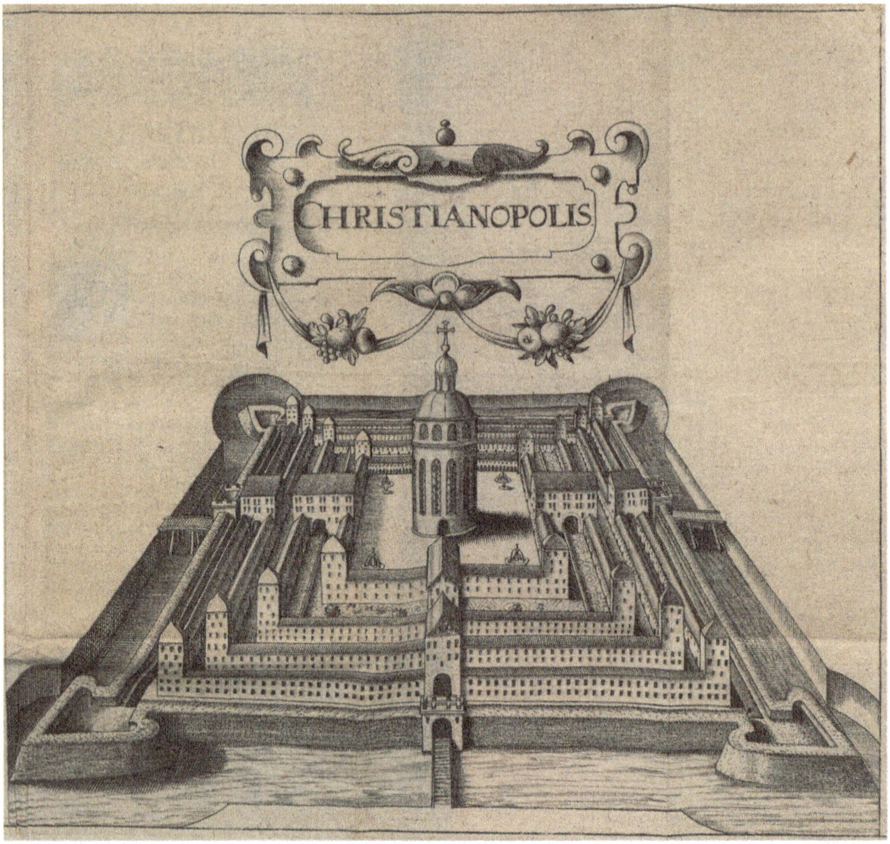

Fig. 14 Johann Valentin Andreae, *Reipublicae christianopolitanae descriptio*, Strasbourg 1619, between p. 24 and 25, Gottfried Wilhelm Leibniz Bibliothek Hannover, T-A 217

of gardens, one general and the other divided into plots corresponding to the houses of the citizens; both are fitted out with more than thousand different sorts of vegetables in such way, that they represent a living herbarium. They are not permitted to confuse the order of the distribution of the plants, which by the skill of the gardener are made to conform to the various zones of the sky, a wonderful and clever combination of colors, representing as it were a painted plate" (ibid.: 268). This ascribes the decisive authority to the gardener—as a botanist (with knowledge of flora) and designer.

The cooperation between practical use and pleasure continues in this garden, going beyond the harmony of botany and aesthetics[7]: "They have here a number of

[7] Cf. Alessandro Tosi in *"Botanical Illustration and the Idea of the Garden in the Sixteenth Century Between Imitation and Imagination"*.

birds in cages and the bees in their hives are very carefully tended. The plants that are for medicine, cooking, or decorative purposes are all in separate plots" (ibid.). The gardens thus become a harmoniously arranged source of enjoyment, health, pleasure and instruction: "Hence they furnish various uses and pleasures,—fragrance, purifying of the air, honey, drugs, harmonious song of birds, and information" (ibid.). It appears as if consideration had even been given to providing a water organ[8]: "There is plenty of water, which is carried by artistically arranged pipes, nor is the music out of harmony with the water. But they avoid all too great expense" (ibid.).

In contrast with Filarete's garden architecture, we have here a more civic version of garden culture for the benefit and devotion of *all* Christian citizens: "Outside the walls they have their very extensive gardens, in which they raise crops for food purposes, for the others [*hortus physicus* and *hortuli civium*] have been planted rather for their elegance" (ibid.) As with everything that one hears of *Christianopolis*, in addition to the real meaning, it also has a second, spiritual and religious meaning. This is particularly true of the gardens, and it demonstrates that the rational and logical, thoroughly planned town, constructed in the imagination, is, at the same time, an extremely edifying town, which is the case neither for Filarete, nor for More, Doni or Campanella. This combination of a trend towards science and spiritualisation lends a special character to *Christianopolis* and its gardens.

> Moreover they learn here to judge the value of human beauty, which is the flower-gathering of a single year. We are born, we grow up, we are in our prime, we droop, and pine away. Out of our death there is again the rise and increase of others. Oh, happy are those who among the wholesome plants learn also to trust in God who feeds the flowers and clothes them without any care on their part, who learn to note down the variety and diverseness of His gifts, and to connect their pleasant odor with God! But why enumerate what man should learn from the creatures of God, when the smallest leaf contains the whole lesson? Rather let us wonder at those who, though they love the earth most of all, neglect entirely that which is the best of the earth, its use and beautiful decoration. Yet they are not willing to seem to burden the earth, though they tread it with crude feet. Let us lament the lost paradise and long for its restoration. For though we look upon natural objects now with faulty vision, when our sight has been restored through the cross, we will behold all things not on the surface, but in their inmost depths (ibid.: 268f).

Nova Solyma or Putting Garden Art into Perspective

Andreae was an ardent admirer and advocate of Johann Arndt (1555–1621), whose four books on *True Christianity* were to become a very popular book in Pietist edification literature. Max Weber, contemplating "all Puritan, baptizing and Pietist

[8] Possibly inspired by Schickhardt's description of the organ in the garden of Pratolino; cf. Simone M. Kaiser and Matteo Valleriani in "The Organ of the Villa d'Este in Tivoli and the Standards of Pneumatic Engineering in the Renaissance".

Christianity", comments: "Quite simply, one believed, starting from the *empirical* comprehension of the divine laws of nature, that one was able to ascend to knowledge of the 'meaning' of the world [...] Empiricism [...] appeared to lead *toward* God, and philosophical speculation appeared to lead away from God" (Weber 2001: ch. 4, 216, fn). In the utopia, *Nova Solyma* (1648), ascribed to the Puritan, poet and politician Samuel Gott (1614-1671), one finds the same idea, but here in a further, interesting connection with garden art.[9]

In the chapter "A Philosophical Garden Party", a dispute is sparked off by the artistic nature of the inviting garden, furnished as it is with statues and an artificial grotto. The question is whether art and invention should be esteemed more highly than nature. The guide and spokesman, Joseph, explains: "[...] when [...] I consider all these Divine arts of Nature, I feel to be encompassed by so many superhuman marvels that every human artifice seems maimed and cobbled [...] The world, indeed which consists by measure, weight and order, contains in itself the exemplar and perfection of every art" (Nova Solyma: 169). Accordingly, natural science represents the privileged path to God via nature: "Now, the special advantage of natural science is to rise from Nature to Nature's God, tracing His footsteps everywhere therein" (ibid.: 172). Secularisation and scientific progress are not necessarily siblings; on the contrary, striving towards God can become the motor of knowledge of the natural world. This—and not the will to rule nature as in Francis Bacon—lends particular importance to natural science in *Christianopolis* and *Nova Solyma*.

In New Jerusalem, too, the garden is integrated into the regular urban environment, whereby its charm is, thanks to the skilfully planted trees, of such a kind "that you seemed to be in the very heart of the country, and all was open to the air and sunshine, green and fertile" (ibid.: 162). This is a new feature, which incites the illusion of natural surroundings. Traces of science are already to be found in the garden's very character: "There were many natural curiosities as well, not collected for mere show or rarity, but in the interest of science" (ibid.: 163f). It is the repeated objection to that which John Dixon Hunt has called "the thrust of garden art", which is new: "[...] its celebration of the superiority of human creation and art over the natural world" (Hunt 2000: 85).[10] Confronted with the works of art in the garden, one reads in *Nova Solyma*, however: "I indeed think that to praise the works of man excessively, and to slight, as so many do, the works of creation and the immense power therein displayed is to show towards the Deity a most unworthy and presumptuous feeling" (ibid.: 163).

With the comment "that Nature is far above all art, and [...] deserves the greater admiration and the greater praise" (ibid.: 168), the "first nature", which in the earlier utopias had been assimilated by the "second nature", makes its return: "[...] there was not a foot of soil to be seen which was not under cultivation or in

[9] Not dealt with in Boesky (1996: 96–100) and Davis (1983: 140–167).

[10] Cf. the chapter of Volker R. Remmert in this volume: "The Art of Garden and Landscape Design and the Mathematical Sciences in the Early Modern Period".

some way put to use for mankind" (Andreae 1916: 143). "There is not a foot of ground that is unproductive" (Campanella 1981: 83). The garden's artfully enhanced natural form must have already obtained a corresponding status in England as an admired creation, to be seen as a contradiction to blossoming religious sensitivity, "whereby it is not the creations of humans that are sublime, it is nature" (Wozniakowski: 54). The architect as the creator of the polity, as he appeared in Filarete and Leon Battista Alberti (1404–1472) (Alberti 1966: I, 10 and 13), would no longer have been conceivable in the *Nova Solyma*.

Conclusion

It is disputed whether the sentence in *Utopia*, *Ciuitas omnis in quatuor aequales partes diuiditur* (More 1965: 136), is to be understood in the sense of a geometric, planned town, based on a square grid. Indisputably, the integration of the gardens in the town planning design (*urbis figuram* [ibid.: 120]) was undertaken at the outset and at the instigation of its mythical founder; in this genre's prototype, they are seen as one of the main elements of a happy social life. A historical and systematic investigation needs to be made into the changes in the concept of gardens and nature in utopian writings up to the eighteenth century and into how these changes compare with the actual development of garden and landscape design. Whereas, for example, the first utopia in the manner of a novel, *Histoire de Sévarambes* (1677/1679) by Denis Veiras (1638–post 1683), clearly owes its garden design to Vaux-le-Vicomte (Fischer 2012: 221), it has been said of its successor, the utopian, Robinson Crusoe-type novel, *Insel Felsenburg* (1736–1751), by Johann Gottfried Schnabel (1692–post 1750), that landscape and garden are presented as an "anti-Versailles" (Höppner 2006–2008: 26). The concepts of nature range from the engineer's conquest of nature (Veiras) to its preformed adjustment to the needs of the utopian settlement's society (Schnabel).

Likewise, the proto-history of utopian gardens in theories of architecture has yet to be analysed systematically. Such an analysis should start with Alberti's *De re aedificatoria libri decem* (finished ca. 1452) and should look beyond Filarete at Francesco di Giorgio Martini's planned gardens in his *Trattato* (1479/1492) (cf. Impelluso 2007: 35–38). Finally, the partially parallel development and antagonism of urban utopias and reform-oriented rural utopias[11]—this utopia conceives of social reconstruction from a rural perspective to the point of "an uncomprising reversion to the 'village'" (Fischer 1981: 270)—also needs to be investigated. The culture of the land and that of the garden moved into focus as a field of conflict, to which urban and rural utopias reacted and responded differently. Urban and castle architecture can thereby become the dividing wall that impedes the longed-for trend towards egalitarian society. Architecture can even be sacrificed to the state of

[11] Johann Eberlin von Günzburg (ca. 1470–1533), *Wolfaria* (1521); Michael Gaismair (1490–1532), *Tiroler lanndsordnung* (1526), [Hans Hergot?], *Von der newen wandlung eynes Christlichen lebens* (1526/1527).

nature, as happens in the very different utopia by Fénelon (1651–1715), "La Bétique" in his *Avantures de Télémaque* (1699). "They find all arts of architecture useless for the simple reason that they build no houses" (Swoboda 1975: 218). Science, architecture and the arts have arrived at their utopian finish—a complete reversal when contrasted with Alberti and Filarete, for whom the human trend towards society has house-building at its very beginning.

Bibliography

Sources

Alberti, Leon Battista. 1966. L'Architettura. Eds. and trans. Giovanni Orlandi and Paolo Portoghesi, 2 vols. Milan: Edizioni il Polifilo.
Andreae, Johann Valentin. 1619. Reipublicae christianopolitanae descriptio. Strasbourg: Lazarus Zetzner.
Andreae, Johann Valentin. 1916. Christianopolis. An Ideal State of the Seventeenth Century. Trans. from the Latin [...] with an historical introduction by Felix Emil Held. New York: Oxford University Press.
Aristoteles. 1965. Politik. Eds. Nelly Tsouyopoulos and Ernesto Grassi. Reinbek: Rowohlt.
Bacon, Francis. 1859. The Works of Francis Bacon. Coll. and eds. James Spedding et al. London: Longmanns et al. 1857-1874 [facsimile reprint, vol.1-14, Stuttgart-Bad Cannstatt: Frommann-Holzboog, 1961-1963], vol. 3.
Campanella, Tommaso. 1981. La Città del Sole: Dialogo Poetico/The City of the Sun: A Poetical Dialogue. Trans. Daniel J. Donno. Berkeley et al.: University of California Press.
Diodorus Siculus. 1933. Library of History. Trans. C. H. Oldfather, vol. I (Loeb Classical Library No. 279). London: William Heinemann; Cambridge, Mass.: Harvard University Press.
Doni, Anton Francesco. 1552. I mondi del Doni, libro primo. Venice: Marcolini.
Doni, Anton Francesco. 1583. Les Mondes celestes, Terrestres et infernaux. Tirez des oeuvres de Doni Florentin, par Gabriel Chappuis Tourangeau. Lyon: Honorati.
Filarete, Antonio Averlino. 1965. Filarete's Treatise on Architecture. Trans. and ed. John R. Spencer, 2 vols. New Haven, London: Yale University Press.
Furttenbach, Joseph. 1988 [1640]. Architectura recreationis. Ed. Detlef Karg. Berlin: Verlag für Bauwesen.
[Gott, Samuel] 1902. Nova Solyma the Ideal City; or Jerusalem Regained. An anonymous romance [...]. Ed. Walter Begley, vol. 1. New York: Scribner's Sons.
More, Thomas. 1518. De optimo reipublicae statu deque nova insula Utopia. Basel: Froben.
More, Thomas. 1965. Utopia. Eds. Edward Surtz and Jack H. Hexter. New Haven, London: Yale University Press.
Plato. 1925. Platons sämtliche Werke in zwei Bänden. Trans. Friedrich Schleiermacher et al. Vienna: Phaidon.
Stiblin, Caspar. 1555. Commentariolus de Eudaemonensium Republica. Basel: Oporinus.
Stiblin, Caspar. 1959. Commentariolus de Eudaemonensium Republica. Ed. Luigi Firpo. Torino: Bottega d'Erasmo.
Swoboda, Helmut. 1975. Der Traum vom besten Staat. 2nd ed. Munich: DTV.

Secondary Literature

Beltramini, Maria. 2001. Le illustrazione del 'Trattato d' architettura' di Filarete: storia, analisi e fortuna. Annali di architettura 13: 25-52.

Boesky, Amy. 1996. Founding Fictions: Utopias in Early Modern England. Athens, Georgia: University of Georgia Press.
Carpo, Mario. 2001. Architecture in the Age of Printing: Orality, Writing, Typography and Printed Images in the History of Architectural Theory. Cambridge, London: MIT Press.
Davis, J. C. 1983. Utopia & the Ideal Society. A Study of English Utopian Writing 1516-1700. Cambridge et al.: Cambridge University Press.
Fabiani Giannetto, Raffaela. 2008. Medici Gardens. From Making to Design. Philadelphia: University of Pennsylvania Press.
Fischer, Hubertus. 1981. Sozialutopien. In Einführung in die deutsche Literatur des 12. bis 16. Jahrhunderts, eds. Winfried Frey et al., vol. 3, 259-287. Opladen: Westdeutscher Verlag.
Fischer, Hubertus. 2012. "Laßt uns das verlorene Paradies bejammern und seine Wiederherstellung herbeisehnen!" – Reisen in die Gärten Utopias. In Reisen in Parks und Gärten – Umrisse einer Rezeptions- und Imaginationsgeschichte, eds. Hubertus Fischer et al., 211-233. Munich: Martin Meidenbauer.
Fischer, Hubertus, Volker R. Remmert, and Joachim Wolschke-Bulmahn. 2011. Wissen und Gärten. Gartenkunst und Naturwissenschaften in der Frühen Neuzeit [...]. In Gartenkunst und Wissenschaft. Diskurs, Repräsentation, Transformation seit dem Beginn der Frühmoderne, eds. Julia Burbulla and Ana-StancaTabarasi-Hoffmann, 271-291. Bern et al.: Lang.
Giordano, Luisa. 1998. On Filarete's 'Libro architettonico'. In Paper Palaces. The Rise of the Renaissance Architectural Treatise, eds. Vaughn Hart and Peter Hicks, 51-65. New Haven: Yale University Press.
Golombek, Lisa. 2008. From Timur to Tivoli: Reflections on 'il giardino all'italiana'. Muqarnas. An Annual on the Visual Culture of the Islamic World 25: 243-254. Leiden: Brill.
Grassi, Liliana. 1987. Diodoro Siculo nel trattato del Filarete: un codice Diodoreo nella Biblioteca degli Sforza?.Aevum 61, 1: 53-58.
Höppner, Stefan. 2006-2008. Neuland. Johann Gottfried Schnabels Insel Felsenburg und die Tradition der utopischen Insel. In Jahrbuch der Johann-Gottfried-Schnabel-Gesellschaft, 9-36. St. Ingbert: Röhrig.
Hubert, Hans W. 2006. Filarete – Der Architekt als Tugendfreund. In Die Virtus des Künstlers in der italienischen Renaissance, eds. Joachim Poeschke et al., 31-54. Münster: Rhema.
Hubert, Hans W. 2003. In der Werkstatt Filaretes: Bemerkungen zur Praxis des Architekturzeichnens in der Renaissance. Mitteilungen des Kunsthistorischen Instituts in Florenz 47: 311-344.
Hunt, John Dixon. 2000. Greater Perfections. The Practice of Garden Theory. Philadelphia: University of Pennsylvania Press.
Impelluso, Lucia. 2007. Gardens in Art. Los Angeles: Getty Publications.
Jellicoe, Geoffrey, and Susan Jellicoe.1988. Die Geschichte der Landschaft. Frankfurt/M et al.: Campus.
Lauterbach, Iris. 1996. The Gardens of the Milanese 'Villegiatura' in the Mid-Sixteenth Century. In The Italian Garden.Art, design and culture, ed. John Dixon Hunt, 127-139. New York: Cambridge University Press.
Rahmsdorf, Sabine. 1999. Stadt und Architektur in der literarischen Utopie der frühen Neuzeit. Heidelberg: Winter.
Ranke, Leopold von. 1884. Universal History: The Oldest Historical Group of Nations and the Greeks. Ed. G. W. Prothero. New York: Charles Scribner's Sons.
Simek, Rudolf. 1992. Erde und Kosmos im Mittelalter. Das Weltbild vor Kolumbus. Munich: Beck.
Tagliolini, Alessandro. 1988. Storia del giardino italiano: gli artistici, l'invenzione, le forme, dall'antichità al XIX secolo. Florence: La Casa Usher, Ponte delle Grazie.
Tausch, Harald. 2003. Im Irrgarten. In Gehäuse der Mnemosyne. Architektur als Schriftform der Erinnerung, ed. Ditto, 79-119. Göttingen: Vandenhoeck & Ruprecht.
Tigler, Peter. 1963. Die Architekturtheorie des Filarete. Berlin: Walter de Gruyter.

Tiller, Elisabeth. 2006. Tugendfreunds 'Trattato'. Antonio Averlino und die Hybridisierung der Register. In Rhetorica: Ordnungen und Brüche, eds. Rita Franceschini et al., 311-324. Tübingen: Narr.

Tönnesmann, Andreas. 2004. Filarete im Dialog: Der Architekt, der Fürst und die Macht. In Dialog und Gesprächskultur in der Renaissance, eds. Bodo Guthmüller et al., 153-164. Wiesbaden: Harrassowitz.

Vercelloni, Matteo, and Virgilio Vercelloni. 2010. Inventing the Garden. Los Angeles: Getty Publications.

Walser, Ernst. 1974. Poggius Florentinus. Leben und Werke. Hildesheim: Olms (1st ed. 1914).

Weber, Max. 2001. The Protestant Ethic and the Spirit of Capitalism. Trans. and introduction Stephen Kalberg. London: Taylor & Francis.

Wozniakowski, Jacek. 1987. Wildnis. Zur Deutungsgeschichte des Berges in der europäischen Neuzeit. Frankfurt/M: Suhrkamp.

Part III

Botany Between Art and Science

Botanical Illustration and the Idea of the Garden in the Sixteenth Century Between Imitation and Imagination

Alessandro Tosi

Abstract
The paper presents some motifs of reflection on the idea and the image of the garden that developed during the course of the sixteenth century and on the relationship between art and science. In particular, the most relevant works in the first age of modern botany (by Brunfels, Fuchs, and Mattioli) that affirm the relevance of images as a method of investigation and identification, are strictly connected to a perception of the garden as a space of organization of knowledge, but also of new cultural and social relations. On one hand, an idea of the garden born from and for science, in which images play a central role in the condivision and communication of knowledge; on the other, the idea of the garden spread in the years 1570–1590 by a flourishing production of prints which underline this evocative, architectural, and emotional dimension. In this encounter between art and science at the end of the century, the garden finds new models of representation and definition ready for further development.

In Brunfels' Garden

As is well known, the visual image and the contribution of the artist to shaping the forms and models of a means of communication that went beyond the verbal constituted fundamental nodal points in early modern European culture. And in the opening years of the sixteenth century, the rise of the scientific illustration accompanied and characterized the "renaissance" of the disciplines connected with the sciences of nature and of man, providing a new and more modern way of seeing,

---

Translated by Lisa Chien.

A. Tosi (✉)
Dipartimento di Civiltà e Forme del Sapere, Università di Pisa, via Trieste 38, I56126 Pisa, Italy
e-mail: alessandro.tosi@unipi.it

H. Fischer et al. (eds.), *Gardens, Knowledge and the Sciences in the Early Modern Period*, Trends in the History of Science, DOI 10.1007/978-3-319-26342-7_9

portraying, cataloguing and, in the end, understanding the natural world and its *historia*.

This phenomenon, which was born out of the long process of the rediscovery and revision of the classical authorities achieved by humanist culture, and the formidable insights of Leonardo and Dürer, may be assigned a precise date and specific context. In 1530, the printer Johann Schott published in Strasbourg the first volume of Otto Brunfels'*Herbarum vivae eicones ad naturae imitationem, summa cum diligentia et artificio effigiatae,* which presented the visual portrayal of the plant world executed *ad vivum* (Fig. 1). In his images, in fact, the artist Hans Weiditz sought to adhere to the rigorous principle of *ad naturae imitationem,* in imitation of nature, revealing a new method of investigating, describing and defining natural data (Landau and Parshall 1994: 247 ff; Swan 1995: 362; Pavord 2005: 161–173; Smith 2008: 15; Touwaide 2008: 43–44; Tongiorgi Tomasi and Willis 2009: 31–36; Kusukawa 2012: 16–19).

In the complex process of preparing the work for publication, the printer succeeded with attentive skill in bringing together a text still anchored in the tradition of the *veteres auctores* with the remarkable naturalism of Weiditz's portrayals of individual specimens. The specific, clearly defined roles and the merits of those who collaborated in the editorial project were by now well recognized in the epigram composed by the humanist Johannes Sapidus (Hans Witz), who would also dedicate verses to the portrait of Hieronymus Bock by David Kandel for the edition of the *Kreuterbuch* (Strasbourg, 1546). The metaphoric reference to a garden that collects seeds from several countries in order to underline Brunfels' botanical knowledge ("... *Quando Ottho velut in quendam Brunfelsius hortum, Huc culta ex vario semina rure tulit*") was united with a traditional reference to Apelles in order to exalt the realism of Weiditz's images, and with the tribute to Schott, who poured considerable money as well as effort into the project.[1]

The scenography of the title page, which can be traced to the close collaboration between the printer and the author and which had already resulted in some visual compositions of the highest quality,[2] conducts the reader into the virtual garden of images and words, illuminating the striking originality of the work and its modernity. Arranged in a refined architectonic frame, with the same elegance and attention to detail that characterize the woodcuts prepared by Weiditz for the text ("*summa cum diligentia et artificio effigiatae*"), the figures of Dioscorides and Apollo underline the continuity of the work with the classical tradition, while the two scenes of the mythical garden of Adonis, with the figures of Venus, Silenus, Mnasilus and Chromis, and of the garden of the Hesperides with Hercules battling

[1] For the reference to Weiditz ("*Ioannes pictor Guidictius*") in the epigram, see also Blunt (1950: 47) and Zucchi (2003: 424).

[2] I need only cite the refined title pages of other works by Brunfels printed by Schott, such as *Vereum Dei multo magis expedit audire...* (1524), *Pandectarum Veteris et Novi Testamenti Libri XII* (1528), or Weiditz's title page of *Almanach ewig werend* (1526); see also Ginzburg (1970: 3 ff).

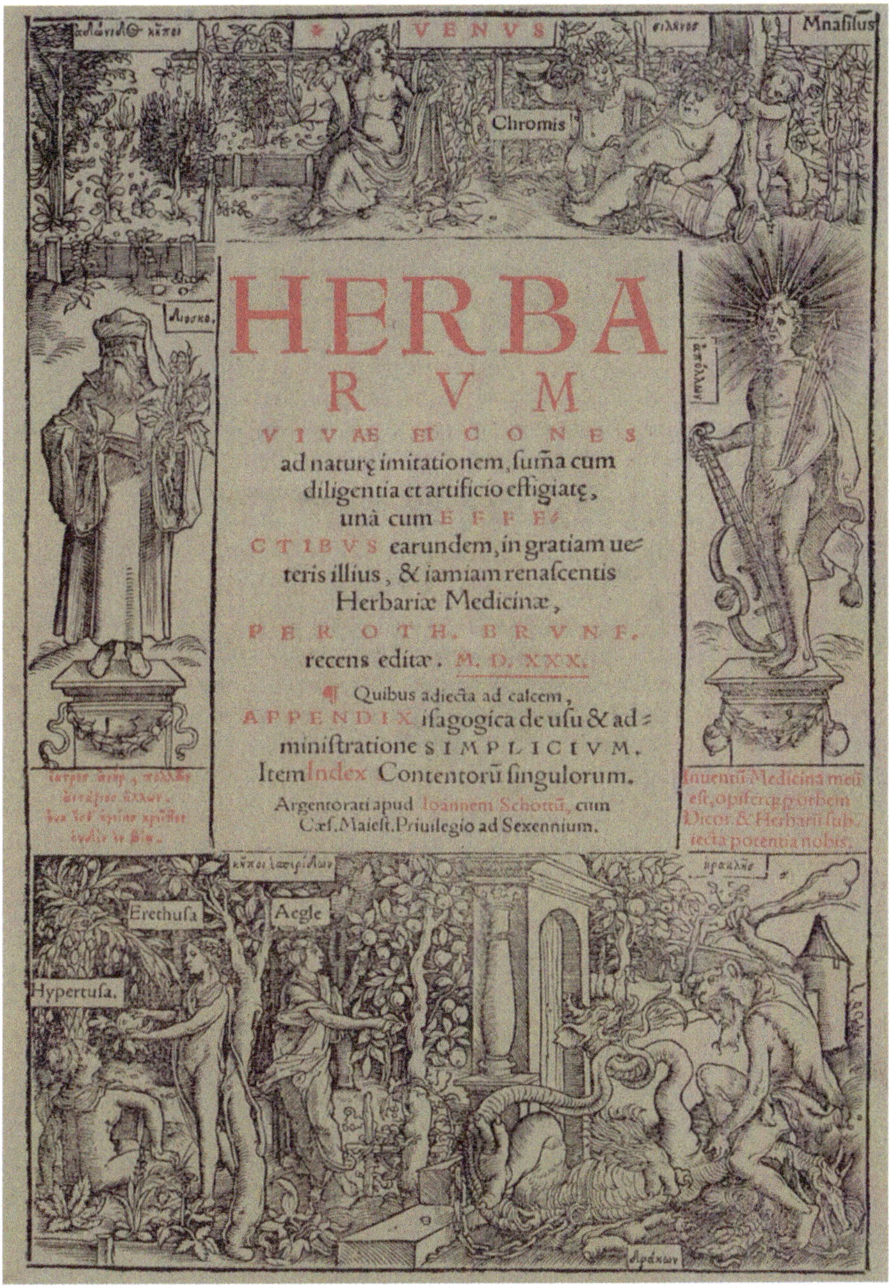
VENVS
Mnaſilus
Chromis
HERBA
RVM
VIVAE EICONES
ad naturę imitationem, ſum̃a cum
diligentia et artificio effigiatę,
unà cum EFFE-
CTIBVS earundem, in gratiam ue-
teris illius, & iamiam renaſcentis
Herbariæ Medicinæ,
PER OTH. BRVNF.
recens editæ. M.D.XXX.
¶ Quibus adiecta ad calcem,
APPENDIX iſagogica de uſu & ad-
miniſtratione SIMPLICIVM.
Item Index Contentorũ ſingulorum.
Argentorati apud Ioannem Schottũ, cum
Cæſ. Maieſt. Priuilegio ad Sexennium.
Inuentũ Medicina meũ
eſt, opiferq; per orbem
Dicor, & Herbarũ ſub-
iecta potentia nobis.
Erethuſa
Aegle
Hypertuſa.

Fig. 1 Otto Brunfels, *Herbarum vivae eicones ad naturae imitationem, summa cum diligentia et artificio effigiatae*, Strasbourg, J. Schott, 1530, title page, private collection

the dragon, form part of a set of deeply rooted symbolical and allegorical practices that have clear links to Dürer's legacy (Tongiorgi Tomasi and Tongiorgi 1984: 53–55; Tongiorgi Tomasi 1988: 374; Zucchi 2003: 439–441; Pavord 2005: 166; Tongiorgi Tomasi and Willis 2009: 32).

The sophisticated title page of *Herbarum vivae eicones*, which was cut in 1529 (the date is inscribed on the plinths bearing the statues of Dioscorides and Apollo), in its complex construction, and in its erudite, classical and literary dimension, borrowed models from the tradition of the printed herbals of the Renaissance and at the same time modernized their iconographic canons, presenting a set of visual and textual coordinates that would have been more accessible to a suddenly expanded readership.

The inclusion of elegant plant motifs and elaborate garlands can be linked to a decorative repertoire often used by printers, as can the refined frame can be attributed to Hans Baldung Grien—who would engrave a fine portrait of Brunfels—which reappears in various texts, including the title page of an edition of *Ortus sanitatis* published in 1517 by the printer Reinhard Beck of Strasbourg (Fig. 2).[3] And the portraits of the *veteres auctores,* in an iconographic model which opens with the frontispiece of the first edition of *Gart der Gesundheit* published by Peter Schoeffer in Mainz in 1485 (New York 2013: 25–26), return on the title page of Brunfels' *Catalogus illustrium Medicorum, sive de primis Medicinae Scriptoribus,* printed by Schott in 1530, in the form of a veritable portrait gallery of eminent physicians from antiquity, framing a scene depicting Galen and Hippocrates giving a lesson on plants. And, moreover, both Schott and Brunfels were well acquainted with the vernacular tradition of the medieval herbal that still flowered in the imagination of the period, documented, for example, in the illustrations to *The Grete Herball,* printed by Peter Treveris in London in 1526, the title page of which depicts gardening practices in the midst of stylized and unnaturally large plants and flowers and anthropomorphic mandrakes, for the pleasure of "reders or practicyens" (Arber 1912: 40–46 and 223; Tongiorgi Tomasi and Willis 2009: 23–27; Elliott 2012: 28). In the edition of *P. Dioscoridae pharmacorum simplicium reique medicae libri VIII* with commentary by Jean Ruel, printed in 1529 by Schott with a preface by Otto Brunfels—who underlined the crucial contribution of the printer—the splendid title page, in fact, harks back to a bookish, fantastical, fabulous world still inhabited by animals and monsters, plants and remarkable remedies, derived from Homer and Pliny, and in which Dioscorides and Cratevas, Helena and Pompeius could appear together (Fig. 3).[4] A significant harbinger, both in terms of concept and composition, but at the same time an obvious counterbalance to the more harmonious and well-ordered title page of *Herbarum vivae eicones,* in which the authority of the classical tradition accompanied a more modern portrayal of the mythical garden.

[3] See Hollstein (1954: 146 and 151) and for his *Portrait of Brunfels* (in *Annotationes…*, Strasbourg, Georgio Ulrichero, 1535) see also Ginzburg (1970: 111–112).

[4] This title page would be the model for the title page of the *Hortus sanitatis* printed in Strasbourg by Mathias Apiarius in 1536, see Pigozzi (2006: 191).

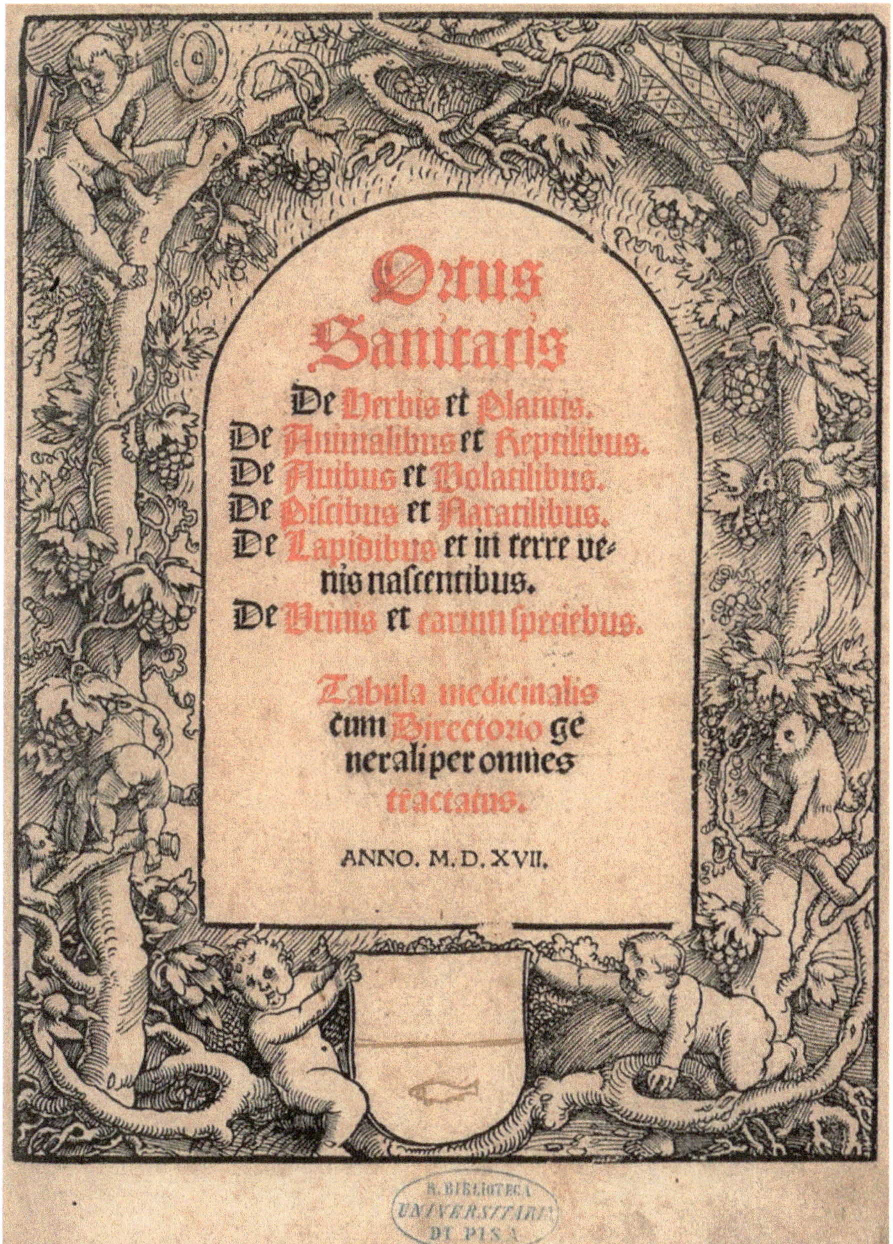

Fig. 2 *Ortus sanitatis*, Strasbourg, R. Beck, 1517, title page (from the ancient Library of the Botanical Garden of Pisa). Pisa, Biblioteca Universitaria, su concessione del Ministero per i Beni e le Attività Culturali e del Turismo

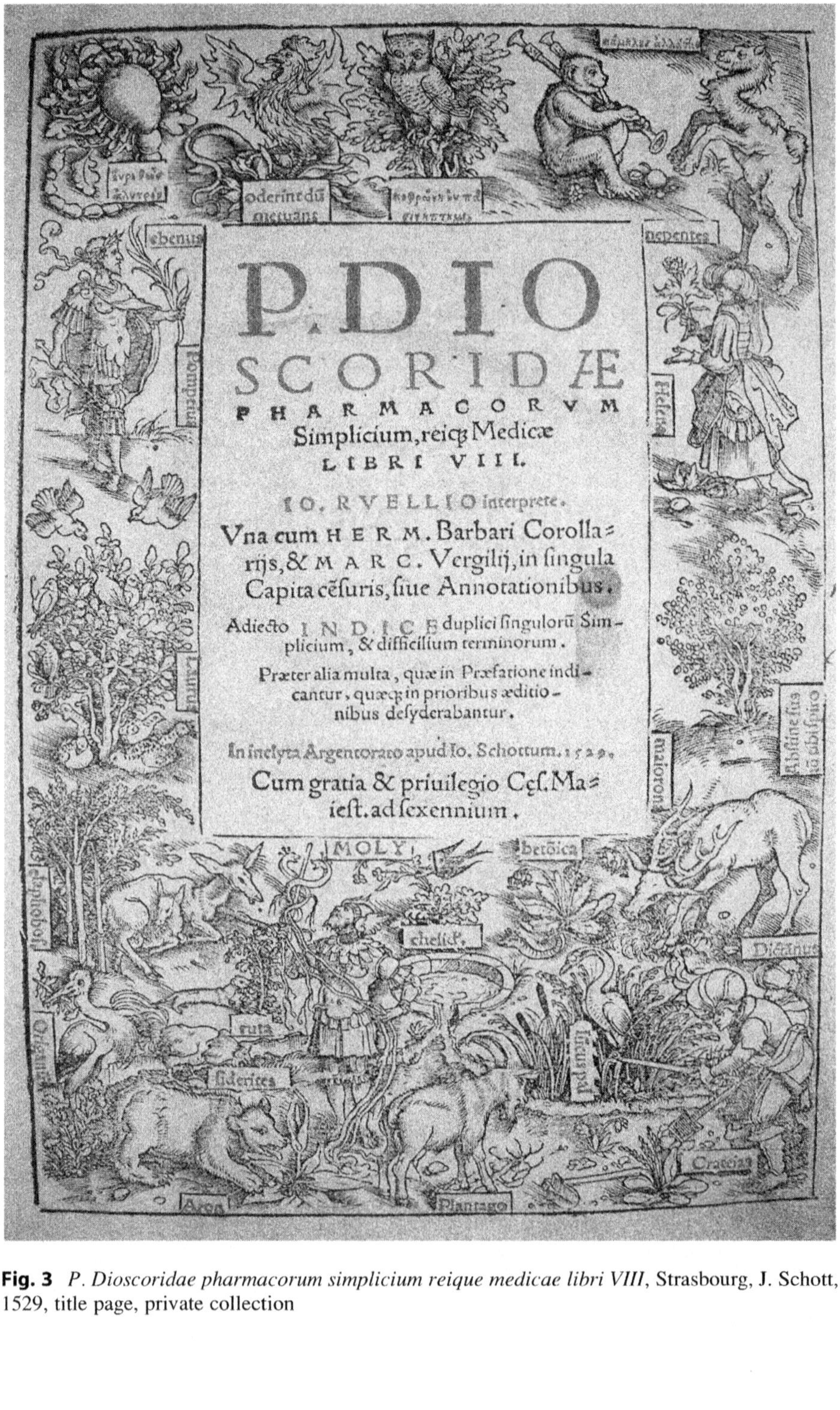
P.DIO
SCORIDÆ
PHARMACORVM
Simplicium, reiq; Medicæ
LIBRI VIII.
IO. RVELLIO interprete.
Vna cum HERM. Barbari Corollarijs, & MARC. Vergilij, in singula Capita cēsuris, siue Annotationibus.
Adiecto INDICE duplici singulorū Simplicium, & difficilium terminorum.
Præter alia multa, quæ in Præfatione indicantur, quæq; in prioribus æditionibus desyderabantur.
In inclyta Argentorato apud Io. Schottum. 1529.
Cum gratia & priuilegio Cęs. Maiest. ad sexennium.

Fig. 3 *P. Dioscoridae pharmacorum simplicium reique medicae libri VIII*, Strasbourg, J. Schott, 1529, title page, private collection

In granting space to the many possible declinations of themes, and to the many different cultural levels for which the complex imagery of the Renaissance herbal provided a fitting vehicle, translating them definitively to a more cultured and refined plane, *Herbarum vivae eicones* constitutes an extremely significant episode in the construction and definition of a new form of visual communication, both in the illustrations by Weiditz and in the title page composed by the author and the printer working in close collaboration. In this sense, the conceptualization and presentation of the plates and the complex architectonic composition of the title page reflect the "dual spirit of the illustration" in the physical space of the book and of the page (Lugli 1985: 62): on the one hand, a spirit that was linked to the emergence of a new form of scientific thought and that, in its close adherence to reality, imposed new rules for the representation and description of natural data; and on the other hand, a spirit inextricably linked to an artistic, literary and philosophical culture which continued and would continue to favour the metaphorical dimension of myth and allegory.[5]

These were twin spirits, both necessary and complementary, spanning imitation and imagination, but adopting the same narrative clarity and descriptive precision: in the naturalistic evidence contained in the illustrations executed *ad naturae imitationem* that heralded profound changes in the sciences of nature, such as in the scrupulous rendition of the fruits and flowers in the gardens of Adonis and Hercules, and in a theatrical title page that provides the perfect translation of the themes and the forms of the Renaissance garden.[6]

Not by chance, Brunfels' title page established a prototype destined to enjoy enduring popularity. In keeping with this model, the image of the garden of Adonis, framed by a gallery of portraits of the physicians and naturalists of antiquity, can be found on the title page of *Spiegel der Artzney*, a work by the physician Lorenz Fries that was edited by Brunfels and printed in Strasbourg by Balthassar Beck in 1532; and above all, on the splendid title page of *De natura stirpium* (Paris, 1536) by the French naturalist Jean Ruel, in which the elegant scene of a garden of delight echoes the humanist dream of Polifilo (Fig. 4).[7]

From this moment onward, the close relationship between art and science would be a determining factor not only in the construction of a more modern approach to the study of natural history, but also in the definition and representation of the garden as a space and repository for knowledge, with its certainties, its naturalistic data, but also the tantalizingly indeterminate nature of its evocative and emotional power. In other words, the route and the development of the botanical illustration in the sixteenth century could add further motifs for reflection about the construction

[5] See also Comito (1979: 23) and Zucchi (2003: 441–442); and, for the frontispieces and title pages as visual documents in the histories of botany and science, see MacDougall (1983: iv), Tongiorgi Tomasi (1988: 372–373) and Remmert (2006: 239–240).

[6] See Conforti (1981: 18–19); and for a general discussion on the idea of the garden in the Renaissance, see Comito (1979), Azzi Visentini (1984) and Keene (2013).

[7] For a description of the title page, attributed to Oronce Finé, see Hocquette (1951: 54).

Fig. 4 Jean Ruel, *De natura stirpium*, Paris, S. de Colines, 1536, title page, private collection

of a visual perception of the Renaissance garden in its various and elaborate declinations, "as an intriguing conceptual system" (Battisti 1972: 4–6).[8]

The Botanical Garden

The processes that led to the revelation and acceptance of the scientific illustration, and the botanical in particular, as a method of knowledge through the union between verbal description and visual representation, are certified in Leonhart Fuchs' *De historia stirpium*, printed in 1542 by Michael Isingrin in Basel. The long title, as well as the *Dedicatory Epistle*, make clear the project and the programme that the author intended to realize. The "lifelike pictures" drawn by Albrecht Meyer, transferred to the woodblock by Heinrich Füllmaurer, and finally realized by the block cutter Veit Rudolph Speckle, "by far the best engraver in Strasbourg", are executed *ad naturae imitationem,* in imitation of nature. And while Brunfels "was the first of all to bring back the correct method of illustrating plants into our Germany, giving others something to imitate", now, the striving for visual clarity was not dictated only by the goal of documenting the natural world by painstaking imitation, revealing a new synthesizing and morphologic approach in a taxonomic definition of scientific data.[9] And the portrait gallery of the *De historia stirpium*—of the *auctor*, the *pictores* and the *sculptor,* in this order and with a significant difference in scale—affirms a definite hierarchy that seems to be a clear declaration of method and practice (Fuchs 1999: 50–51; Tongiorgi Tomasi 2000: 141; Zucchi 2003: 453–455).

Not only this, but the title included a reference to the garden:

> ... this compendium will enable you to learn, with great pleasure, all about the subject with the greatest possible savings in time and expense, far from all risks, as if you found yourself in a living and most pleasant garden (... *eam tibi materiam universam summo & impensarum & temporis compendio, procul discrimine omni, tanquam in vivo iocundissimoque viridario, magna cum voluptate, hinc cognoscere licebit*).

A garden of words, therefore, but also and above all a garden of images that was no longer the verbal garden presented by Euricius Cordus in his *Botanologicon* (1534), or by Hieronymus Bock in his *Kreütter Buch* (1539) (Ogilvie 2006: 145; Swan 2011: 186–191). On the title page, significantly, we find only the emblem with the 'palm' of Michael Isingrin, "the most painstaking printer of Basel": there are no images of mythological gardens, no scenes of herbalists portrayed like hobbits in search of the secrets of herbs, as were still widely used in the *Orti sanitatis,* and even published by the printer Christian Egenolph of Frankfurt, who

[8] Now in Battisti (2004: 9–11). See also MacDougall (1972).

[9] See Ivins (1985: 44–46), Landau and Parshall (1994: 253–254), Kusukawa (1997), Ogilvie (2006: 194 ff), Tongiorgi Tomasi and Willis (2009: 37–45), Dackerman (2011: 142), Kusukawa (2012: 107 ff). The quotation is from the *Dedicatory Epistle*, in Fuchs (1999: 208).

incurred the severe disapproval of Fuchs and a formal accusation of plagiarism by Johann Schott.[10] The artist is guided by the naturalist, who sets the criteria for a representation subject to a scientific method within the conceptual space offered by the book, criteria that do not permit any straying into the intellectual and illusory dimensions of myth or vernacular tales. Every possible temptation to fall back on the classical repertoire of symbols and allegories seemed to be prohibited with the same rigor that was imposed on the realization of the plates in which "we have purposely and deliberately avoided the obliteration of the natural form of the plants lest they be obscured by shading and other artifices that painters sometimes employ to win artistic glory" (Fuchs 1999: 213).

Fuchs' idea of the garden, for which there was as yet no model in the iconography to date, was well expressed in his *Dedicatory Epistle* addressed to Joachim II, Margrave of Brandenburg, which closes with a statement regarding the motives and aims associated with the planting of a garden:

> For never would you be going to such trouble and expense in constructing gardens with so many trees and growing plants if pleasure and a particular love for medicine were not attracting you.

And therefore, with a reference to the motives and aims of his paper and painted garden:

> Moreover there is good hope that our gift of *this garden* will be most pleasing by far to Your Highness, not only because you can find in it almost every variety of plant most elegantly pictured, but also because you can pluck from it remedies for every kind of disease (Fuchs 1999: 218–219).

In a wholly new and yet immediately recognizable typology, Fuchs' idea of the garden fits perfectly with that of a botanical garden (Prest 1981: 6; Ogilvie 2006: 151 ff).

At the same time, in fact, the first botanical gardens were created in association with prestigious academic institutions—in Pisa in 1543–1544, then in Padua, Florence and eventually the rest of Europe—certifying the garden as a place of knowledge and science, but also as a center for the production of botanical images, a veritable artists' workshop. The metaphor by Fuchs, who had also studied botany in his garden in Tübingen, was invited by the grand duke Cosimo I of the Medici to become director of the botanical garden of Pisa and to teach botany at the University (where Vesalius was invited to teach anatomy), and was sent images of plants by Luca Ghini, "that most learned man, and deserving of eternal memory" (Fuchs 1999: 185; cf. Garbari et al. 2002: 7, 12–14, 101 and 108–110), acquires remarkable relevance and significance. The garden, like the collection, is an encyclopaedia and a space of knowledge, both metaphorical and real, a living and most pleasant garden

[10] See Eamon (1994: 108 ff) and Kusukawa (2012: 87–90). For the emblem of Isingrin, see Fuchs (1999: 49).

of images and paper, and a living and most pleasant garden of flowers and plants, in which aesthetic and emotional pleasure arises from knowledge.

It is no coincidence that the same idea and metaphor of the garden was broadened and further developed by the physician and naturalist Pietro Andrea Mattioli of Siena, author of one of the most celebrated and widely read botanical works of the period, his *Commentarii* on Dioscorides. The first edition was printed in Venice by Vincenzo Valgrisi in 1544 without any illustrations but, beginning with the first Latin edition in 1554, Mattioli made certain that his text, which offered a profound revision of the *materia medica* of Dioscorides, was always accompanied by illustrations depicted in imitation of nature ("*ad naturae imitationem expressas*"), to conceive a paper garden easily accessible in every season.[11]

The 1555 Italian edition, the first to be printed in folio "with true portraits of the plants and of the animals" ("*con i veri ritratti delle piante & de gli animali*") presented what was, for the time, in terms of both quantity and quality, a truly extraordinary repertoire of images realized with "inestimable art, genius and patience" ("*con arte, ingegno, & patientia inestimabile*") by the "most refined painter" ("*gentilissimo dipintore*") Giovanni Liberale da Udine:

> ... so that those who are not able to go out into the world to learn about simples, who have no skilled men who can show them these things, may have from me a garden, where they can at any time of the year without the need to cultivate, see them, and know them well.[12]

As in the works of Fuchs, the title page presents only the printer's mark of Vincenzo Valgrisi: no images of mythological gardens, no reference to the Garden of the Hesperides, quite unlike the title page of the *Cruydeboeck* then recently published by Rembert Dodoens (1554), in which the scene of the "*Hesperidum Horti*" was clearly inspired by the title page in *Herbarum vivae eicones*.[13] And no images drawn from the repertoires made popular by the printer Christian Egenolph in Frankfurt for the work of Adam Lonicer (Fig. 5), or by the printer Symon Cock in

---

[11] Mattioli (1554), *Epistola Nuncupatoria* addressed to Ferdinand I of the Hapsburgs (Gorizia, January 1554): "Adiecimus insuper quam plurimas plantarum et animalium imagines, quam fieri potuit diligentius, ad naturae imitationem expressas, magnis cum laboribus, tum sumptibus: non alia mehercle ratione, quam ut ijs, qui terras peragrare non possunt, nec praeceptores habent, quasi hortulum exhiberemus, in quo omni tempore, nullo cultu adhibito, vivas fere stirpium effigies spectarent." On Mattioli and the editions of his works, see Ferri (1997).

[12] Mattioli (1555), *Epistola* addressed to the cardinal Cristofano Madruzzo, prince of Trento (Gorizia, 20 January 1555): "acciò che coloro, che non possono andare per lo mondo alla cognitione de semplici, né hanno huomini periti, che glieli dimostrino, habbiano da me un giardino, ove possano in qual si voglia tempo dell'anno senza veruna coltura vederle, & conoscerle bene". On Mattioli's "paper garden", see Findlen (1999: 382).

[13] The same title page would also adorn the editions published in Anvers (1557, 1563) and London (1578): see Antwerp (1993: 102–103), Swan (1995: 363–364) and Ogilvie (2006: 87–89).

DE HISTORIA, NATV-
RA AC VIRIBVS ARBORVM ET FRVTICVM,
Et primo,
DE HORTO ET IIS QVAE IN EO
naſcuntur, plantis, & arboribus.

DE SOLI, FVNDI, AC TERRAE
adparatione, paſtinatione, ſtercoratione,
atque cultu.

PRINCIPIO igitur, quando de arboribus & cultu hortorum tradere propoſitum nobis eſt, etiam de ſoli conditione & qualitate aliquid præmittendum uidetur. Terra igitur ea, quæ in locis ſqualidis & ſiccis uaſta incultaq; iacet, ſolis ardore plerunq; ita torretur & induratur, ut humorem tranſmittere facilè non poſsit, neque imbribus aut rore reficiatur. Deinde ut plurimum ita quoque conſolidatur, ut neq; frumenta, neq; quicquid aliud in ea ſeritur aut plantatur, protuberare facilè, aut ad frugem peruenire poſsit. Quare ſi qua eiuſmodi fuerit, ea principiò fodiendo, paſtinando, aut etiam arando diligenter ac multum ſubigi præpariq; debet, ut hoc pacto duritiem ac feritatem illam priſtinam exuat, cultuq; uelut manſuefacta miteſcat.

Quod ſi qua autem in tantum ſolida ac ſpiſſa erit, ut ſubinde in glebas maiores ac duriores concreſcat, ea omnino ſurculis aut fuſtibus

Fig. 5 Adam Lonicer, *Naturalis historiae opus novum*, Franckfurt, C. Egenolph, 1551, private collection

Antwerp in *Den Groten herbarius* (1547), the last Dutch edition of the *Ortus sanitatis*.[14]

In the case of the paper and painted garden of Mattioli, the references are even more overt and recognizable. In his opening *premessa* addressed "to learned readers", Mattioli lists a series of exemplary gardens, beginning with the "*amplissimo*" botanical garden in the "*floridissima città di Padova*", implanted by the Venetian Senate and by the University "*per commodo publico, & ornamento della medicina*". He also lists the Garden of Simples, designed by the illustrious botanist and scholar Luca Ghini, one of the most authoritative figures in the natural sciences in the early modern age, and transplanted by Cosimo I of the Medici in

> the most ancient city of Pisa ... where by the good offices of his patron, today flourish many rare plants that have never been seen before elsewhere in Italy, for the public convenience and improvement of doctors, scholars and all others who may delight in such a facility.[15]

And in the wake of these illustrious examples, and with the same ends in mind, others gardens constructed by "kind and virtuous men" in Padua (the garden of Filippo Pasqualigo, owner of a celebrated painted herbal), Venice, Ferrara, Florence, and Rimini: gardens for physicians, herbalists, collectors, virtuosi, dilettantes and poets, places where the guiding principle was the notion of sharing, offering the opportunity to

> partake of all the rare plants to be found there; thwarting the handful of other envious and miserly persons, who have built gardens into which not only are scholars not allowed to enter ... but in which they would not give even a single leaf to anyone, not for any great prize, let alone out of kindness & courtesy, in order to be able to vaunt their being the only ones to have this & that other rare plant in prison.[16]

To the geographical map of the *respublica botanica*, animated by such figures as Luca Ghini, Gabriele Falloppia, Bartolomeo Maranta and Ulisse Aldrovandi, and which Mattioli would amplify with more precise references, we can add other gardens designed "for the public convenience and improvement", such as the "public garden ... abounding in all the honoured and useful plants ... that are sought after by a great apothecary in order to safeguard the City" imagined by the

[14] See de Nave and Imhof (1993: 92); and for the *Naturalis historiae* of Adam Lonicer, printed by Egenolph in 1551, see Tongiorgi Tomasi and Tongiorgi (1984: 57–58).

[15] Mattioli (1555), *Epistola*: "nell'antichissima città di Pisa [...] dove per opera del suo promotore verdeggiano hoggi molte rare piante, che altrove non si sono in Italia fin hora vedute, a commodo & ornamento publico de i medici, de gli scholari, & d'ogni altro, che di quella facultà si diletti". On the botanical gardens of Pisa and Padova, see Garbari et al. (2002) and *The Botanical Garden of Padova* (2013).

[16] Mattioli (1555), *Epistola*: "partecipare con tutti delle piante rare, che vi si trovano; a confusione d'alcuni altri invidiosi et avari, che hanno fatto giardini, nei quali non solamente non lasciano entrare i virtuosi [...] ma non ne darebbono pure una foglia ad alcuno per ogni gran premio, non che per gentilezza & cortesia, per potersi vantare che essi soli hanno questa & quella altra rara pianta in prigione".

Florentine apothecary Stefano Rosselli.[17] And we may, in addition, cite the letter that Alessandro Grifoni wrote from Pisa to Ulisse Aldrovandi in November 1569, requesting information on the name and properties "of that plant which has such a beautiful flower, in the garden of the most reverend Paleotti", that Grifoni also had occasion to admire "in a beautiful garden belonging to the most illustrious gentleman from Piombino, which he prizes for its beauty, but without knowing its name or anything else about it".[18]

Mattioli did not mention in his list the garden of Pietro Antonio Michiel, which may be considered the most emblematic example of the immediate acceptance and application of these principals and ideals. In addition to collaborating on the renovation of the "useful" and "public" Botanical Garden of Padua, the Venetian patrician was the owner of a celebrated garden at an estate in San Torvaso in Venice, and commissioned a considerable *corpus* of botanical illustrations from the artist Domenico Dalle Greche. In his three gardens—his private garden, his painted garden, and the public Botanical Garden ("in my own small garden with pots containing plants from the East, as in that of Our Lords in Padua, and also in the garden that *semper viret depinto* ...") –[19] Michiel adhered with enthusiam to the goals that were being pursued by the most authoritative naturalists in Europe. Writing in 1554 to Ulisse Aldrovandi about the sycamore, he went so far as to criticize Mattioli's text, which in his view presented a 'virtual' garden whose "many gaps" could be ascribed to a lack of direct knowledge of gardening:

> ...It is no novelty to me that with a twig or a dry leaf, and through discourse it is possible to do good things. But it requires labor similar to mine who with diligence and even great obsessiveness has sacrified himself in raising, nourishing, and watching over plants from beginning to end, and when they are to be found in full fruition with all of their qualities, presenting them to the artist, and he with his brush and I with my pen glean from them all of the useful qualities possible. I have gone through this almost unthinkable process of reasoning, for which you must forgive me....[20]

[17] See Tosi (1989: 21); and for Mattioli's "*respublica botanica*", see Findlen (1999: 380 ff).

[18] Ibidem: 63, letter to Aldrovandi in Biblioteca Universitaria di Bologna, ms. 38^2, IV, c. 335r.: "... ne ho veduto qua copia infinita in un bellissimo giardino dell'illustrissimo signor di Piombino, che in quello la tiene questo signore per bellezza, né sa il nome, né tampoco altro di quella".

[19] "sì nel piccolo mio con li vasi delle cose de levante, come in quello di le Signorie Nostre a Padova, et anchora nel giardino che semper viret depinto ...": letter to Ulisse Aldrovandi, from Venice, 14 October 1553, in Biblioteca Universitaria di Bologna, ms. 38^2, I, cc. 175r.-175v., quoted in De Toni (1910: 39–41).

[20] Letter to Ulisse Aldrovandi, from Venice, 10 April 1554, in Biblioteca Universitaria di Bologna, ms. 38^2, I, cc. 176r.-176v., quoted in De Toni (1910: 41–43): "... non mi è cosa nova che con uno ramusculo over una foglia secha, et parlare per relatione non è impossibile far cose buone, ma el ci vole una faticha simile alla mia che con diligenza et summo impazzio ho havuto suferenza in alevare le piante, nutrirle, et vedere da suo principio fino alla fine, et poi quando si ritrovano in sue tutte qualità et producione presentarle al pitore, et lui con il penelo et io con la pena cavarne tutte interamente le utilità che sia posibile. Io son trascorso in questo ragionamento per la impensata, ma mi perdonarete...". See also Palmer (1985: 153), Palmer (1988: 59), Kusukawa (2000: 107), Findlen (2005: 60) and Egmond (2010: 102).

The "unthinkable process of reasoning", i.e., the intuitive understanding on the part of Michiel of the relationship between the word and the image, between the pen and the brush, striving to glean all of the "useful qualities" of the plants, is extremely significant. Setting aside his criticism of Mattioli's text, this intuition came to Michiel in the physical space of the garden, and it was promptly documented in some of the most decorative illustrations to be found in *Cinque libri di piante*.[21]

Mattioli corrected his omission, and in the 1559 edition of the *Discorsi* included the garden of Michiel—"noteworthy for the foreign plants to be found there, and also for the fountains and grotesques of the greatest artistic value fashioned with admirable artifice"—in his list of the Italian gardens modeled on the botanical gardens of Pisa and Padova.[22] The eminent reputation of the Venetian "in the science of plants" and the fame of his "most pleasant" garden were also affirmed by Luigi Anguillara in his *Semplici* and by Conrad Gessner in his *De Hortis Germaniae*, works themselves that appeared, in a significant coincidence, in 1561[23] (Anguillara 1561: 34 and 40–51; Gessner 1561: 239).

In the story of the close ties between gardens and botanical illustration that, from this moment onward, would continue to unfold with new modalities and strategies, Mattioli and his artists contributed further important chapters. The painted garden of the Italian physician became notably enriched after his arrival at the Habsburg court in Prague in 1555, where Giorgio Liberale eventually joined him, drawing "large images of plants and animals, as beautifully wrought as nature herself could have done" (Tongiorgi Tomasi 1997: 370). And Liberale participated, together with the German Wolfgang Meyerpeck and other artists, in vivacious publishing activities; between 1562 and 1563, Mattioli's work would be published in a Czech translation (*Herbarz*) prepared by the renowned astronomer and humanist Tadéas Hájek, who served as physician and surgeon to Massimiliano II and Rudolf II, and in a German translation (*New Kreuterbuch*) published by Georg Handsch (Lenzi 1997: 91; Tongiorgi Tomasi and Willis 2009: 45–56).

Mattioli's little garden ("*l'orticello del Matthioli*"), first shown to scholars by the naturalist and his artist Giorgio Liberale in a significant portrait painted midway through the century, was now a "vastly larger and amplified garden, whose doors were perpetually open to everyone", as Mattioli noted with pride in the Italian edition of the *Discorsi* printed by Vincenzo Valgrisi in Venice in 1568.[24] And while, in the 1530s, Euricius Cordus cited a "Dioscorides minor" and the

---

[21] See Azzi Visentini (1984: 109–115, 2008: 154–155). On the celebrated painted herbal of Michiel (Venice, Biblioteca Nazionale Marciana), see also Michiel (1940); Di sana pianta (1988: 157–159) and Tongiorgi Tomasi (2002b: 47).

[22] Mattioli (1559), preface: "notevole sì per le piante peregrine che vi si ritrovano, come ancho per gl'acquedotti & groteschi di grandissimo valore con mirabil artificio ivi fabricati". See also Findlen (2005: 60).

[23] On Gessner, see Ogilvie (2006: 151–155).

[24] Mattioli (1568), letter to the Princess Giovanna d'Austria (Insbruck, 1 april 1568): "...cresciuto & ampliato giardino, le porte del quale staranno in perpetuo aperte a ciascuno"; and also: "tutto quel poco di tempo, che m'è avanzato fra tanti travagli, disconci, & intrighi, che apportano i

Brunfels' volumes as bibliographical instruments in botanizing, now the "Mattioli with figures" had become an indispensable tool for herborizing, and also an outstanding intellectual achievement, as is certified in the splendid *Portrait of a Naturalist*, once conserved in the Sanford Collection.[25]

The *respublica botanica* therefore gradually expanded into an emerging *respublica eruditorum,* in a network that included and linked naturalists, physicians and herbalists in Italy, but also in Germany, the Low Countries and the other intellectual centers of Europe, from Ulisse Aldrovandi to Francesco Calzolari, Giacomo Antonio Cortuso, Carolus Clusius ... (Olmi 1991; Egmond 2007, 2010). This nascent scientific community found in the image a uniquely effective channel of scientific communication, and in the garden—as in the collection—the perfect setting for the sharing of pleasure and knowledge.

Images of Gardens

Therefore, at the very moment when Mattioli's painted garden had reached the height of its success, actual gardens were constructed for the Hapsburgs in Vienna, Prague, and Ambras; and at the Castle of Ambras, in addition to his celebrated *Wunderkammer*, the archduke Ferdinand II of Tyrol created in the same period a large collection of prints, one of the first and most important examples in northern Europe (Baroni 2011: 5; Parshall 1994; cf. Gothein 2006: 613 ff. Habsburg 2008).

This coincidence is extremely significant, because it is from this moment that the prints produced by some of the painters, engravers and publishers in the Low Countries began to recount and visualize the garden, presenting a renewed idea of its form and concept in the usual frame of myth and allegory, but also in the renovated spaces of man. It was an account marked by the invention of Frans Floris, whose *Vertumnus deus hortorum*, engraved by Cornelis Cort and printed by Hieronymus Cock in 1565, presents a garden as the background for the athletic god surrounded by flowers; and, above all, by Pieter Bruegel's *Spring,* drawn in the same year, engraved by Pieter van der Heyden and published by Hieronymus Cock in 1570, an extraordinary and often quoted visual document in the history of the garden in the sixteenth century, which must be read in conjunction not only with a

negotii & le faccende delle Corti, l'ho veramente speso tutto nel coltivare con ogni studio & in ampliare questo mio principiato giardino". See also Findlen (1999: 382, Crisciani (1999: 7). For the *Portrait of Pietro Andrea Mattioli* by Giorgio Liberale (1550: private collection), see Tosi (1997: 378–380) and Tosi (2007: 33–35).

[25] On the portrait, with the edition of Mattioli opened to the page with the description of *Melanthium,* now identified as *Portrait of Francesco I de' Medici* and attributed to Francesco Morandini called il Poppi (Christie's, sale 5964, lot 35, 4 December 2012), see also Nicolson (1955: 214). On Cordus see Ogilvie (2003: 30–31) and Swan (2011: 189). Writing to Ulisse Aldrovandi in 1567 when he was on an herborizing expedition near Camaldoli, Sebastiano Soavi requested "uno Mathiolo con le figure perché noi habbiamo il Fuchiscio, ma non serve molto" ("a copy of Mattioli with the illustrations because we have the [work by] Fuchs, but it is not very useful"); see Tosi (1989: 180) and Findlen (1999: 388).

Fig. 6 Pieter Bruegel—Pieter van der Heyden, *Spring,* 1570, engraving, private collection

consolidated figurative and allegorical tradition, but also with the illustrations in the works printed by Christian Egenolph or—showing 'how to dress, sow and set a garden'—in the "pleasant treatise" published in London in 1563 by Thomas Hill (Fig. 6).[26]

What seems to have unfolded therefore, beginning in the 1560s, was a correspondence and cross-fertilization of themes and motifs in a visual culture in which art and science succeeded in defining the scenarios of the modern "theatre of knowledge" between books and prints, gardens and collections. On the one hand, the image would serve as an instrument of knowledge and scientific communication for the construction of a natural history that would be transmitted through the extraordinary success of the works by the great founders of modern botany, and also through replicas, plagiarized editions, and the reuse of original illustrations. One need only consider the 1568 edition of *A New Herball* by William Turner, with

[26] For the *Vertumnus* by Floris and Cort, see Hellerstedt (1986: 22–23), Strauss and Shimura (1986: 195), Tongiorgi Tomasi and Tosi (1990: 65). For Bruegel's *Spring*, see Hellerstedt (1986: 8–9), Clayton (1990: 44–45), Tongiorgi Tomasi and Tosi (1990: 60), Aikema and Brown (2000: 572), Tongiorgi Tomasi (2004: 113) and Sellink (2007: 147–150). On the work printed by Hill, *A most briefe and pleasaunt treatyse, teatchynge howe to dress, sowe and set a Garden* (Thomas Hill, London 1563), see Giorgetta (2010: 84–91).

its illustrations drawn in large part from Fuchs ("... that English herbal which contains plants not painted by others, but not however that it is a book made by the hand of a master, and thus it was described by someone who had seen it", as Giovanni Vincenzo Pinelli wrote to Ulisse Aldrovandi from Padova in 1572),[27] and the works of Antoine Du Pinet, Rembert Dodoens, Pierre Pena, Matthias de L'Obel, Carolus Clusius and Leonhart Thurneisser, produced by Italian, French, English and German printers, or by the *Officina Plantiniana* of Cristophe Plantin in Anvers. And then, and above all, a natural history recounted by scientists and painters, with plants and animals painted "from life", together with "their conjoint stories", as Ulisse Aldrovandi wrote in 1577 to Francesco I of the Medici after having seen and admired the grand duke's collection of botanical and zoological paintings by Jacopo Ligozzi, masterpieces of art and science born in the Medici gardens and collections, between *vaghezza* and *utilità,* pleasure and utility.[28]

On the other hand, the image would also be a customary and familiar container for allegories, metaphors and symbols that, at the end of the sixteenth century and in the unifying theme of the garden, would embrace a new aesthetic, conceptual and practical dimension, between pleasure and utility, delight and sociability. To this end, contextually joined with the motifs to be found in the paintings of Lodovico Pozzeserrato (Lodewijck Toeput) and Lucas van Valckenborg, as well as in tapestries and the decorative arts, the wonderfully imaginative and successful architectural compositions of Hans Vredeman de Vries from his 1577 *Theatrum Vitae humanae,* engraved by Hieronymus Wierix, to his *Hortorum Viridariorumque elegantes et multiplicis formae ad architectonicae artis normam affabre,* printed by Philips Galle in Anvers in 1583,[29] constitute extraordinary testimony. Not only this, but the theme of the garden recurs with significant frequency in the engravings produced by the most celebrated Flemish artists of the period.

The variations on the themes of the four seasons and the twelve months of the year drawn by Maarten de Vos, Hans Bol, and Joos de Momper, which were translated into engravings during the 1580s and 1590s by Philips Galle, Adriaen Collaert, Crispijn de Passe, Julius Goltzius or Johan Sadeler, provided the possibility of imagining landscapes that opened on formal gardens manicured by high-spirited gardeners or animated by scenes of well-dressed visitors engaged in conversation, in which the figures of Venus and Flora alternated with aristocratic

[27] In Biblioteca Universitaria di Bologna, ms. 38^2, I, c. 76r., quoted in De Toni (1919: 309–310): "Quell'herbario inglese intendo ch'habbia piante non dipinte da altri, ma non però che sii libro fatto per man di maestro, et cosi l'intende chi l'ha veduto".

[28] On Ligozzi, see Olmi (1992: 61–64) and Tongiorgi Tomasi (2002a). For the letter to Francesco I, see Tosi (1989: 231–247).

[29] See Tongiorgi Tomasi and Tosi (1990: 71–72), De Jong (2000, 2004: 62–64, 2008); see also Gothein (2006: 621–624); and, on Lodovico Pozzeserrato (Limentani Virdis 1988).

Fig. 7 Hans Bol - Johan Sadeler I, *Spring*, 1580, engraving, private collection

ladies and young lovers absorbed in "the pleasure of plucking flowers" (Fig. 7).[30] Beyond the allegorical context, the inclusion of such realistic details as gardening tools, the tasks of the gardener, and the layout of the flower beds in gardens by now familiar, translated themselves into iconographic motifs that should be read in parallel, in terms of their divergences as well as their correspondences, with the visualization of projects for the layout and partitioning of flower beds and for the horticultural practices in the botanical gardens (Fig. 8). An important documentation on this aspect is offered by the drawings of garden and agricultural tools and projects for flower-beds connected to the Garden of Simples of Pisa, and by images of the layout of the botanical garden of Padua presented in *L'Orto dei semplici di Padova* published by Girolamo Porro in 1591 (Tongiorgi Tomasi 1983).

[30] This is the title of an engraving by Crispijn de Passe from the series *Deliciarum juvenilium libellus*: see Veldman (2001: 14–15, 21–23 and 137–146). For the inventions by Maarten de Vos (*Ver*, c. 1587–1588, engraved by Philips Galle; *Ver*, c. 1588, engraved by A. Collaert; *April*, engraved by C. De Passe; *Ver*, engraved by Julius Goltzius, and later by Nicolas de Bruyn), see Hollstein (1996: 279–281, 283 and 286). For Hans Bol's *Spring*, from a drawing made in 1579 and engraved in 1580 by Johan Sadeler I, see Tongiorgi Tomasi and Tosi (1990: 61) and de Ramaix (2003: 44); for his beautiful drawing *Castle and garden* (1580, Firenze, Gabinetto Disegni e Stampe degli Uffizi), then printed by Johan Sadeler, see Tongiorgi Tomasi and Tosi (1990: 72–73), de Ramaix (2003: 42). For the *Months* engraved by Adriaen Collaert from drawings by Joos de Momper and Hans Bol, see Hollstein (1949: 206–207).

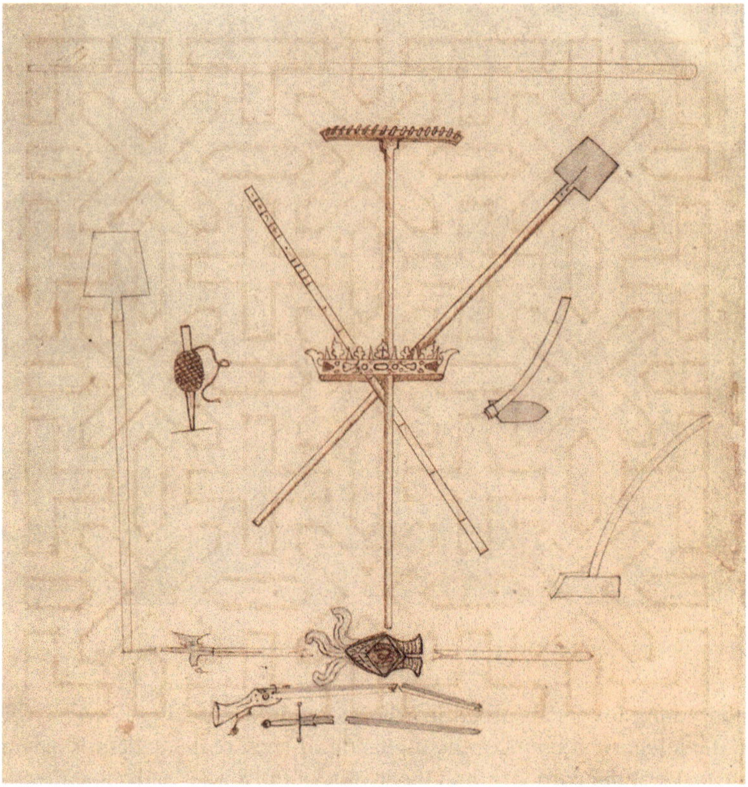

Fig. 8 Garden tools. From *Libro di Compartimenti di giardini*, sixteenth century (from the ancient Library of the Botanical Garden of Pisa). Pisa, Biblioteca Universitaria, su concessione del Ministero per i Beni e le Attività Culturali e del Turismo

Motifs should also be read in parallel with the increasingly dense network of ties that bound together artists and scientists in the spaces of knowledge, in the garden as in the book, with boundaries that became ever more blurred between different levels of interpretation and visual communication. One need only think of the symbols and emblems that Joachim Camerarius drew *ex re herbaria*, or the *Herbario nuovo* of Castore Durante, printed in 1585 with woodcuts by Leonardo Parasole from drawings by his wife Isabella Cattani, derived from the works of Mattioli and Fuchs, "one of the most interesting herbals to be produced in Italy at the end of the sixteenth century" (Fig. 9).[31] The work by this physican and herbalist of Rome was conceived as an updated and expanded version of the works of Mattioli, "the great and illustrious Mattioli, with whom you tilt as an equal" ("*il chiaro gran Matthiol, con cui giostrate al paro*"), as Ercole Filarete recited in

[31] See Tongiorgi Tomasi and Willis (2009: 169–177). On the work by Camerarius (*Symbolorum et Emblematum ex re herbaria desumtorum centuria una*, 1590), see Vignau-Wilberg (2000: 46–47).

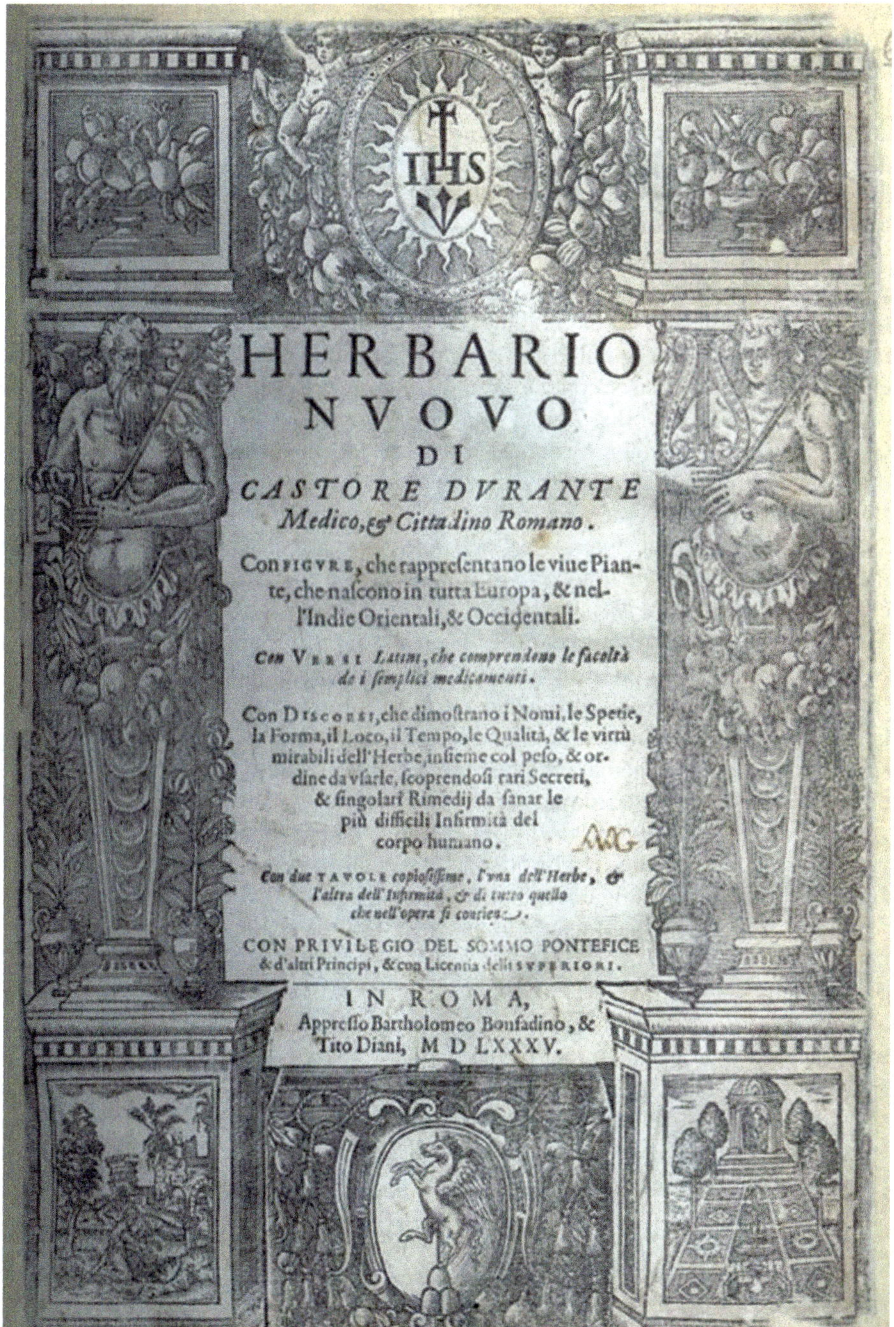

IHS

HERBARIO
NVOVO
DI
CASTORE DVRANTE
Medico, & Cittadino Romano.

Con FIGVRE, che rappresentano le viue Piante, che nascono in tutta Europa, & nell'Indie Orientali, & Occidentali.

Con VERSI Latini, che comprendono le facoltà de i semplici medicamenti.

Con DISCORSI, che dimostrano i Nomi, le Spetie, la Forma, il Loco, il Tempo, le Qualità, & le virtù mirabili dell'Herbe, insieme col peso, & ordine da vsarle, scoprendosi rari Secreti, & singolari Rimedij da sanar le più difficili Infirmità del corpo humano.

Con due TAVOLE copiosissime, l'vna dell'Herbe, & l'altra dell'Infirmità, & di tutto quello che nell'opera si contiene.

CON PRIVILEGIO DEL SOMMO PONTEFICE & d'altri Principi, & con Licentia delli SVPERIORI.

IN ROMA,
Appresso Bartholomeo Bonfadino, & Tito Diani, M D LXXXV.

Fig. 9 Castore Durante, *Herbario nuovo*, Roma, Bonfadino e Diani, 1585, title page, private collection

one of the many epigrams from the poetical gallery that introduces the reader to the herbal. The elegant architectonic frame on the title page, derived from the prototype of Brunfels, is now embellished not only with the figures of Dioscorides and Apollo, but also with two realistic scenes, one depicting an herborizing expedition conducted in the countryside near Rome, and the other a garden with flower beds laid out in geometric forms and a small temple dedicated to Apollo, as if to unite the landscapes of Gherardo Cibo born on the pages of Mattioli with the layouts for the garden of Padua printed by Girolamo Porro.[32]

At the end of the century, therefore, the idea of the garden created and developed with the contribution of botanists and artists seems to have been further enriched by new motifs that testify to the complexity of its definition and interpretation. The process that had been initiated by Brunfels and continued with a multiplicity of episodes and protagonists concludes and is summed up in the title pages of two celebrated botanical works, *The Herball, or Generall Historie of Plantes* by John Gerard, printed in London by John Norton in 1597, and the edition of the *Opera omnia* by Pietro Andrea Mattioli, edited by the Swiss physician and botanist Caspar Bauhin and printed in Frankfurt by Nicolas Basseus in 1598 (Fig. 10).[33]

These two editorial enterprises, separated by the space of one year, reused illustrations, in skilful patchworks, taken from the 'botanical bestsellers' of the century: for Gerard's *The Herball*, the same woodblocks copied from Fuchs, Mattioli, Dodoens, or Matthias de L'Obel just recently employed by the printer Nicolas Basseus for the *Neuw Kreuterbuch* of Tabernaemontanus (Jacob Dietrich of Betrzabern); and for Mattioli's *Opera Omnia,* a collection of plant illustrations borrowed from previous editions. The two engraved title pages also proposed the same compositional schema, with an architectonic frame that brought together the portraits of Flora lifted from the prints of the period, floral garlands, and the portraits of the *veteres auctores*. But above all, they presented a quite similar image—printed in counterpart by Nicolas Basseus—of a garden, a modern, familiar and wholly recognizable garden, not by chance derived from a print engraved by Adriaen Collaert.[34] This was an image that was easily comprehensible, as the Garden of the Hesperides and the Garden of Adonis had been during the first half of the century, but which replaced the meaning and importance of myth with the meaning and importance of modernity; indeed, an image that was well suited to evoke the poetical enchantment of the gardens of Theobalds and the pleasures of Lord Burghley, the patron of Gerard to whom the botanist dedicated *The Herball* (Comito 1971: 497 ff. Corbett 1977; Strong 1979: 52). But at the same time, it was an image designed to evoke a model of a botanical garden, such as those of Stuttgart

[32] See Lugli (1985: 62). On Gherardo Cibo, see now Mangani and Tongiorgi Tomasi (2013).

[33] On Gerard's *The Herball,* see de Nave and Imhof (1993: 124), Tongiorgi Tomasi and Willis (2009: 79–84) and Knight (2009: 69 ff). On Mattioli's *Opera omnia* edited by Bauhin (a second edition would be printed in Basel, Joannis Konig, 1674), see Gentilini (1985: 103–104).

[34] See Corbett (1977: 227). For the print, engraved by Adriaen Collaert from a drawing by Hans Bol for the series of the months (*April*), see also Hellerstedt (1986: 13) and Gothein (2006: 621 and 624).

Fig. 10 Pietro Andrea Mattioli, *Opera quae extant omnia, hoc est Commentarii in VI Libros Pedacii Dioscoridis Anazarbei de Materia medica...*, Frankfurt, N. Basseus, 1598, title page (from the ancient Library of the Botanical Garden of Pisa). Pisa, Biblioteca Universitaria, su concessione del Ministero per i Beni e le Attività Culturali e del Turismo

or Montbéliard cited by Caspar Bauhin in the *Epistola Dedicatoria* written in Basel and dedicated to the duke Frederick of Württemberg[35]; or perhaps those frequented by Mattioli, the quintessential *modern author* portrayed, beneath portaits of the ancients, as a physician and herbalist engaged in healing the sick and preparing remedies.

Therefore, it is an image that speaks to us of "the pleasure of plucking flowers", as in a print engraved by Crispijn de Passe or in a page written by Justus Lipsius[36]; and an image that speaks to us of the pleasure of plucking "remedies for every kind of disease", as Fuchs was careful to remind the reader of the *De historia stirpium* in his *Dedicatory Epistle* to Joachim of Brandenburg. It evokes an idea of the garden as a place for the exchange of emotions and information, between delight and sociability, pleasance and knowledge, in the complex construction of a *texture poétique*—to borrow the expression coined by Michel Conan (Conan 2004)—in which the interwoven development of art and science played, and would continue to play, a fundamental role.

Bibliography

Sources

Anguillara, Luigi. 1561. Semplici. Venezia: Valgrisi.
Gessner, Conrad. 1561. De Hortis Germaniae. Strasbourg: Rhiel.
Mattioli, Pietro Andrea. 1554. Commentarii in libros sex Pedacii Dioscoridis Anazarbei de Medica Materia. Venetiis: in officina Erasmiana, apud Vincentium Valgrisium.
Mattioli, Pietro Andrea. 1555. I Discorsi. Venezia: Valgrisi.
Mattioli, Pietro Andrea. 1559. I Discorsi. Venezia: Valgrisi.
Mattioli, Pietro Andrea. 1568. I Discorsi. Venezia: Valgrisi.
Mattioli, Pietro Andrea. 1598. Petri Andreae Matthioli Medici Caesareiet Ferdinandi Archiducis Austriae Opera quae extant omnia, hoc est Commentarii in VI Libros Pedacii Dioscoridis Anazarbei de Materia medica... Francofurti: ex officina Nicolai Bassaei.

Secondary Literature

Aikema, Bernard, and Beverly Louise Brown (eds.). 2000. Il Rinascimento a Venezia e la pittura del Nord ai tempi di Bellini, Dürer, Tiziano, exhib. cat. Milano: Bompiani.
Arber, Agnes. 1912. Herbals: their origin and evolution. A chapter in the history of botany 1470-1670. Cambridge: Cambridge University Press.

[35] Caspar Bauhin, *Epistola Dedicatoria*, in Mattioli (1598): "duos hortos Simplicibus exoticis refertissimos, si qui alii hodie sunt, vel fuerint olim, praetor I.C.T. excoli curat; alter Stutgardiae, multis miraculo; Montisbeligardi alter exoticarum fertilitate notissimus, quem Iohannes Bauhinus, Celsitudinis Tuae Archiater, optimus frater meus, magna cura, indefessis laboribus, ex I.C.T. mandato, indies auget & illustrat".

[36] On Lipsius, see Swan (2005: 114–120).

Azzi Visentini, Margherita. 1984. L'Orto botanico di Padova e il giardino del Rinascimento. Milano: Il Polifilo.

Azzi Visentini, Margherita. 2008. "... ma spesse volte fa bisogno all'architetto accomodarsi più alla volontà di coloro che spendono che a quello che si devrebbe osservare". Il ruolo della committenza nei giardini veneti del Rinascimento. In Delizie in villa. Il giardino rinascimentale e i suoi committenti, eds. Gianni Venturi and Francesco Ceccarelli, 147-184. Firenze: Olschki.

Baroni, Alessandra. 2011. I 'libri di stampe' dei Medici e le stampe in volume degli Uffizi. Firenze: Olschki.

Battisti, Eugenio. 1972. Natura Artificiosa to Natura Artificialis. In The Italian Garden, ed. David R. Coffin, 1-36. Washington: Dumbarton Oaks.

Battisti, Eugenio. 2004. Iconologia ed ecologia del giardino e del paesaggio. Ed. Giuseppa Saccardo Del Buffa. Firenze: Olschki.

Blunt, Wilfrid. 1950. The Art of Botanical Illustration. London: Collins.

Clayton, Virginia Tuttle. 1990. Gardens on paper. Prints and drawings 1200-1900. Washington: University of Pennsylvania Press.

Comito, Terry. 1971. Renaissance Gardens and the Discovery of Paradise. Journal of the History of Ideas 32, 4: 483-506.

Comito, Terry. 1979. The Idea of the Garden in the Renaissance. Hassocks: The Harvester Press.

Conan, Michel. 2004. Essais de poetique des jardins. Firenze: Olschki.

Conforti, Claudia. 1981. Mitologia classica e pagana nel giardino mediceo fiorentino dall'Umanesimo al Manierismo. In Giardini italiani. Note di storia e conservazione, 13-21. Roma: Ministero per i beni culturali e ambientali.

Corbett, Margery. 1977. The engraved title-page to John Gerarde's Herball or General Historie of Plantes, 1597. Journal of the Society for the Bibliography of Natural History 8: 223-230.

Crisciani, Chiara. 1999. Il 'giardino di carta' di Pietro Andrea Mattioli. In Scienziati a Siena, 7-18. Siena.

Dackerman, Susan (ed.). 2011. Prints and the pursuit of knowledge in early modern Europe. New Haven, London: Yale University Press.

De Jong, Erik A. 2000. Gärten auf Papier. Hans Vredeman de Vries und sein Hortorum Viridariorumque elegantes et multiplicis formae von 1583. In Gärten und Höfe der Rubenszeit im Spiegel der Malerfamilie Brueghel und der Künstler um Peter Paul Rubens, ed. Ursula Harting, 37-49. München: HirmerVerlag.

De Jong, Erik A. 2004. Graceful ornament or formal mockery? Dutch topiary between taste and trend. In Topiaria. Architetture e sculture vegetali nel giardino occidentale dall'antichità a oggi, ed. Margherita Azzi Visentini, 57-70. Treviso: Fondazione Benetton Studi Ricerche.

De Jong, Erik A. 2008. A Garden Book made for Emperor Rudolf II in 1593: Hans Puechfeldner's Nutzliches Khunstbuech der Gartnereij. In The Art of Natural History: Illustrated Treatises and Botanical Paintings, 1400-1850, eds. Therese O'Malley and Amy R.W. Meyers, 187-203. New Haven, London: Yale University Press.

De Nave, Francine, and Dirk Imhof (eds.). 1993. Botany in the Low countries (end of the 15th century – ca. 1650), exhib. cat. Antwerp: Snoeck-Ducaju & Zoon.

De Toni, Giovanni Battista. 1910. Contributo alla conoscenza delle relazioni del patrizio veneziano Pietro Antonio Michiel con Ulisse Aldrovandi. Memorie della R. Accademia di scienze, lettere ed arti in Modena 3, 9: 21-70.

De Toni, Giovanni Battista. 1919. Lettere di Giovanni Vincenzo Pinelli bibliofilo del secolo XVI. Archivio di storia della scienza 1 (1910-1920): 297-312.

Di sana pianta 1988. Di sana pianta. Erbari e taccuini di sanità. Le radici storiche della nuova farmacologia, exhib. cat. Modena: Panini.

Eamon, William. 1994. Science and the secrets on nature. Books of secrets in medieval and early modern culture. Princeton: Princeton University Press.

Egmond, Florike. 2007. Clusius and friends: cultures of exchange in the circles of European naturalists. In Carolus Clusius: towards a cultural history of a Renaissance naturalist, eds.

Florike Egmond, Paul Hoftijzer and Robert Visser, 9-48. Amsterdam: Koninklijke Nederlandse Akademie van Wetenschappen.
Egmond, Florike. 2010. The world of Carolus Clusius: natural history in the making. 1550-1610. London: Pickering & Chatto.
Elliott, Brent. 2012. The world of the Renaissance herbal. In Locus amoenus: gardens and horticulture in the Renaissance, ed. Alexander Samson, 24-41. Chichester: Wiley-Blackwell.
Ferri, Sara (ed.). 1997. Pietro Andrea Mattioli (Siena 1501-Trento 1578). La vita e le opere. Perugia: Quattroemme.
Findlen, Paula. 1999. The formation of a scientific community: natural history in sixteenth-century Italy. In Natural particulars. Nature and the disciplines in Renaissance Europe, eds. Anthony Grafton and Nancy Siraisi, 369-400. Cambridge and London: The MIT Press.
Findlen, Paula. 2005. The market and the world. Science, culture, and collecting in the Venetian Republic. In Il collezionismo a Venezia e nel Veneto ai tempi della Serenissima, eds. Bernard Aikema, Rosella Lauber and Max Seidel, 55-68. Venezia: Marsilio.
Fuchs, Leonhart. 1999. The Great Herbal of Leonhart Fuchs: De historia stirpium commentarii insignes, 1542. Eds. Frederick G. Meyer, Emily Emma Trueblood and John L. Heller, vol. 1. Stanford: Stanford University Press.
Garbari, Fabio, Lucia Tongiorgi Tomasi, and Alessandro Tosi. 2002. Giardino dei semplici. Garden of Simples. Pisa: Plus.
Gentilini, Anna Rosa (ed.). 1985. Faenza. Bibliotheca Botanica. Erbario e libri dal Cinquecento al Settecento del naturalista Lodovico Caldesi (1821-1884), exhib. cat. Imola: Santerno.
Ginzburg, Carlo. 1970. Il nicodemismo. Simulazione e dissimulazione religiosa nell'Europa del '500. Torino: Einaudi.
Giorgetta, Franco. 2010. Hortus librorum, liber hortorum. L'idea di giardino dal XV al XX secolo attraverso le fonti a stampa. Cremona: Il Polifilo.
Gothein, Marie Luise. 2006. Storia dell'arte dei giardini. Firenze: Olschki. (1st ed.: Geschichte der Gartenkunst, 1914).
Habsburg 2008. Habsburg: das Haus Habsburg und die Gartenkunst. Worms: Werner. (Die Gartenkunst, 20, 2, 2008).
Hellerstedt, Kahren Jones. 1986. Gardens of Earthly Delight. Sixteenth and Seventeenth-Century Netherlandish Gardens, exhib. cat. Pittsburgh: The Frick Art Museum.
Hocquette, Maurice. 1951. Histoire générale des jardins. Jardins flamands et lillois. Lille: Bibliothéque Universitaire.
Hollstein, F.W.H. 1949. Dutch and Flemish etchings, engravings and woodcuts, ca. 1450-1700, 4. Amsterdam: Hertzberger.
Hollstein, F.W.H. 1954. German engravings, etchings and woodcuts, ca. 1400-1700, 2. Amsterdam: Hertzberger.
Hollstein, F.W.H. 1996. Hollstein's Dutch and Flemish etchings, engravings and woodcuts, 1450-1700, 44. Ed. Christiaan Schuckman. Rotterdam: Sound & Vision.
Ivins, William M. Jr. 1985. Prints and Visual Communication. Cambridge: The MIT Press. (1st ed. 1953)
Keene, Bryan C. 2013. Gardens of the Renaissance, exhib. cat. Los Angeles: J. Paul Getty Museum.
Knight, Leah. 2009. Of books and botany in early modern England: sixteenth-century plants and print culture. Aldershot: Ashgate.
Kusukawa, Sachiko. 1997. Leonhart Fuchs on the importance of pictures. Journal of the History of Ideas 58, 3: 403-427.
Kusukawa, Sachiko. 2000. Illustrating nature. In Books and the Sciences in History, eds. Marina Frasca-Spada and Nick Jardine, 90-113. Cambrige: Cambridge University Press.
Kusukawa, Sachiko. 2012. Picturing the book of nature. Image, text, and argument in Sixteenth Century human anatomy and medical botany. Chicago: The University of Chicago Press.
Landau, David, and Peter Parshall. 1994. The Renaissance Print. 1470-1550. New Haven and London: Yale University Press.

Lenzi, Maria Ludovica. 1997. Dal "Regno di Iatria" alla corte di Praga. In Pietro Andrea Mattioli (Siena 1501-Trento 1578). La vita e le opera, ed. Sara Ferri, 83-104. Perugia: Quattroemme.
Limentani Virdis, Caterina. 1988. I piaceri della villa nel Pozzoserrato. In Toeput a Treviso: Ludovico Pozzoserrato, Lodewijk Toeput, pittore neerlandese nella civiltà veneta del tardo Cinquecento, eds. Stefania Mason Rinaldi and Domenico Luciani, 71-77. Asolo: Acelum.
Lugli, Adalgisa. 1985. Una biblioteca di immagini. In Bibliotheca Botanica. Erbario e libri dal Cinquecento al Settecento del naturalista Lodovico Caldesi (1821-1884), ed. Anna Rosa Gentilini, exhib. cat., 49-70. Imola: Santerno.
MacDougall, Elisabeth. 1972. Ars Hortulorum: Sixteenth Century Garden Iconography and Literary Theory in Italy. In The Italian Garden, ed. David R. Coffin, 37-59. Washington: Dumbarton Oaks.
MacDougall, Elisabeth. 1983. Introduction. In Hortus librorum: early botanical books at Dumbarton Oaks, iii-v. Washington: Dumbarton Oaks.
Mangani, Giorgio, and Lucia Tongiorgi Tomasi (eds.). 2013. Gherardo Cibo. Dilettante di botanica e pittore di 'paesi'. Arte, scienza e illustrazione botanica nel XVI secolo. Ancona: Il lavoro editoriale.
Michiel, Pietro Antonio. 1940. I Cinque libri di Piante. Ed. Ettore De Toni.Venezia: Ferrari.
New York 2013. The Renaissance Herbal, exhib. cat. New York: The New York Botanical Garden.
Nicolson, Benedict. 1955. The Sanford Collection. The Burlington Magazine 97, 628: 207-214.
Ogilvie, Brian W. 2003. The many books of nature: Renaissance naturalists and information overload. Journal of the history of ideas 64, 1: 29-40.
Ogilvie, Brian W. 2006 The science of describing. Natural history in Renaissance Europe. Chicago and London: The University of Chicago Press.
Olmi, Giuseppe. 1991. 'Molti amici in varii luoghi': studio della natura e rapporti epistolari nel secolo XVI. Nuncius. Annali di storia della scienza 6, 1: 3-31.
Olmi, Giuseppe. 1992. L'inventario del mondo. Catalogazione della natura e luoghi del sapere nella prima età moderna. Bologna: Il Mulino.
Palmer, Richard. 1985. Medical botany in Northern Italy in the Renaissance. Journal of the Royal Society of Medicine 78: 149-157.
Palmer, Richard. 1988. La botanica medica nell'Italia del Nord durante il Rinascimento. In Di sana pianta. Erbari e taccuini di sanità. Le radici storiche della nuova farmacologia, exhib. cat., 55-60. Modena: Panini.
Parshall, Peter. 1994. Art and the theater of knowledge: the origins of print collecting in northern Europe. Harvard University Art Museums Bulletin 2, 3: 7-36.
Pavord, Anna. 2005. The naming of names. The search for order in the world of plants. New York: Bloomsbury.
Pigozzi, Marinella (ed.). 2006. Filtri di salute, d'amore, di morte, di conoscenza. I trattati di botanica della Biblioteca Comunale Passerini-Landi, exhib. cat. Piacenza: Tp.le.co.
Prest, John. 1981. The Garden of Eden. The Botanic Garden and the re-creation of Paradise, New Haven and London: Yale University Press.
Ramaix, Isabelle de. 2003. The Illustrated Bartsch. Johan Sadeler I. 70, 4. New York: Abaris books.
Remmert, Volker R. 2006. 'Docet parva pictura, quod multae scripturae non dicunt'. Frontispieces, their functions, and their audiences in seventeenth-century mathematical sciences. In Transmitting knowledge: words, images, and instruments in early modern Europe, eds. Sashiko Kusukawa and Ian Maclean, 239-270. Oxford: Oxford University Press.
Sellink, Manfred. 2007. Bruegel. The complete paintings, drawings and prints. Ghent: Ludion.
Smith, Pamela H. 2008. Artisanal knowledge and the representation of nature in sixteenth-century Germany. In The Art of Natural History: Illustrated Treatises and Botanical Paintings, 1400-1850, eds. Therese O'Malley and Amy R.W. Meyers, 15-31. New Haven and London: Yale University Press.
Strauss, Walter L., and Tomoko Shimura (eds.). 1986. The Illustrated Bartsch. Netherlandish artists: Cornelis Cort. 52. New York: Abaris books.

Strong, Roy. 1979. The Renaissance Garden in England. London: Thames and Hudson.
Swan, Claudia. 1995. Ad vivum, near het leven, from the life: defining a mode of representation. Word & Image 11: 353-372.
Swan, Claudia. 2005. Art, science, and witchcraft in early modern Holland: Jacques de Gheyn II (1565-1629). Cambridge: Cambridge University Press.
Swan, Claudia. 2011. Illustrated Natural History. In Prints and the pursuit of knowledge, ed. Susan Dackerman, 186-191. New Haven, London: Yale University Press.
The Botanical Garden of Padova 2013. The Botanical Garden of Padova. Venice: Padova University Press.
Tongiorgi Tomasi, Lucia. 1983. Projects for Botanical and Other Gardens: a 16th-Century Manual. Journal of Garden History 3, 1: 1-34.
Tongiorgi Tomasi, Lucia. 1988. Image, symbol and word on the title pages and frontispieces of scientific books from the sixteenth and seventeenth centuries. Word & Image 4, 1: 372-382.
Tongiorgi Tomasi, Lucia. 1997. Il problema delle immagini nei "Commentarii". In Pietro Andrea Mattioli (Siena 1501-Trento 1578). La vita e le opere, ed. Sara Ferri, 369-376. Perugia: Quattroemme.
Tongiorgi Tomasi, Lucia. 2000. L'immagine naturalistica: tecnica e invenzione. In Natura-cultura. L'interpretazione del mondo fisico nei testi e nelle immagini, eds. Giuseppe Olmi, Lucia Tongiorgi Tomasi and Attilio Zanca, 133-151. Firenze: Olschki.
Tongiorgi Tomasi, Lucia. 2002a. The Flowering of Florence: Botanical Art for the Medici. In The Flowering of Florence: Botanical Art for the Medici, eds. Lucia Tongiorgi Tomasi, and Gretchen A. Hirschauer, exhib. cat., 15-107. Washington: National Gallery of Art.
Tongiorgi Tomasi, Lucia. 2002b. L'illustrazione botanica nell'Orto patavino e le tavole di Baldassarre Cattrani. Un caso esemplare di fine Settecento. In L'orto rappresentato. Scienza, didattica e immagine a Padova tra Sette e Ottocento, ed. Luciana Sitran Rea, 43-69. Cittadella: Biblos.
Tongiorgi Tomasi, Lucia. 2004. Tradizione iconografica, simbologia e scienza in alcuni progetti di arte topiaria tra Cinquecento e Seicento. In Topiaria. Architetture e sculture vegetali nel giardino occidentale dall'antichità a oggi, ed. Margherita Azzi Visentini, 104-113. Treviso: Fondazione Benetton Studi Ricerche.
Tongiorgi Tomasi, Lucia, and Alessandro Tosi (eds.). 1990. "Flora e Pomona". L'orticoltura nei disegni e nelle incisioni dei secoli XVI-XIX, exhib. cat. Firenze: Olschki.
Tongiorgi Tomasi, Lucia, and Paolo Tongiorgi. 1984. La nascita della moderna iconografia botanica. In Immagine e natura. L'immagine naturalistica nei codici e libri a stampa delle Biblioteche Estense e Universitaria. Secoli XV-XVII, exhib. cat., 53-59. Modena: Panini.
Tongiorgi Tomasi, Lucia, and Tony Willis. 2009. An Oak Spring Herbaria. Herbs and herbals from the fourteenth to the nineteenth centuries. A selection of the rare books, manuscripts and works of art in the collection of Rachel Lambert Mellon. Upperville: Oak Spring Garden Library.
Tosi, Alessandro. 1989. Ulisse Aldrovandi e la Toscana. Carteggio e testimonianze documentarie, ed. Alessandro Tosi. Firenze: Olschki.
Tosi, Alessandro. 1997. In Matthioli effigiem. In Pietro Andrea Mattioli (Siena 1501-Trento 1578). La vita e le opere, ed. Sara Ferri, 377-389. Perugia: Quattroemme.
Tosi, Alessandro. 2007. Portraits of men and ideas. Images of science in Italy from the Renaissance to the nineteenth century. Pisa: Plus.
Touwaide, Alain. 2008. Botany and humanism in the Renaissance: background, interaction, contradictions. In The Art of Natural History: Illustrated Treatises and Botanical Paintings, 1400-1850, eds. Therese O'Malley and Amy R.W. Meyers, 33-61. New Haven, London: Yale University Press.
Veldman, Ilja M. 2001. Profit and pleasure. Print books by Crispijn de Passe. Rotterdam: Sound & Vision.
Vignau-Wilberg, Thea. 2000. Devotion and observation of nature in art around 1600. In Natura-cultura. L'interpretazione del mondo fisico nei testi e nelle immagini, eds. Giuseppe Olmi, Lucia Tongiorgi Tomasi and Attilio Zanca, 43-55. Firenze: Olschki.
Zucchi, Luca. 2003. Brunfels e Fuchs: l'illustrazione botanica quale ritratto della singola pianta o immagine della specie. Nuncius. Annali di storia della scienza 18, 2: 411-465.

Gardens on Canvas and Paper: Cataloguing Botanical Abundance in Late Medici Tuscany

Irina Schmiedel

Abstract

What is the purpose of portraying a branch from a pear tree in bloom and fruit, of listing hundreds of citrus varieties or of dissecting a huge truffle? When Cosimo III (1642–1723), the penultimate grand duke of the Medici dynasty, commissioned artists and botanists alike to create comprehensive visual and textual catalogues of Tuscan flora, aesthetics and curiosity seemed to go hand in hand with a demonstration of knowledge and prosperity. This article deals with the collection of botanical paintings by Bartolomeo Bimbi (1648–1730), housed in the Casino della Topaia close to the city of Florence, and with the mainly handwritten work of the grand ducal botanist Pier Antonio Micheli (1679–1737), who, among other things, compiled a list of fruit varieties brought to Cosimo's table. Recording both botanical diversity and specialty, these 'gardens on canvas and paper' simultaneously acted as representatives of taste and erudition, of possession and the geographical reach of the grand ducal influence. One of the questions to address is whether such issues of conspicuous consumption fostered or rather constrained 'scientific progress', like the taxonomic discourse or the study of obviously less prestigious lower plants that Micheli could only pursue by extending his 'network' far beyond the borders of Tuscany.

Una Rama di Pere, chiamate di Gerusalemme, o pure d'Inghilterra staccata per Maraviglia dall'Albero piantato in un Orto d'una Casa vicina alla Nonziatina in Firenze, che del Mese di Settembre 1720: era con i fiori, e Frutti, come si vede.

I. Schmiedel (✉)
Institut für Kunstgeschichte und Musikwissenschaft (IKM), Johannes Gutenberg-Universität Mainz, Germany
e-mail: irina_schmiedel@gmx.de

H. Fischer et al. (eds.), *Gardens, Knowledge and the Sciences in the Early Modern Period*, Trends in the History of Science, DOI 10.1007/978-3-319-26342-7_10

> A marvellous branch of pears, called 'di Gerusalemme', or 'd'Inghilterra' taken from a tree that was planted in the garden of a house close to the [Convento della] Nonziatina in Florence, and that in the month of September 1720 was bearing flowers and fruit, as one can see.

Thus reads the inscription in the upper right corner of an oil painting by Bartolomeo Bimbi from Settignano (1648–1730), depicting—"come si vede"—a branch from a pear tree that was both flowering and bearing fruit at the same time (Fig. 1). The branch is set against stony ground, with only a slight allusion to trees and cloudy skies in the background. In this setting, the branch leaps out from the painting, and the viewer is tempted to reach out and pluck one of the faint green fruits placed to the fore, while the little white blossoms gleam in the depths of the canvas. So what is the purpose of the above cited inscription, with its almost ironic undertone (stating something that is plain to see)? The text stands in for those parts of the story that are not directly revealed by the painted flowers and fruits. It informs the viewer of two synonymic denominations—"Pere, chiamate di Gerusalemme, o pure d'Inghilterra"—the reason why the branch was taken from its tree—"per Maraviglia"—as well as where and when the event in question occurred. Image and text explain and complement one another. By using both

Fig. 1 Bartolomeo Bimbi: Pere di Gerusalemme, 1720, oil on canvas, 50 × 63.7 cm, Florence, Museo di Storia Naturale, Sezione Botanica. © Museo di Storia Naturale dell'Università degli Studi di Firenze, Sezione Botanica

pictorial and textual data, Bimbi records and conveys information on the wondrous "Rama di Pere".

All Kinds of Fruit: Bimbi's Paintings and Micheli's Lists for Cosimo III

Far from being a single remarkable work of art, Bimbi's painting was part of an extensive collection of 'botanical portraits' for Cosimo III (1642–1723), the penultimate grand duke of the Medici dynasty.[1] The vast collection of works depicting various fruits and flowers was housed in the Casino della Topaia, one of Cosimo's favourite retreats close to the famous Castello villa on the western outskirts of Florence.[2] Bimbi's biographer, Francesco Saverio Baldinucci (1663–1738), states that the grand duke wanted this place to be adorned with paintings representing the things growing outside 'after life' ("al vivo").[3] This was not only to recall the forms and colours of the various species when the original plants were gone; it also involved the desire to document and disseminate their proper names by means of copies and inscriptions. Therefore, every remarkable fruit was sent directly to the artist to be portrayed ("a farne fare il ritratto") before being sent to the Casino della Topaia for display.[4] Baldinucci includes descriptions of some of Bimbi's works for

[1] Bimbi's botanical and zoological paintings for Cosimo III and some other members of the Medici family and court have been the subject of numerous research projects and expositions, especially within the last 30 years; see Baldini et al. (1982), Mosco (1985) and Mosco (1988), Chiarini (1997), Meloni Trkulja and Tongiorgi Tomasi (1998), Savoia and Strocchi (2001), Tongiorgi Tomasi and Hirschauer (2002), Casciu and Nepi (2008), Nepi and Signorini (2008), Hubert (2008) and Casciu (2009).

[2] For information on the Casino della Topaia, which was built in the middle of the sixteenth century under Cosimo I and restructured in the 1690s under Cosimo III, see Lapi Ballerini (2003: 90–91), Acanfora (2007) and Spinelli (2008).

[3] On the issue of 'after life' or 'ad vivum' representations of natural things, see Swan (1995).

[4] "Avendo poi la stessa Altezza Reale di Cosimo III fabbricato un bel casino in luogo detto 'la Topaia' e posto sopra l'eminenze delle famose vigne della Real Villa di Castello—per dar luogo al suo riposo, in caso che avesse voluto far gita intorno alle medesime—sì come questo luogo era ed è ripieno di tutte le sorti di frutte, d'agrumi, d'uve e di fiori che finora si sono potute trovare, sì di naturali, come anche di stravaganti e bizzarri aborti della natura, così volle che restasse tutto adorno di quadri rappresentanti al vivo le stesse cose. [E ciò] non solo perché, in caso della mancanza delle piante originali, restassero sempre vive nella memoria di ciaschedun dilettante le forme e colori d'ogni specie de' frutti loro, ma eziandio perché noti si facessero distintamente i veri nomi, già variamente confusi, delle medesime, colle loro copie e inscrizioni. Per qual cosa non capitò mai frutta forestiera e stravagante che Sua Altezza Reale non la mandasse subito a farne fare il ritratto al Bimbi, per collocarsi poi in detto casino, col dovuto e destinato ordine, al luogo suo." (Baldinucci 1975 (1725–1730): 247).

la Topaia, including one showing a table with baskets full of different pear varieties ripening in summer, autumn and winter, respectively.[5]

Actually, it seems as if pear trees were the predominant tree type in the garden. The long rectangular orchard in front of the rather small and flat building is often referred to as the "Giardino de' Peri", as can be seen in plans of the house and gardens held in the *Biblioteca Nazionale Centrale* in Florence and in the *Österreichische Nationalbibliothek* in Vienna (Casciu and Nepi 2008: 27 and 45).[6] The painting that most likely corresponds to Baldinucci's description of Bimbi's pear varieties shows 114 cultivars arranged neatly in baskets, on plates or directly on the table.[7] Each different variety is labelled with a little number that corresponds with a list of names on a large cartouche in the front. And, perhaps not surprisingly, among the winter fruits, at the very right edge, one can see a pair of nicely shaped green pears, the "Di Gerusalemme" variety that usually ripens in November if not in September, as documented by Bimbi around 20 years later when he painted the amazing branch with both blossom and fruit. Almost all the species depicted, labelled and named in Bimbi's large canvas (as in 11 other similar paintings of different fruit varieties[8]) are also recorded in the contemporary *Lista di tutte le Frutte*... compiled by the Florentine botanist Pier Antonio Micheli (1679–1737) for the grand duke.[9] The "Di Gerusalem" entry in the *Lista*... does not give any additional information (Micheli ms. 25, c. 284v),[10] but in other instances Micheli provides a very detailed description of the enlisted cultivars

[5] "Per mostrare poi la bella e copiosa diversità di frutte che nell'amene campagne e vigne di Castello produce l'industria e la natura, fece fare dal nostro pittore, il Serenissimo Granduca, due gran quadri. [...] Nell'altro figurò sopra una simil tavola, più canestre piene di diverse specie di pere stative, autunnali e vernine [...]" (Baldinucci 1725–1730 (1975): 248–249). The first of the two paintings described by Baldinucci would have been dedicated to apples.

[6] A telling example of Cosimo's 'pearmania' is found in Antonio Targioni Tozzetti's *Cenni storici sulla introduzione di varie piante nell'agricoltura ed orticoltura toscana* from 1853. The author reports on a variety of pears from Portugal that was introduced to la Topaia, spending the huge sum of "100 doppied'oro" (around 1800 florins, Baldini 2001: 21): "Della pera *Dorice* di Portogallo, ne furono fatti venire i nesti da Cosimo III, per propagarla nella villa della Topaia, ed avendo speso per acquistarla 100 doppie d'oro, fu perciò in seguito chiamata pera cento doppie, ed ancora con tal nome si conosce; più anche con quello di pera Ducale o del Duca." (Targioni Tozzetti 1853: 154–155).

[7] Bartolomeo Bimbi: Pears, 1699, 169 × 227 cm, Poggio a Caiano, Museo della Natura Morta; see Casciu and Nepi (2008: 48) and Casciu (2009: 74–75).

[8] For pictures and the complete records of Bimbi's twelve series of fruit varieties, see Casciu (2009: 58–61, 72–83 and 104–111).

[9] Some technical information on the *Lista*... is found in Ragazzini (1993: 39–41). Micheli later incorporated the fruits of the *Lista*... into the ten-volume *Enumeratio quarundam Plantarum sibi per Italiam, et Germaniam observatarum iuxta Tournefortii methodum dispositarum*. Biblioteca Botanica dell'Università di Firenze, Fondo Micheli, ms. 40–49. Many of the lists and descriptions (especially of fruit) in the *Enumeratio*... are accompanied by soberly sketched watercolour drawings. On the different volumes of the *Enumeratio*... see Ragazzini (1993: 63–78). For a short consideration of Bimbi's and Micheli's pears, see Schmiedel (2013: 20–21).

[10] Neither does it in the *Enumeratio*... (Micheli ms. 49, 159v).

that correspond well to Bimbi's pictorial rendering of the respective fruits. With respect to the external appearances and ripening times, the 'painted' and 'written' catalogues coincide, but Micheli also deals with questions of smell and taste, growth and—especially if a species is not recorded in Bimbi's large canvas—the provenance of cultivars. For example, the "Bianchetta Agostina", which does not appear in the painting, is afforded an elaborate description, including the detail that it is said to have been found in an estate beyond Impruneta close to Florence: "A 4. d'Agosto era matura, in un podere del Bartoli Lanciaio, all'Olmo, luogo di la dall Impruneta. Se il tempo della maturazione non guastasse, la crederei la Pera Canina J.B., che è la Pera Can. H. Cathol. 181." (Micheli ms. 25, c. 109r–110r). Quoting and evaluating contemporary botanical literature is a very typical phenomenon that not only highlights the erudition of the author, but also gives an idea of how difficult it was to identify and name plants. Micheli states that if it were not for the early ripening period, the variety in question would most likely have been identified as Jean Bauhin's "Pira Canina" (Bauhin and Cherler 1650–1651: vol. 1, 51) that also appears in Francesco Cupanis *Hortus Catholicus* (Naples 1696) on page 181. Apart from the obvious formal differences between Bimbi's and Micheli's (to varying degrees) pictorial and textual catalogues of fruit, in terms of content, the involvement of the latter in the early modern botanical discourse seems to be the main distinction. But there are some multifaceted relations that characterise and connect the work of the artist on the one hand and the botanist on the other.[11]

Like Bimbi's paintings for la Topaia, the *Lista di tutte le Frutte...* reflects the grand duke's ardent interest in botany and gardening. As Micheli's student and biographer Giovanni Targioni Tozzetti puts it in § VII *Come il Serenissimo Granduca Cosimo III concepisse qualche stima del Micheli* of his extensive *Notizie della Vita...*, he was introduced to Cosimo III by his *archiatrus*, Giuseppe del Papa (1648–1735). Following this introduction, the botanist and the grand duke often conversed about the different species of fruit that intrigued the prince. Cosimo had been told of the importance of fruit as a means to a long and healthy life, which is why he had procured grafts and scions from far and wide. These were then cultivated and propagated with great care in the princely gardens and manors close to the city of Florence, and it was Micheli's role to compile a catalogue of the most exquisite species and varieties of foreign fruit and vine that grew in the grand ducal estates and vineyards. This catalogue became the *List of all the fruits that are put upon the table of the serene Grand Duke of Tuscany over the year*. Micheli later enlarged his list, adding new kinds of fruit, vine and citruses that he found in other estates, orchards and gardens in Florence and its

[11] Such connections and a possible collaboration between Bimbi and Micheli have been highlighted before; see, for instance, Nepi and Signorini (2008).

surroundings.[12] Targioni Tozzetti's description of his master's engagement at the princely court, the origin of his *Lista*... and its successive extension resulting in the *Enumeratio quarundam Plantarum*... (see footnote 9), agree with Baldinucci's consideration of the la Topaia collection of botanical paintings and its background.

The Glory of Hercules: Territorial and Intellectual Statements

While the la Topaia orchard might have been dedicated to pears, the garden of Castello, situated just a few hundred meters below the casino, was famous for its different cultivars of citrus fruit. When the garden was established under Cosimo I (1519–1574), Vasari wrote in his vita for the sculptor and architect Niccolò Tribolo that the design of the garden, which provided protection from the *tramontana* and other winds, proved the perfect setting for the cultivation of oranges.[13] Citruses had been cultivated in the Medicean gardens of the Careggi and Fiesole villas since the early fifteenth century. The 'golden apples' soon found their way into the family's coat of arms, as was formulated in a panegyric poem to the dynasty's first grand duke Cosimo I.[14] The abundance of the Castello *limonaia* and the traditional Medicean interest in *agrumi* were also recorded by Bartolomeo Bimbi, who, apart from a couple of portraits showing single remarkable specimens, created four large canvases of 116 varieties of citrus fruit altogether.[15] The diverse fruits are depicted (mostly in pairs, as in Bimbi's painting of pear varieties) as if growing from four single espaliers upon which they have been grafted. Little tags with numbers and cartouches with names at the bottom of each canvas provide identification of the

[12] "Più che altro i discorsi del Granduca col Micheli erano sulle varie specie di frutte, delle quali esso principe era molto dilettante, perchè stategli lodate come un mezzo per viver lungamente sano, e perciò ne aveva d'ogni parte fatte venire le barbatelle, o i nesti, e le faceva con gran premura coltivare e moltiplicare nei suoi Reali Giardini, e nelle sue possessioni più vicine alla città. Da ciò prese occasione il Micheli di fare un catalogo delle più pregiabili specie o varietà di esse frutte, e delle viti forestiere che si coltivavano nelle Regie Possessioni e vigne, e lo intitolò: *Lista di tutte le frutte, che giorno per giorno, dentro all'anno, son poste alla mensa del serenissimo Granduca di Toscana*; ma successivamente lo andò ampliando coll agiunta di nuovi frutti, di uve, e di agrumi che gli riuscì osservare in altri poderi, orti e giardini di Firenze e del suo contado." (Targioni Tozzetti 1858: 39–40).

[13] "[...] e questo giardino aveva a essere tutto pieno d'aranci, che vi avrebbono avuto ed averanno, quanto che sia, commodo luogo, per essere dalle mura e dal monte difeso dalla tramontana ed altri venti contrari." (Vasari 1973 (1568): vol. 6, 75). For the early history of the Castello garden, see Acidini Luchinat and Galletti (1992: 41–62 and 67–77); on its iconography with a close consideration of Vasari's description, see Bertsch (2012: 205–212).

[14] On citrus cultivation and symbolism under the Medici, see Galletti (1996). In his examination of the *Sala di Leone X* frescoe cycles in the Medicean villa in Poggio a Caiano (in this particular case, the Hercules and Fortuna lunette), Julian Kliemann highlighted the connection between *palle* and *mala medica*, also referring to Borgaeus' poem *Ad Cosmum Medicem*, reproduced in the Appendix (Kliemann 1976: 83–85 and 259–265).

[15] For the portraits of single specimens, see Casciu and Nepi (2008: 122–131). For Bimbi's series of citrus varieties, all dating to 1715, see Casciu (2009: 104–111).

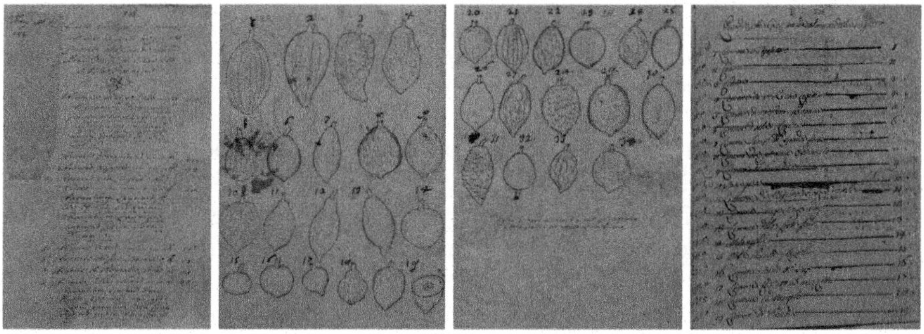

Fig. 2 Lists and ink drawings of Bartolomeo Bimbi's citrus fruit, Pier Antonio Micheli: *Enumeratio...*, vol. 9, Florence, Biblioteca Botanica dell'Università, ms. Micheli 48, c. 206r, 210v, 211r, 215r. © Biblioteca Botanica dell'Università degli Studi di Firenze

painted species. It is interesting to note that Micheli actually used Bimbi's paintings of citruses as a reference for his own work (Fig. 2). In the ninth volume of his *Enumeratio...*, among his notes on citrus fruit, there are four corresponding sets of schematic ink drawings documenting the species depicted by Bimbi (no. 19, the "Limone di Paradiso", for instance, is sliced both in the painting and in the drawing to show the small amount of flesh inside). The names of the citrus varieties are also copied from Bimbi, and as the title reveals, we are looking at the "Lista di tutti gl'Agrumi, che son dipinti ne' quattro Quadri del Casino della Real Villa di Castello e della Topaia" (Micheli ms. 48, c. 206r–218v).[16]

The cultivation of both pears and citruses was very popular in early modern orchards and ornamental gardens.[17] However, the instances of painted single and serial portraits, lists, descriptions and drawings were not limited to pears and citrus fruit; indeed, both Bimbi and Micheli documented and examined a wide variety of Tuscan flora in their service to Cosimo's passion for flowers, fruit, etc. The historian Riguccio Galluzzi described, in his 1781 monumental work on the Medici reign, that the grand duke had suffered from serious bilious complaints, which led him to adopt a strict vegetarian diet, "un rigoroso vitto pitagorico", prescribed by his *archiatrus* Francesco Redi (1626–1697), predecessor of del Papa:

> After Cosimo III had been cured from the life-threatening illness that had troubled him he followed the prudent advice of Redi and totally changed his diet [...] So botany, gardening

---

[16] Micheli's *Lista di tutti gl'Agrumi...* further allows the drawing of conclusions about the original installation of Bimbi's paintings inside the casino (Casciu 2008: 34).

[17] On the cultivation of citruses and pears in early modern Tuscany, see Baldini et al. (1982: 17–44 and 101–122). The literature on the cultural history of citrus fruit is vast; see, for example Tagliolini and Azzi Visentini (1996) and Doosry et al. (2011). Pears, however, were a very sought after genus of fruit as well. Leaving the Italian context, Abbé Gobelin's plea for the species-rich and more resistant pears in favour of the fashionable "Orangers, Citronniers & Grenadiers" is very telling (Gobelin 1677: 145–165). On this topic, see also Schnapper (1988: 51–54).

and fruit cultivation became one of the prince's greatest passions, and he had the rarest and finest plants, fruits and flowers from India, America and Africa brought to his gardens not shying away from spending huge sums to obtain them. Presumably, the Prince's adherence to his strict diet played a role in his long and healthy life.[18]

Galluzzi's explanation not only refers to Cosimo's health problems as a possible catalyst for his growing interest in botany, but also underlines the representative aspects connected with the prince's costly passion. The possession of a specific plant implied knowledge about it and promoted the prestige of the relevant garden and its owner.

When, in 1723, Michelangelo Tilli (1655–1740), director of the *Giardino dei Semplici* in Pisa, published the *Catalogus plantarum horti Pisani*, the grand duke's long life was coming to an end. The garden itself had been founded almost 200 years before by Cosimo I as part of the Pisan university reform, when it had become one of the first botanical gardens in Europe.[19] In Francesco Redi's words, the Pisan *Giardino dei Semplici* and its Florentine counterpart fulfilled an intellectual function, rather than merely being a site of aesthetic interest and pleasure (though the truth may lie somewhere in between[20]). The physician and member of the famous *Accademia del Cimento* further interweaves the aspects of knowledge and representation in a subtle rhetorical manner when, in his *Esperienze intorno a diverse cose naturali...*, he states that:

[...] if it was not the least of Hercules's glorious deeds to transport the citruses from the African gardens of the Hesperides to Greece, so it still shines among the achievements of my Serene Lord, who nobly provides the gardens of Florence and Pisa with every foreign

[18] "Dopo che [Cosimo III] si trovò sanato dalla mortale malattìa che lo avea travagliato, seguitando i prudenti consigli del Redi variò totalmente il sistema nel regime della sua vita [...] Quindi è che la botanica, la giardineria, e la coltivazione delle frutte divennero una delle più forti passioni di questo Principe, che non risparmiò tesori per avere nei suoi giardini le piante, i frutti, ed i fiori più rari e deliziosi dell'Indie, dell'America e d'Affrica. Questo metodo continuato con estremo rigore, e con la maggiore esattezza fu certamente quello che li produsse una vita lunga e sana [...]" (Galluzzi 1781: vol. 4, 260).

[19] Extensive information on the Pisan *Giardino dei Semplici* and its history is found in Garbari et al. (1991). For a shorter impression and the general functions of early modern botanical gardens, see Tongiorgi Tomasi (2005).

[20] That the gardens incorporated both the concepts of *docere* and *delectare* is not least revealed by numerous travel accounts and letters. John Evelyn, for instance, mentioned "a noble palm tree, and very fine waterworks" (Evelyn 1901: 88), while his compatriot John Raymond stated that the "Physitians garden [...] is more for use then delight; although there be good walks, & water-works that well washt us, yet for the most'tis cover'd with simples, outlandish Plants and the like. Joyning to it is a gallery very commodious for Medecinall things, it abounding with all curiosities of Nature, as foreign creatures, Stones, Mineralls, and whatsoever strange the farthest Indies produce" (Raymond 1648: 21–22).

plant; and not just for a vain and curious delight, but for the sole benefit of those who investigate and describe the diverse characters and properties of the plants.[21]

The display of botany as an evolving modern science in this case is underpinned by traditional allegoric and representative values. As an essential part of early modern scientific practice, such allegoric and representative contents often helped to legitimise and explain the research and its results: for instance, by making a connection with typical figures of princely identification such as Hercules, who, as part of his twelve labours, stole the golden apples of the Hesperidean nymphs that were guarded by the fierce dragon Ladon.[22]

However, in the case of Tilli's catalogue of the Pisan botanical garden from 1723, this is not so evident. It contains alphabetical lists of about 5000 plant species and 50 engraved plates that show a selection of the registered specimens. The catalogue also owes much to Tilli's *aiuto custode*, who was Pier Antonio Micheli. In fact, there are so many mentions of his name in the 175 pages, it is hard to keep count. Even the *Hortus Pisanus* plays on the relation between possession, knowledge, and representation, especially with the choice of 'eye-catcher' on the title page (Fig. 3). The vignette underneath the prominent dedication to Cosimo III shows the personification of a royal Pisa and the four continents around her offering their botanical treasures to the sovereign. The Pisan coat of arms on her stony throne, the personification of the river Arno to the left and the silhouette of the famous cupola and leaning tower in the background leave no doubt as to the setting. The scenery of the continents paying homage to a civic ruler is not unusual, and in Tilli's case seems to be inspired by Jan Commelijn's catalogue of the Amsterdam botanical garden from 1697.[23] Nevertheless, it perfectly confirms Targioni Tozzetti's, Galluzzi's and Redi's accounts of Cosimo III as a wealthy amateur procuring plants from all parts of the world to adorn his gardens, enrich his table and advance botanical knowledge. It could also be argued that the volume's lists, brief descriptions and engravings cataloguing the abundance of and knowledge about all the different plants were representative of the geographical reach of the grand ducal influence.

---

[21] "[...] se tra le glorie di Ercole non fu la minore l'aver trapiantati i Cedri nella Grecia dagli orti Affricani delle Esperidi, così tra le glorie del Sereniss. mio Signore rifulge ancora quella di far nobilmente mantener provveduti d'ogni pianta straniera i giardini di Firenze, e di Pisa, non già per un vano, e curioso diletto, ma per lo solo benefizio di coloro, che investigano, e scrivono le diverse nature, e proprietà delle piante." (Redi 1741 (1671): 33). The cited passage is preceded by the detailed description of an exotic fruit from Brazil that was cultivated in the Tuscan gardens among many other species (Redi 1741 (1671): 31–33).

[22] On the importance of the 'Hesperides myth', see Segre (1996) and Yasmin Doosry's introductory article in Doosry et al. (2011: 27–67) (*Die goldenen Äpfel der Hesperiden. Antike Mythen und ihre bildlichen Spuren*).

[23] Both Tilli and Micheli were in close contact with Caspar Commelijn, director of the garden in Amsterdam and nephew of Jan Commelijn. With their letters, they also exchanged seeds and plants, as one can see from numerous entries in Tilli's catalogue. Apart from that, throughout the *Hortus Pisanus*, the Commelijns' publications are often cited as references.

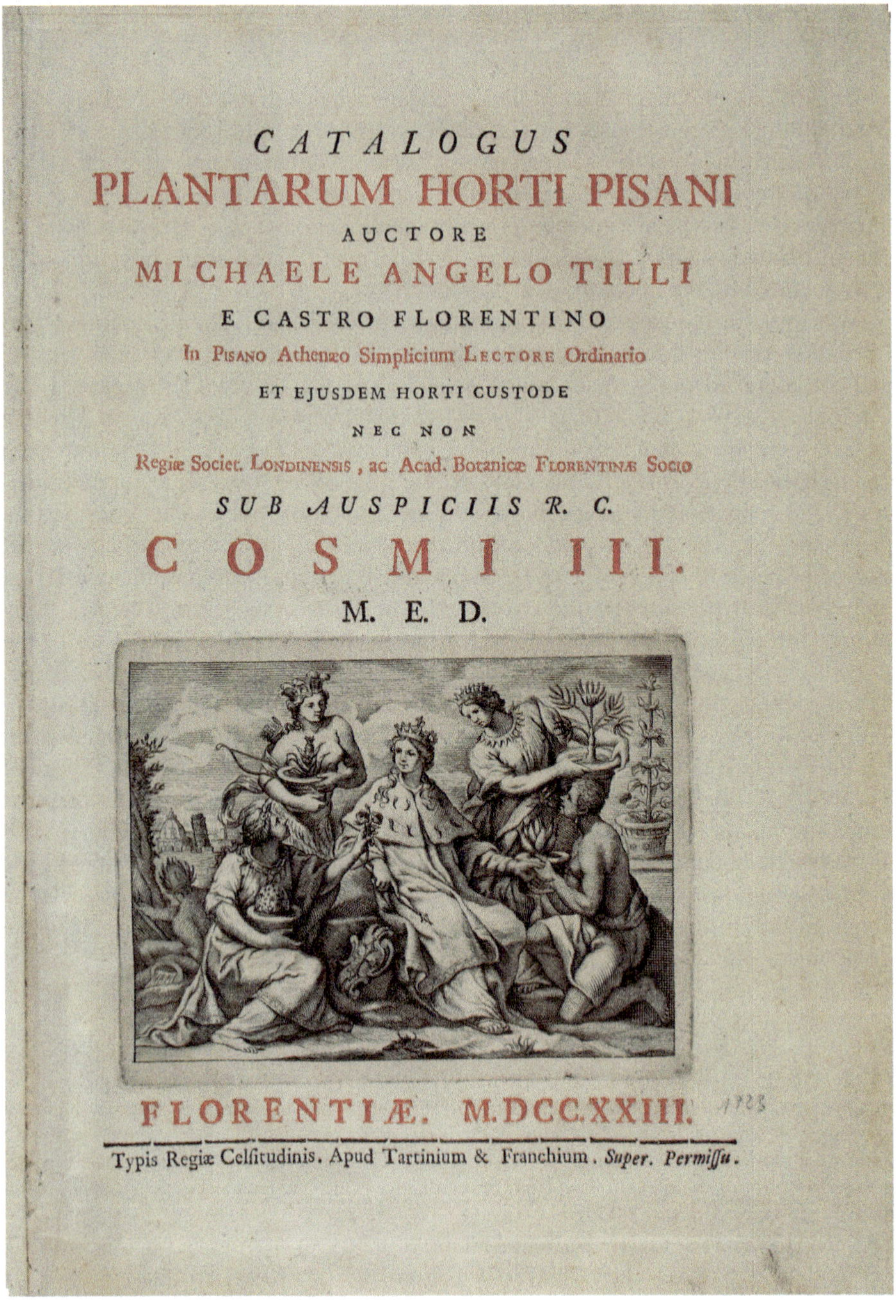

CATALOGUS
PLANTARUM HORTI PISANI
AUCTORE
MICHAELE ANGELO TILLI
E CASTRO FLORENTINO
In Pisano Athenæo Simplicium Lectore Ordinario
ET EJUSDEM HORTI CUSTODE
NEC NON
Regiæ Societ. Londinensis, ac Acad. Botanicæ Florentinæ Socio
SUB AUSPICIIS R. C.
COSMI III.
M. E. D.

FLORENTIÆ. M.DCC.XXIII.
Typis Regiæ Celſitudinis. Apud Tartinium & Franchium. *Super. Permiſſu.*

Fig. 3 Title Page, Michelangelo Tilli: *Catalogus plantarum horti Pisani*, Florence 1723. © Universitätsbibliothek Erlangen-Nürnberg

Variety and Wonder: Traditions of Collecting and Display

Both Bimbi's paintings for la Topaia and Micheli's handwritten legacy also act as inventories of (partly exclusively Tuscan) botanical abundance that transmit specific information, as well as serving, to some degree, the purposes of princely representation. Baldinucci writes that many amateurs and specialists, locals and foreigners, came to la Topaia to see Bimbi's works, and that these contributed to the fame of the whole site, with its prestigious villas and gardens.[24] It is not easy to trace who actually visited the casino to see the collection of paintings. There was an annual festival in honour of San Fiacre, patron saint of the gardeners, to whom the small chapel inside the casino was dedicated, that may have attracted a selected circle of visitors.[25] Pier Antonio Micheli was, of course, one of the local specialists who, in all probability, saw and studied Bimbi's paintings, as can be seen from his drawings and lists that are based on the artist's four large canvases of citrus fruit. Apart from his extensive *Lista di tutte le frutte...*, and its elaboration in the *Enumeratio...*, Micheli left a huge number of handwritten notes (Ragazzini 1993). We find preparatory lists, descriptions and drawings of plants (mostly fungi[26]) for his *Nova plantarum genera* (Florence 1729), plant lists of the Florentine botanical garden, the *Giardino delle Stalle*, that were published posthumously by Giovanni Targioni Tozzetti,[27] numerous drawings, lists and descriptions, for example, of native orchids and olives,[28] and much more. Among the vast collection of manuscripts in the *Biblioteca Botanica* in Florence, there are also two inconspicuous volumes that Targioni Tozzetti describes as Micheli's reference works when entering the service of the grand duke in 1706.[29] Possibly compiled around the

[24] "E quindi è che quella fran [sic!] fama che gode questo fortunato paese di Castello, lascia in dubbio se deva attribuirsi più all'amenità e rara dovizia del medesimo che al valore di questo nostro singolare pittore, che tanto ornamento gl'apportò colle sue pitture: per vedere le quali non vi è dilettante e professore o nazionale o forestiero, che a questo luogo, molto di buona voglia e con somma curiosità, non si porti, e con altrettanto di sodisfazione d'averlo veduto non se ne parta." (Baldinucci 1725-30 (1975): 249–250).

[25] On San Fiacre, whose relics were translocated from Meaux in France to la Topaia in 1696 and whose marble statue in the semicircle at the front side of the orchard was erected in the same year in the course of the casino's restructuring, see Hubert (2008: 216–219) and Spinelli (2008: 24–25).

[26] Throughout the seventeenth and eighteenth centuries, fungi were commonly considered to be plants (Dörfelt and Heklau 1998).

[27] Including a detailed preface describing the history of the garden and an appendix to Micheli's lists of plants by Targioni Tozzetti, Antonio Cocchi's *Discorso sopra l'Istoria Naturale* and eight engraved plates of the garden's ground plan and some plant species, the *Catalogus plantarum horti Caesarei Florentini* was published in 1748.

[28] On the *Orchidum agri florent. Icones* and *Oleorum Agri Florent: Descr et Icones*, see Ragazzini (1993: 34–37). Micheli's handwritten treatise on olives has been published by Baldini and Ragazzini (1998).

[29] "Queste operette fruttarono al Micheli il conseguimento d'un annua provvisione di scudi ottanta, sulla Cassa delle decime ecclesiastiche, conferitagli dal Granduca con benigno rescritto sotto il dì 18 ottobre 1706 (cioè nella sua età di soli 27 anni), col titolo di aiuto del custode del Giardino de' Semplici di Pisa, e coll'obbligo ingiuntogli verbalmente di cercar piante per esso Giardino, e per quello di Firenze." (Targioni Tozzetti 1858: 50).

same time as the *Lista*..., the two volumes register and arrange plants and other things of natural history following different systems of order. The *Ristretto della Toscana illustrata* lists and explains all the natural things of interest (animals, plants, minerals, etc.) occurring in Tuscany that begin with the letter A (Micheli ms. 4).[30] In the *Corollarium Institutionum Rei Herbariae, iuxta Tournefortianum methodum dispositum*..., Micheli obviously adopted the French botanist Joseph Pitton de Tournefort's method of classification for his inventory of plants observed in the Tuscan environs and princely gardens (Micheli ms. 12).[31]

The symbolic contents of Bimbi's and Micheli's 'image and text catalogues' clearly transcend the *agrumi* topos with Hercules's golden apples, the so-called *mala medica*, as the famous *palle* from the Medici coat of arms. Both the artist and the botanist highlight and expose the (native and imported) botanical plenty of Tuscany, which often involves the desire to identify and define (new) species. What Londa Schiebinger in another context describes with the slogan 'naming and knowing' (Schiebinger 2007) can similarly be related to the Tuscan case. The factual and intellectual authority over things, in this case, plants, is not only closely related to their actual possession but also to the practice of naming and thus communicating selected information to the public. This practice is well documented by garden catalogues such as Tilli's *Hortus Pisanus*. Bimbi and Micheli also take a similar approach in their paintings, lists and drawings. Both pictorial and textual means are used to convey information about names, appearances or the provenance of plants. While Bimbi uses his artistic skills to create an 'after life' documentation and representation of native fruits, vegetables and flowers, Micheli, with his botanical expertise, accumulates a vast collection of data mostly arranged in clear thematic lists as exemplified above.

Though the described 'catalogues' may all be interpreted as expressions of the same range of interests in the late Medici grand duchy, one can easily identify different connotations and traditions. In his 12 large canvases of fruit varieties, Bimbi provides a striking testimony to the extraordinary variety of cultivars, but in most of his other paintings, he focuses on single exceptional specimens, like the miraculous "Rama di Pere" (Fig. 1). Baldinucci describes some of those "stravaganti e bizzarri aborti della natura" that today, for the most part, belong to

[30] The full title reads *Ristretto del primo volume della Toscana illustrata o vero sia Istoria generale, nella quale si dimostra tutte le cose naturali, che in essa, e nell'Isole e Monti suddiacenti spontaneamente nascono, come Animali, Alberi, Erbe, Pietre, Metalli, e Terre, de' Monti e Selve, Spelonche o Antri, de' Fiumi, Laghi, Stagni, Paludi, Bagni, Acquitrini e Fonti, e generalmente di tutte le acque, con l'etimologia e spiegazione dei nomi, non solo volgari, come greci e latini, con lo scoprimento di molte cose non da altri autori osservate*; see Targioni Tozzetti (1858: 40–42) and Ragazzini (1993: 11–12).

[31] On the *Corollarium Institutionum Rei Herbariae, iuxta Tournefortianum methodum dispositum, in quo plantae plusquam bis mille omnino novae reperiuntur, nec hucusque ab alio auctore descriptae, quae in Etruria, et in montibus, et in insulis adjacentibus, tum vero in amoenis viridariis Regiae Celsitudinis Cosmi III, Magni Ducis Etruriae, observatae fuerunt*, see Targioni Tozzetti (1858: 48–49) and Ragazzini (1993: 23–24).

the botanical collections of the *Museo di Storia Naturale* in Florence.[32] Among the extraordinary specimens depicted by Bimbi for la Topaia was an enormous ball-shaped truffle (Fig. 4). Apart from its portrayal, Cosimo wanted the huge fungus to be carefully dissected.[33] Bimbi's painting shows the entire truffle on a silver plate to the left of its two halves. The result of the truffle's dissection, which was possibly carried out by Giuseppe del Papa, is not only documented by Baldinucci but also described in an inscription on the painting itself:

> In the Princedom of Castel Leone, belonging to the Reverend Highness Prince Cardinal Francesco Maria de' Medici, in the month of October of the year 1706 this huge mass was found in the ground where the black odorous truffles grow. And as this globe in its external appearance had all the qualities of a real truffle: like shape, colour, texture and also smell, it was sent as it was to His Royal Highness Cosimo III Grand Duke of Tuscany. This huge truffle weighed 4 ½ libbre [ca. 1.5 kg]. Its bark was almost petrified, it was as thick as two fingers and therefore this truffle was thought to be very, very old. It may also be for this reason that the truffle's internal substance was not uniform, soft and mushy like ordinary truffles, but all unequal, earthen, perforated and separated by many, many tunicae [membranes] similar to fine leather in white, red and other colours. Some of these tunicae were infested by little red worms. These were all the side-effects of the strange change of substance that had occurred over the course of many years.[34]

In combination with the prominent pictorial presentation of the huge truffle on the canvas, such an inscription instantly evokes early modern collections of natural

[32] For an impression of Bimbi's paintings of remarkable botanical specimens and the collections of the museum, see Casciu and Nepi (2008) and Raffaelli (2009).

[33] "Fu poi portato in Firenze un grossissimo tartufo, di forma sferica e grossa circa un palmo di diametro, trovato ___. Il che saputosi da Sua Altezza Reale, fattolo portare a sé, volle che il Bimbi con tutta diligenza lo ritraesse, per accomodarlo nella Topaia fra i sopradetti ed altri frutti di straordinaria grandezza: in che fu, con somma diligenza e prontezza, ubbidito. Volle poi lo stesso Serenissimo Granduca che se ne facesse notomia. E partito in più pezzi, fu riconosciuto essere non già un solo tartufo ma un gruppo di moltissimi tartufi d'ogni grandezza, conglutinati insieme con una certa sorta di terra nera talmente soda che pareva calcina. Separati e partiti poi alcuni di questi, furon trovati dello stesso odore, colore e sapore degl'altri. Ma perché questi avevano dentro di sé molti vermicciuoli rossi, fu proibito a tutti il mangiarne—a secondo del parere dell'Eccellentissimo Dottor del Papa—e indi ordinato che si buttasse ogni frammento nel fiume, dopo essere stati dipinti dal Bimbi tutti i pezzi nella forma che furono ritrovati." (Baldinucci 1725–30 (1975): 248).

[34] "Nel Principato di Castel Leone, posseduto dall'Altezza Reverendissima del Signor Principe Cardinale Francesco Maria de' Medici, nel mese di ottobre dell'Anno 1706 fù trovato questa gran massa di materia in quel terreno, ove si producono i Tartufi neri odorosi; E perché questo globo nell'esterna sua apparenza aveva tutte le qualità di vero Tartufo, cioè figura, colore, consistenza, e odore eziandio; Fu perciò come tale mandato à Firenze a Sua Altezza Reale di Cosimo III. Gran Duca di Toscana. Questo smisurato Tartufo pesò libbre 4½ La sua corteccia era quasi affatto impietrita, ed era grossa due dita traverse, e ciò fù creduto procedere dall'esser questo Tartufo molto, e molto antico, e per tal cagione è credibile esser parimente proceduto, che la sua interna sostanza non fusse uniforme, tenera, e pastosa, siccome è quella de' Tartufi ordinari; ma tutta ineguale, terrea, traforata, e d'ogn'intorno divisa da molte, e molte tuniche, simili a un sottile quoio, e bianche, e rosse, e di altri colori, ad alcune delle quali tuniche erano uniti molti piccioli vermicciuoli rossi, effetti tutti dell'essersi in progresso di molti Anni stranamente alterata la sua sostanza."

Fig. 4 Bartolomeo Bimbi: Tartufo di Castel Leone, 1706, oil on canvas, 89 × 121.5 cm, Florence, Museo di Storia Naturale, Sezione Botanica. © Museo di Storia Naturale dell'Università degli Studi di Firenze, Sezione Botanica

curiosities, so-called *Kunst- und Wunderkammern*.[35] Bimbi's *stravaganti e bizzarri* also stand in the tradition of chronicles and single-sheet prints illustrating and interpreting abnormal natural occurrences.[36] It is commonly known that the preternatural played an important role throughout sixteenth and seventeenth century natural sciences. Browsing the Bolognese naturalist Ulisse Aldrovandi's (1522–1605) vast collection of prints and drawings and his (posthumous)

[35] Samuel Quiccheberg for instance, in his *Inscriptiones vel tituli theatri amplissimi* from 1565, mentioned paintings, drawings, etc., representing remarkable specimens of animals and plants as typical elements of the 'theatrum' (Roth 2000: 44–45 and 54–57). The literature on *Kunst- und Wunderkammern* and early modern collection practices has become extremely vast within the last 20 years; see, for example, Impey and MacGregor (1985), Bredekamp (1993), Grote (1994), Findlen and Smith (2002) and Felfe and Lozar (2006).

[36] This is well illustrated by Hans Hubert, who, in his article on the la Topaia collection, included two prints from the Erlangen university library, showing a very fertile bunch of wheat and a two-headed calf, as a comparison to Bimbi's paintings of similar subjects (Hubert 2008: 215–216). For an insight into the variety of botanical single-sheet prints, see Stopp (2001). On chronicles documenting and communicating natural anomalies, see the case of the *Theatrum Europaeum* (Frankfurt 1634–1738), exemplified by Schramm (2003: 24–27). For a short general overview of the phenomena, see Schenda (1963).

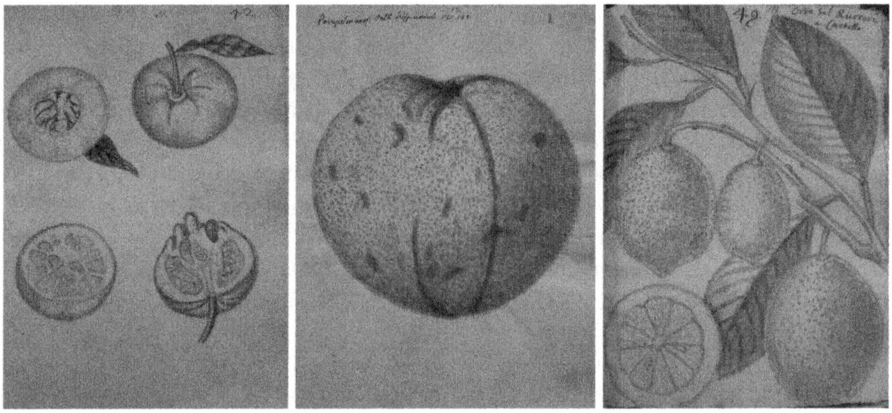

Fig. 5 Watercolour drawings of citrus fruit, Pier Antonio Micheli: *Enumeratio*..., vol. 9, Florence, Biblioteca Botanica dell'Università, ms. Micheli 48, c. 55r, 13r, 185r. © Biblioteca Botanica dell'Università degli Studi di Firenze

publications, for instance, one comes across an enormous variety of different species, as well as many bizarre specimens, both of animals and plants.[37]

Even though the inscription in another of Bimbi's paintings mentions that Micheli brought an enormous turnip to the grand duke,[38] it seems that the botanist was more interested in completing series of plant species and establishing classificatory order among the botanical chaos than in the exposure of single remarkable specimens. Apart from nature herself, Micheli drew his information from many different sources, like the publications of Jean Bauhin or Francesco Cupani, as indicated in the beginning. Furthermore, he often refers to his contemporary and forerunner in the matter of plant classification, Joseph Pitton de Tournefort (1656–1708), and to Ulisse Aldrovandi's *Dendrologiae*..., published in 1668.[39] It has been shown that Micheli referred to Bimbi's paintings when he copied the *agrumi* from the la Topaia canvases (Fig. 2). In addition, when taking a closer look at Micheli's examination of citrus fruit in the *Enumeratio*..., one can easily make out copies from Giovan Battista Ferrari's and Johann Christoph Volkamer's *Hesperides* publications next to drawings that seem related to direct observations (Fig. 5).[40] The lists and descriptions also depend on Micheli's own experience,

[37] On wonder and the preternatural as epistemological categories, see Daston and Park (1998). Extensive information on Aldrovandi's collection of drawings and a rich bibliography is given by Alessandrini and Ceregato (2007). For Aldrovandi's consideration of 'botanical monsters', see Baldini (2000).

[38] Bartolomeo Bimbi: Giant Cauliflower and Turnip, 1706, 88 × 118 cm, Florence, Museo di Storia Naturale, Sezione Botanica; see Casciu and Nepi (2008: 90–93).

[39] These, of course, are only a few examples among many. A more complete impression of Micheli's sources can be gained when looking at the list of authors in the *Nova plantarum genera*.

[40] The model of the first drawing is clearly found in Ferrari (1646: 405) (*Aurantium foetiferum*). The relevant engraving of the *Pompelmus* (as in the manuscript (ms. Micheli 48, c. 14r) followed

like the list of "Agrumi osservati dal Micheli nello Stato Veneto nel 1736" (Micheli ms. 48, c. 222r), and on works by other authors, for instance, the "Distribuzione metodica degl'Aranci [de Limoni/de Cedri] cavati dall'Opera del Clarici [...]" (Micheli ms. 48, c. 10v, 132r, 221r).[41] Towards the end, in a short *Catalogue of some very distinguished species of Agrumi with their precise qualities and variations...*,[42] Micheli notes that to describe all the other more ordinary qualities of citruses, he needed a book with images of fruits and leaves printed in Amsterdam, most likely referring to Jan Commelijn's *Nederlantze Hesperides* (Amsterdam 1676) that, for some reason, was not available to him.[43]

Progress and Discoveries? The Range of Princely Patronage

In the above-mentioned cases, Micheli benefits from works of 'science', 'art' and their 'in-betweens' to create his comprehensive account of citrus fruit. The entanglement of art and science in different fields of natural history was, of course, nothing unusual.[44] Micheli's extensive inventories of the plant world and Bimbi's splendid oil paintings, with Cosimo III's passion for botany as catalyst, show how an aesthetical as well as intellectual documentation of fruits, flowers and the like could serve traditional representative purposes. Apart from composing beautiful landscapes and still lifes, Bimbi integrates 'hard facts' like denominations, information on measure, weight and date or first-hand observation into his works. It is plain to see that the collection's purpose was both to delight and to educate its visitors. As for Micheli's botanical drawings, lists and descriptions, like many of his contemporaries, he noticed and incorporated the useful qualities of works of art and visual delight at the same time as acknowledging the accelerating scientific progress and specialisation going on around him. Riguccio Galluzzi, at the end of his account on Cosimo III's life and reign, comes back to the grand duke's great passion and to Pier Antonio Micheli:

by the image of a sliced fruit) appears in Volkamer (1708: 182a and 182b). On Ferrari's and Volkamer's publications, see Freedberg (1992), Baldini and Freedberg (1997) and the articles by Iris Lauterbach in Doosry et al. (2011: 237–263) (*Johann Christoph Volkamers Hesperidenwerk*) and in this volume.

[41] Paolo Bartolomeo Clarici's *Istoria e cultura delle piante* [...] *con un copioso trattato degli Agrumi* was published in Venice 1726. For information on Clarici and his treatise, see the relevant entry by Maria Letizia Strocchi in *Dizionario Biografico degli Italiani*, vol. 26 (1982), online access via http://www.treccani.it/enciclopedia.

[42] "Catalogo di alcune Specie più distinte di Agrumi con sue più precise qualità e diversità [...]" (Micheli ms. 48, c. 226r).

[43] "Vi sarebbe da descrivere tutte le altre qualità più ordinarie, ma vi vorrebbe un volume, quale si ritrova alle Stampe in Amsterdam con i ritratti de' Frutti e foglie." (Micheli ms. 48, c. 228r).

[44] The literature on this topic is at least as broad as the respective fields of research. For an insight into some fundamental theories and questions, see Smith (2008).

> He was immensely fond of botany and followed a pythagorean diet for many years, which contributed both to his health and also to his great delight in plants. So we owe him [Cosimo III] the progress and the discoveries made by Pier Antonio Micheli in the science of botany that largely contributed to the discipline's further perfection in the future.[45]

Before concluding, it is worth questioning this last statement. A closer look at the procedures and results of cultural and scientific patronage in Tuscany around 1700 reveals that dealing with plants in different contexts involved an examination and documentation of both botanical diversity and specialty. But scholarship, economy and prosperity were still closely linked at this time to aspects of symbolism and traditional interests in curiosity and wonder. As for Micheli, who was a non-university trained autodidact, he surely benefited from Cosimo's interest in botany when entering his service in 1706.[46] Ten years later, he co-founded and became the 'botanical leader' of the *Società Botanica Fiorentina*, an institution consisting mainly of wealthy amateurs, which, in 1718, was acknowledged by the grand duke and entrusted with the maintenance of the Florentine *Giardino delle Stalle*.[47] Those circumstances helped to improve Micheli's working conditions and provided a regular income. As mentioned before, his primary concerns were the issue of plant classification and the examination of lower plants like fungi, mosses and lichens—activities that fitted Galluzzi's ideas of 'progress and discoveries'. Micheli, however, had a hard time getting the results of his classificatory research published. The grand ducal printing press obviously gave priority to Tilli's *Hortus Pisanus*, and Micheli was asked to suspend his own work in order to bring forward the Pisan catalogue project.[48] A few years later, he succeeded in having his *Nova plantarum genera* printed without any noteworthy support from Cosimo's son Gian Gastone de' Medici (1671–1737), but with the help of a long list of subscribers from Florence, the rest of the Italian peninsula and abroad. Whether things would have turned out differently if Cosimo III had lived some five years longer we do not know. Nevertheless, Micheli's project had been delayed in the first place, and in terms of structure and content, the book differs not only from Bimbi's paintings and Tilli's garden catalogue but also from the majority of Micheli's own lists of plants and other things of natural history. The *Nova plantarum genera* is not so much about the enormous abundance of species and varieties but a coherent classification

---

[45] "Amò singolarmente la Botanica perchè addetto da lungo tempo al vitto pitagorico trovava in essa ciò che interessava la sua salute, e tutto il pascolo per la delizia. Quindi è che ad esso sono dovuti i progressi e le scoperte fatte in questa scienza da Pietro Antonio Micheli che tanto hanno contribuito per perfezionarla nell'avvenire." (Galluzzi 1781: vol. 4, 409).

[46] For a shorter account of Micheli's life and work than Targioni Tozzetti (1858), see Negri (1938) and Casciu and Nepi (2008: 116–121).

[47] On the *Società Botanica Fiorentina*, see Arrigoni (1988: 166–183).

[48] This is recorded by Giovanni Targioni Tozzetti in § XIX *Opera de' nuovi generi di piante* (Targioni Tozzetti 1858: 248–251), though, of course, Micheli's vita, as written by his student, may not always be judged historically objective. On Tilli's and Micheli's book projects, see Tongiorgi Tomasi (2000: 122–127) and Schmiedel (2013).

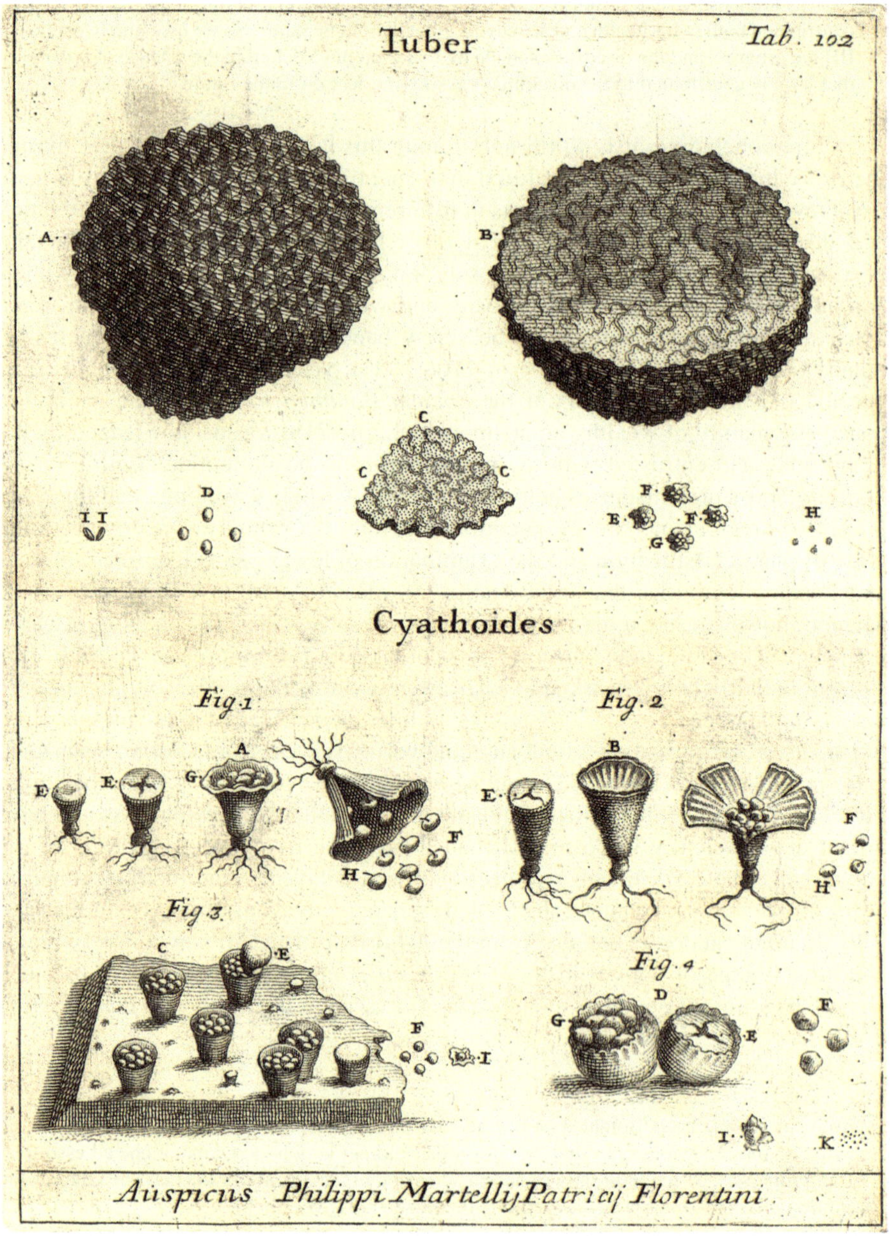

Fig. 6 *Tuber* and *Cyathoides*, Pier Antonio Micheli: *Nova plantarum genera*, Florence 1729. © ETH-Bibliothek Zürich

of the plant world into classes and genera. The 'nova genera' are exemplified in 108 engraved plates that, apart from a meticulous rendering of each genus' characteristics, also state the names of Micheli's various supporters (Fig. 6).

Looking at the body of work from another perspective, the engraving dedicated to the genus *Tuber* resembles Bimbi's pictorial interpretation of the "smisurato Tartufo". But unlike Micheli's ink drawings of Bimbi's *agrumi*, mentioned above, it is very unlikely that the print was inspired by the painting; however, the formal similarities once more underline the 'scientific' contents of Bimbi's paintings. Of course, one can hardly compare the la Topaia collection with Micheli's book on plant classification, the functions and messages of which are quite different. Nonetheless, the case of the *Nova plantarum genera*, the exceptional collection of botanical paintings for Cosimo III, Micheli's extensive handwritten lists and Tilli's *Hortus Pisanus* together exemplify how botanical abundance could be 'recruited' to serve purposes of princely representation.[49] The cataloguing of variety and certain particularities seems to have dominated over the support of scientific 'up-to-date issues' like plant classification. In contrast to his French colleague Tournefort, whose publications were sponsored by the *Académie Royale des Sciences* and thus by the French King,[50] the grand ducal botanist had to pursue other strategies to succeed in this field.

Taken together, this episode from the late Medici grand duchy highlights the opportunities and constraints of patronage in relation to gardening and botanical knowledge. While Bimbi, the painter of still lifes, found his niche documenting the splendour of Tuscan flora, the botanist Micheli, lacking the support of a state-sponsored academy of sciences, took it upon himself to create a vast network of international specialists and also established one of the first popular societies dedicated to botany. Such tendencies of both specialisation and popularisation reflect many of the typical characteristics of enlightenment science. But, at the same time, the large-scale campaign of aesthetically as well as intellectually cataloguing plants to render homage to an absolutist ruler is reminiscent of baroque traditions of art and science mixing and mingling to teach the mind, delight the eyes and convey political messages.

Bibliography

Sources, Primary Literature

Baldinucci, Francesco Saverio. 1975 (1725-30). Vite di artisti dei secoli XVII-XVIII. Biblioteca Nazionale Centrale di Firenze. Fondo Palatino. Ms. 565. Ed. Anna Matteoli. Rome: De Luca.

[49] On the socio-political connotations of nature and her products in the service of a ruler, see, for the case of France under Louis XIV; Mukerji (1997) and Hyde (2005). For the specific case of the Sun King's vegetable garden, the *Potager du Roi*, see Chandra Mukerji's article "The Power of the Sun-King at the Potager du Roi" in this volume.

[50] Tournefort's three volume *Élémens de botanique*, containing more than 400 engravings by the *peintre miniaturiste au Jardin du Roi* Claude Aubriet, was published in 1694 by the *Imprimerie Royale* in Paris. An enlarged Latin version entitled *Institutiones rei herbariae* appeared in 1700. On Tournefort and his work, see Becker et al. (1957) and Hamonou-Mahieu (2010: 85–109).

Bauhin, Jean, and Johann Heinrich Cherler. 1650-1651. Historia plantarum universalis, nova, et absolutissima. 3 vols. Yverdon.
Commelijn, Jan. 1697. Horti medici Amstelodamensis rariorum plantarum historia. Amsterdam: P. & J. Blaeu.
Cupani, Francesco. 1696. Hortus Catholicus. Naples: Francesco Benzi.
Evelyn, John. 1901. The Diary of John Evelyn. New York and London: M. W. Dunne.
Ferrari, Giovanni Battista. 1646. Hesperides, sive de Malorum aureorum cultura et usu Libri Quatuor. Rome: Hermann Scheus.
Galluzzi, Riguccio. 1781. Istoria del Granducato di Toscana sotto il governo della Casa Medici. 5 vols. Florence: Cambiagi.
Gobelin, Abbé. 1677. Le Jardinier Royal qui enseigne la maniere de planter, cultiver, & dresser toutes sortes d'Arbres. First published 1661. Paris: Charles de Sercy.
Micheli, Pier Antonio. 1729. Nova plantarum genera iuxta Tournefortii methodum disposita. Florence: Bernardo Paperini.
Micheli, Pier Antonio. ms. 4. Ristretto del primo Volume della Toscana Illustrata [. . .]. Biblioteca Botanica dell'Università di Firenze.
Micheli, Pier Antonio. ms. 12. Corollarium Institutionum Rei Herbariae iuxta Tournefortianum methodum dispositum [. . .]. Biblioteca Botanica dell'Università di Firenze.
Micheli, Pier Antonio. ms. 25. Lista di tutte le frutte, che giorno per giorno dentro all' Anno sono poste alla mensa dell' A. R.e del Ser.mo Gran Duca di Toscana [. . .]. Biblioteca Botanica dell'Università di Firenze.
Micheli, Pier Antonio. ms. 48. Petri Ant. Michelii M. Ducis Etruriae Botanici Enumeratio quarundam Plantarum sibi per Italiam, et Germaniam observatarum iuxta Tournefortii methodum dispositarum Tomus IX. Biblioteca Botanica dell'Università di Firenze.
Micheli, Pier Antonio. ms. 49. Petri Ant. Michelii M. Ducis Etruriae Botanici Enumeratio quarundam Plantarum sibi per Italiam, et Germaniam observatarum iuxta Tournefortii methodum dispositarum Tomus X. Biblioteca Botanica dell'Università di Firenze.
Raymond, John. 1648. An Itinerary contayning a Voyage made through Italy in the yeare 1646, and 1647. Illustrated with divers figures of Antiquities. London: Humphrey Moseley.
Redi, Francesco. 1741 (1671). Esperienze intorno a diverse cose naturali, e particolarmente a quelle, che ci son portate dall'Indie. In Opere di Francesco Redi, vol. 2, ed. Raffaele Gessari, 1-73. Naples: Angelo Carfora.
Targioni Tozzetti, Antonio. 1853. Cenni storici sulla introduzione di varie piante nell'agricoltura ed orticoltura toscana. Florence: Tipografia Galileiana.
Targioni Tozzetti, Giovanni. 1858. Notizie della vita e delle opere di Pier Antonio Micheli botanico fiorentino. Ed. Adolfo Targioni Tozzetti. Florence: Le Monnier.
Tilli, Michelangelo. 1723. Catalogus plantarum horti Pisani. Florence: Typis Regiae Celsitudinis.
Vasari, Giorgio. 1973 (1568). Le opere di Giorgio Vasari, con nuove annotazioni e commenti di Gaetano Milanesi. Ed. Paola Barocchi, 9 vols. Florence: Sansoni.
Volkamer, Johann Christoph. 1708. Nürnbergische Hesperides, oder Gründliche Beschreibung der Edlen Citronat- Citronen- und Pomerantzen-Früchte [. . .]. Nürnberg.

Secondary Literature

Acanfora, Elisa. 2007. Le decorazioni della villa La Topaia. In Fasto di corte. La decorazione murale nelle residenze dei Medici e dei Lorena. L'età di Cosimo III de' Medici e la fine della dinastia (1670-1743), ed. Mina Gregori, 126-131. Florence: Edifir.
Acidini Luchinat, Cristina, and Giorgio Galletti. 1992. Le Ville e i Giardini di Castello e Petraia a Firenze. Ospedaletto: Pacini.
Alessandrini, Alessandro, and Alessandro Ceregato (eds.). 2007. Natura picta. Ulisse Aldrovandi. Bologna: Editrice Compositori.

Arrigoni, Tiziano. 1988. Per una storia delle istituzioni scientifiche nella Toscana del Settecento. Atti e memorie dell'Accademia Toscana di Scienze e Lettere 'La Colombaria' 53: 115-203.
Baldini, Enrico. 2000. Prodigi, simulacri e mostri nell'eredità botanica di Ulisse Aldrovandi. In Natura – Cultura. L'interpretazione del mondo fisico nei testi e nelle immagini, eds. Giuseppe Olmi, Lucia Tongiorgi Tomasi and Attilio Zanca, 215-243. Florence: Olschki.
Baldini, Enrico. 2001. Agrumi, frutte e uve alla corte dei Granduchi di Toscana. In Le Belle Forme della Natura. La pittura di Bartolomeo Bimbi (1648-1730) tra scienza e 'maraviglia', eds. Daniela Savoia and Maria Letizia Strocchi, 20-32. Bologna: Abacus.
Baldini, Enrico, and David Freedberg. 1997. Citrus Fruit. London: Harvey Miller.
Baldini, Enrico, and Stefania Ragazzini. 1998. Le varietà di ulivo dell'agro fiorentino: manoscritto inedito di Pietro Antonio Micheli. Florence: Accademia dei Georgofili.
Baldini, Enrico et al. (eds.). 1982. Agrumi, frutta e uve nella Firenze di Bartolomeo Bimbi pittore mediceo. Florence: Consiglio Nazionale delle Ricerche.
Becker, Georges et al. 1957. Tournefort. Paris: Muséum National d'Histoire Naturelle.
Bertsch, Christoph. 2012. Villa Garten Landschaft. Stadt und Land in der florentinischen Toskana als ästhetischer und politischer Raum. Berlin: Gebr. Mann.
Bredekamp, Horst. 1993. Antikensehnsucht und Maschinenglauben. Die Geschichte der Kunstkammer und die Zukunft der Kunstgeschichte. Berlin: Wagenbach.
Casciu, Stefano. 2008. Allestimenti di nature morte nelle ville medicee al tempo di Cosimo III: il caso della villa di Topaia. In Stravaganti e bizzarri. Ortaggi e frutti dipinti da Bartolomeo Bimbi per i Medici, eds. Stefano Casciu and Chiara Nepi, 27-43. Florence: Edifir.
Casciu, Stefano (ed.). 2009. Villa Medicea di Poggio a Caiano: Museo della Natura Morta, catalogo dei dipinti. Livorno: Sillabe.
Casciu, Stefano, and Chiara Nepi (eds.). 2008. Stravaganti e bizzarri. Ortaggi e frutti dipinti da Bartolomeo Bimbi per i Medici (exhibition catalogue). Florence: Edifir.
Chiarini, Marco (ed.). 1997. Il giardino del granduca: natura morta nelle collezioni medicee. Torino: Edizioni Seat.
Daston, Lorraine, and Katherine Park. 1998. Wonders and the Order of Nature. 1150-1750. New York: Zone Books.
Doosry, Yasmin, Christiane Lauterbach, and Johannes Pomeranz (eds.). 2011. Die Frucht der Verheißung. Zitrusfrüchte in Kunst und Kultur (exhibition catalogue). Nürnberg: Germanisches Nationalmuseum.
Dörfelt, Heinrich, and Heike Heklau. 1998. Die Geschichte der Mykologie. Eine Übersicht von den Anfängen bis zur Gegenwart. Schwäbisch Gmünd: Einhorn.
Felfe, Andreas, and Angelika Lozar (eds.). 2006. Frühneuzeitliche Sammlungspraxis und Literatur. Berlin: Lukas.
Findlen, Paula, and Pamela Smith (eds.). 2002. Merchants & Marvels. Commerce, Science, and Art in Early Modern Europe. New York et al.: Routledge.
Freedberg, David. 1992. Ferrari on the Classification of Oranges and Lemons. In Documentary Culture. Florence and Rome from Grand-Duke Ferdinand I to Pope Alexander VII, eds. Elizabeth Cropper, Giovanna Perini and Francesco Solinas, 287-306. Bologna: Nuova Alfa Ed.
Galletti, Giorgio. 1996. Agrumi in casa Medici. In Il giardino delle Esperidi. Gli agrumi nella storia, nella letteratura e nell'arte, eds. Alessandro Tagliolini and Margherita Azzi Visentini, 197-215. Florence: Edifir.
Garbari, Fabio, Lucia Tongiorgi Tomasi, and Alessandro Tosi. 1991. Giardino dei Semplici. L'Orto Botanico di Pisa dal XVI al XX secolo. Pisa: Pacini. English edition: 2002. Garden of Simples. Pisa: Plus.
Grote, Andreas (ed.). 1994. Macrocosmos in microcosmo: die Welt in der Stube. Zur Geschichte des Sammelns 1450 bis 1800. Opladen: Leske und Budrich.
Hamonou-Mahieu, Aline. 2010. Claude Aubriet: artiste naturaliste des Lumières. Paris: Comité des travaux historiques et scientifiques.
Hubert, Hans W. 2008. 'Cosmic Delight': Bartolomeo Bimbi and the Representation of Nature at the Court of Cosimo III de' Medici. In The Art of Natural History: Illustrated Treatises and

Botanical Paintings 1400-1850, eds. Therese O'Malley and Amy R.W. Meyers, 205-225. New Haven: Yale University Press.
Hyde, Elizabeth. 2005. Cultivated Power. Flowers, Culture, and Politics in the Reign of Louis XIV. Philadelphia: University of Pennsylvania Press.
Impey, Oliver, and Arthur MacGregor (eds.). 1985. The Origins of Museums. The Cabinet of Curiosities in Sixteenth- and Seventeenth-Century Europe. Oxford: Clarendon Press.
Kliemann, Julian. 1976. Politische und humanistische Ideen der Medici in der Villa Poggio a Caiano. Untersuchungen zu den Fresken der Sala grande. Bamberg: PhD Dissertation Heidelberg.
Lapi Ballerini, Isabella. 2003. Le ville medicee. Florence: Giunti. English edition: 2011. The Medici Villas. Florence: Giunti.
Meloni Trkulja, Silvia, and Lucia Tongiorgi Tomasi (eds.). 1998. Bartolomeo Bimbi. Un pittore di piante e animali alla corte dei Medici. Florence: Edifir.
Mosco, Marilena (ed.). 1985. Natura viva in Casa Medici. Dipinti di animali dai depositi di Palazzo Pitti con esemplari del Museo Zoologico La Specola (exhibition catalogue). Florence: Centro Di.
Mosco, Marilena (ed.). 1988. Floralia. Florilegio dalle collezioni fiorentine del Sei-Settecento (exhibition catalogue). Florence: Centro Di.
Mukerji, Chandra. 1997. Territorial ambitions and the gardens of Versailles. Cambridge et al.: Cambridge University Press.
Negri, Giovanni. 1938. Pier Antonio Micheli (1679-1737). Nuovo Giornale Botanico Italiano 45: LXXX-CVII.
Nepi, Chiara, and Maria Adele Signorini. 2008. 'Forme e colori d'ogni specie di frutti': Bartolomeo Bimbi e la tradizione pomologica e botanica alla corte medicea. In Miti, arte e scienza nella pomologia italiana, ed. Enrico Baldini, 101-123. Rome: Consiglio Nazionale delle Ricerche.
Raffaelli, Mauro. 2009. Il Museo di Storia Naturale dell'Università degli Studi di Firenze. Le collezioni botaniche. Florence: Firenze University Press.
Ragazzini, Stefania. 1993. I manoscritti di Pier Antonio Micheli conservati nella Biblioteca Botanica dell'Università di Firenze. Milan: Giunta.
Roth, Harriet. 2000. Der Anfang der Museumslehre in Deutschland. Das Traktat 'Inscriptiones vel tituli theatri amplissimi' von Samuel Quiccheberg. Berlin: Akademie-Verlag.
Savoia, Daniela, and Maria Letizia Strocchi (eds.). 2001. Le belle forme della natura. La pittura di Bartolomeo Bimbi (1648-1730) tra scienza e 'maraviglia' (exhibition catalogue). Bologna: Abacus.
Schenda, Rudolf. 1963. Die deutschen Prodigiensammlungen des 16. und 17. Jahrhunderts. Archiv für Geschichte des Buchwesens 4: 637-710.
Schiebinger, Londa. 2007. Naming and Knowing: The Global Politics of Eighteenth-Century Botanical Nomenclatures. In Making Knowledge in Early Modern Europe. Practices, Objects, and Texts, 1400-1800, eds. Pamela Smith and Benjamin Schmidt, 90-105. Chicago et al.: University of Chicago Press.
Schmiedel, Irina. 2013. Zwischen Repräsentation und Klassifikation: botanische Bücher und Bilder für die letzten Medici. Imprimatur NF XXIII: 9-38.
Schnapper, Antoine. 1988. Le géant, la licorne et la tulipe. Collections et collectionneurs dans la France du XVIIe siècle. Paris: Flammarion.
Schramm, Helmar. 2003. Kunstkammer – Laboratorium – Bühne im Theatrum Europaeum. Zum Wandel des performativen Raums im 17. Jahrhundert. In Kunstkammer, Laboratorium, Bühne. Schauplätze des Wissens im 17. Jahrhundert, eds. Helmar Schramm, Ludger Schwarte and Jan Lazardig, 10-34. Berlin et al.: de Gruyter.
Segre, Ada. 1996. Alla ricerca dell'Orto delle Esperidi. In Il giardino delle Esperidi. Gli agrumi nella storia, nella letteratura e nell'arte, eds. Alessandro Tagliolini and Margherita Azzi Visentini, 19-40. Florence: Edifir.

Smith, Pamela. 2008. Art is to Science as Renaissance is to Scientific Revolution? The Problematic Algorithm of Writing a History of the Modern World. In Renaissance Theory, eds. James Elkins and Robert Williams, 427-445. New York et al.: Routledge.

Spinelli, Riccardo. 2008. La Topaia, 'villa della fruteria', il restauro di Giovan Battista Foggini e Cosimo III de' Medici. In Stravaganti e bizzarri. Ortaggi e frutti dipinti da Bartolomeo Bimbi per i Medici, eds. Stefano Casciu and Chiara Nepi, 19-26. Florence: Edifir.

Stopp, Klaus. 2001. Botanische Einblattdrucke und Flugschriften vor 1800. 2 vols. Stuttgart: Hiersemann.

Swan, Claudia. 1995. Ad Vivum. Naer het Leven. From the Life: Defining a Mode of Representation. Word and Image 11: 353-372.

Tagliolini, Alessandro, and Margherita Azzi Visentini.1996. Il giardino delle Esperidi. Gli agrumi nella storia, nella letteratura e nell'arte. Florence: Edifir.

Tongiorgi Tomasi, Lucia. 2000. Naturalistic Illustrations and Collections in Tuscany in the Eighteenth Century. In Science and the Visual Image in the Enlightenment, ed. William R. Shea, 111-136. Canton: Science History Publications.

Tongiorgi Tomasi, Lucia. 2005. Gardens of Knowledge and the 'République des Gens de Sciences'. In Baroque Garden Cultures. Emulation, Sublimation, Subversion, ed. Michel Conan, 85-129. Washington DC: Dumbarton Oaks Research Library and Collection.

Tongiorgi Tomasi, Lucia, and Gretchen Hirschauer. 2002. The Flowering of Florence. Botanical Art for the Medici (exhibition catalogue). Washington DC: National Gallery of Art.

Reconstructing Order: The Spatial Arrangements of Plants in the Hortus Botanicus of Leiden University in Its First Years

Gregory Grämiger

Abstract

The University of Leiden featured one of the first botanical gardens north of the Alps, which was fully equipped in 1594. The first three persons responsible for the garden—Carolus Clusius, Dirk Outgaertsz. Cluyt and Pieter Pauw—had to distribute the plants and herbs into the different garden beds. The logic behind the spatial order they chose gives insight into various approaches on plant classification in the early modern period. It is possible to reconstruct and analyze the order of plants using catalogues from different years. These catalogues show the changes in the distribution of plants under the prefecture of several natural historians. The catalogues themselves are an interesting historical source, showing a way to transfer a garden and its plants into a printed counterpart. In the early modern period, the classification of plants was not yet strictly defined. Also, in the botanical garden of the University of Leiden, we do not find a single, stern systematic approach to ordering the plants, but rather a number of different ones. Plants were classified according to their virtues and to empirically comprehensible criteria, but also according to their beauty or even to the chapters of an herbal. The reconstruction of the distribution of the plants shows those different approaches to ordering the flora and casts light on the development of botany.

G. Grämiger (✉)
ETH Zürich, Institut für Geschichte und Theorie der Architektur, Stefano-Franscini-Platz 5, Postfach 118, 8093 Zürich, Switzerland
e-mail: gregory.graemiger@gta.arch.ethz.ch

H. Fischer et al. (eds.), *Gardens, Knowledge and the Sciences in the Early Modern Period*, Trends in the History of Science, DOI 10.1007/978-3-319-26342-7_11

Introduction

In 1610, four copper engravings were published in Leiden.[1] They presented the newest facilities of the town's still young university. The series featured images of the library, the anatomical theatre, the fencing school and the botanical garden. The prints were proclaiming that the University of Leiden possessed the most important and newest research and teaching typologies of its time.[2] Indeed, the botanical garden of Leiden University, which will be the focus of this paper, opened in 1594 as one of the first institutionalized botanical gardens in the world, a new garden typology only introduced from 1545 onwards (Fig. 1).[3]

The botanical garden displayed a *book of nature* quite different from those found in the library, allowing students to examine the plants first hand and to compare the resulting empirical insights with the legacy stored in the books.[4] The garden was also closely linked to the anatomical theater. Anatomical demonstrations took place during winter times in the newly established anatomical theatre (Huisman 2009). There, the human body and its diseases could be shown to the students of medicine, while, during the summer, the garden was full of remedies against the discussed bodily disorders. Therefore, all those spaces depicted in the series of copper engravings built a closed system of relationships.

The engraving of the garden was republished several times. Prints from the year 1611 feature an accompanying text either in German, Latin or French. This text gives a good idea of how the *hortus botanicus* was perceived and used in its time. The text claims that the shown picture is "a realistic image of the herbs garden",[5] which was "established with great industry and with high costs for the usefulness and teaching of all students of medicine, but also for pharmacists, doctors and all other admirers of plants".[6] The garden was therefore a public space used by different people. Aside from that, the text says that "the herbs are laid out in a beautiful, graceful and convenient order".[7] The aspect of this mentioned order will be the main topic of this paper. Furthermore, the text proclaims that "a lot of strange and foreign herbs, plants and flowers of different sorts and colours can be found"

[1] The images were engraved by Willem Isaacsz. van Swanenburg after a drawing by Jan Cornelisz. van 't Woudt and published by Andries Clouck; see Boon (1984: 27–28).

[2] For a detailed discussion of these institutions of the University of Leiden, see Grämiger (2014). For a general introduction about anatomical theatres, botanical gardens and collections of curiosities, see Findlen (2006).

[3] For early modern botanical gardens, see Tongiorgi Tomasi (2005); for the botanical garden in Leiden, see Veendorp and Baas Becking (1938) and Tjon Sie Fat (1991).

[4] For the "book of nature", see Jorink (2010).

[5] "die grundtliche ond ware abconterfeytung des gemeinen Kreutergartens".

[6] "mit sonderlicher embsigkeyt un grossen kosten / zu nutz und onterweisung deren / so in der Artzney studieren / auch Apothekern / Wundartzten / ond allen andern Studenten und liebhabern der Kreuter".

[7] "in schoner/zierlicher und bequemer ordnung/auffgerichtet worden".

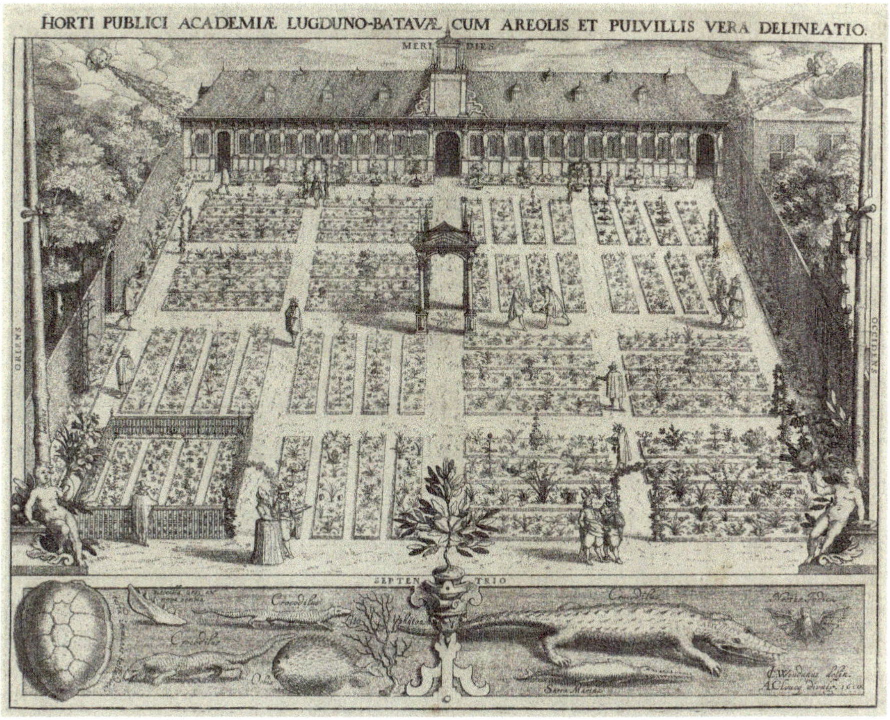

Fig. 1 The botanical garden of Leiden University, engraving by Willem Isaacsz. van Swanenburg after a drawing by Jan Cornelisz. van 't Woudt, 1610 (Amsterdam, Rijksmuseum, Object number RP-P-1893-A-18089)

and that they were given by God to preserve health,[8] clearly showing that the botanical garden was, in fact, a medical one. Additionally, it claims that "day-to-day new growths, plants and herbs arrive in the garden",[9] not only indigenous but also many exotic varieties, reflecting the wish of the prefectures to collect all plants of the earth. The garden was "not only an adornment and ornament for the university"[10] but also a pleasure ground—a *locus amoenus*—for all visitors. Furthermore, the newly established gallery at the back of the garden is mentioned, which—as is also proclaimed in the text—featured different curiosities from the East and West Indies, building up an early and typical *Wunderkammer*.[11] Some of these specimens are shown at the bottom of the engraving. With the specimens of the

[8] "Aldah viel setzame frembde Artzneykreuter/pflantzen und blumen / von mancherley art unfarben/gefunden werden; welche ons Gott d'allmechtige zu erhaltung unser leibes gesundtheit gnediglich mitgetheilt hat."

[9] "teglich mehr und mehr an frembden gewechsen/pflantzen und kreutern".

[10] "das er nicht allein der Universitet ein grosz ornament und zierd ist".

[11] On the gallery in Leiden, see de Jong (1991).

garden and the gallery, all the main categories of nature could be found: minerals, plants and animals, showing the effort to put the whole natural world on display.

The plants in the garden had to be spatially distributed inside the single garden beds. But how were they organized? Which specimens were set next to each other? How was this order useful for empirical research, the teaching of students and in showing the wonders of nature to those visiting the garden? Using diverse sources, it is possible to reconstruct the garden in terms of its architecture and dimensions. Furthermore, different surviving catalogues give us the exact location of each plant. Hence, it is possible to deduce the distribution and therefore the classification of the plants for different years.

Index Stirpium (1594)

The first catalogue of the botanical garden—titled *Index stirpium*—is a handwritten inventory, featuring all plants which grew in the newly established garden in 1594.[12] The first page of the catalogue features a schematic plan of the garden and its layout, showing and naming all its parts (Fig. 2). The garden was divided by two main axes into four so-called *quadrae*. These four squares were surrounded by a small stripe, encircling the whole of the garden. The *quadrae* were internally further divided by two main axes into another four parts. The smaller parts and the main layout were similar in appearance, connecting the small with the big, hence forming a geometrical unity.

Each *quadra* contained stripes called *areae*. They were turned in each *quadra*, forming a system like a windmill rotating around the centre. The result is a non-hierarchical layout of the garden and its plots. As can be seen in the plan, there were three *quadrae* containing 4 by 4 stripes, and one with only 4 by 3 stripes. The small stripes made it easy to cultivate and inspect the plants. These stripes contained the actual plots for the plants, the so-called *areolae*. Each stripe had—dependent on their *quadra*—18, 26 or 32 single garden beds. At the bottom of the page, there is a calculation, showing that a total of 1400 plots could be found in the four quadrants. The subsequent pages of the catalogue feature the name and location of each plant in a grid-like structure, each grid showing a stripe or *area*. The roman digits number and define each bed. This grid-like structure therefore gives us an image of a stripe.

The garden divided into four similar-looking parts was a common layout in the medieval and early modern periods and could be found all over Europe (Tongiorgi Tomasi 2005: 91–93). In secondary literature concerning early botanical gardens, it is often claimed that the quadripartite layout of the garden was used to arrange the plants after the four main virtues according to Galen, the four continents, or that

[12] Leiden, University Library, shelfmark Ms. AC.1, Nr. 101, fol. 117A–T; images of the catalogue can be found online in the Digital Special Collections of Leiden University Library; a typographically simplified transcription provides (Molhuysen 1913–1924: Deel 1, Bijlage 291, 317*–334*).

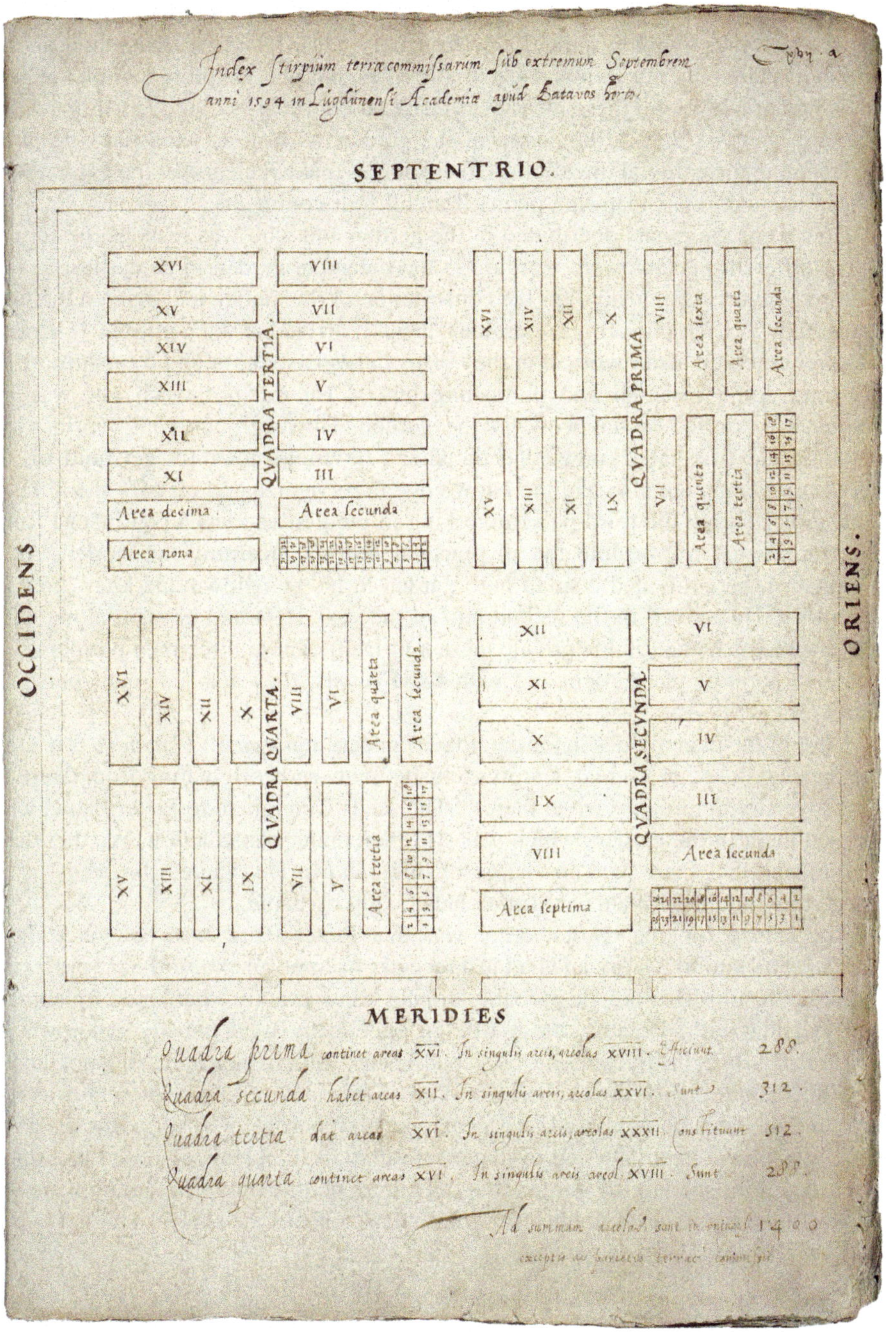

Fig. 2 Index stirpium, schematic floorplan of the garden, 1594 (Leiden, University Library, Ms. AC1, Nr. 101, fol. 117Ar)

they were distributed following magical-hermetical or natural-philosophical-cosmological topics (Müller-Wille 2004, especially 591). However, the actual distribution of plants did not follow such simple rules, but was far more complex. Only on a symbolic level did these ideas manifest. Our engraving of the garden therefore shows a compass dial in the centre of the garden, as well as wind-blowing *putti* representing the cardinal directions. Other sources make it clear that the garden was indeed understood as featuring plants from all four continents.

How were the plants distributed in the garden and who was responsible for it? Three different persons were working in the botanical garden in its earliest years. Pieter Pauw was responsible for the construction of the garden and planted the first seeds in it. The famous scholar Carolus Clusius[13] arrived in the winter of 1593 and was from then on also involved in the *hortus botanicus*, although he spent most of his time doing research and publishing books. He did not teach nor was he responsible for the actual work in the garden. However, with his world-wide network and first-hand knowledge of plants, he could not only get hold of all necessary plants, but also had the knowledge to recognize and classify them. Dirk Outgaertsz. Cluyt, the third person working in the garden, was a pharmacist from The Hague and responsible for all practical needs concerning the garden. As a pharmacist, he cultivated a small herb garden in his hometown and knew how to care for it. He arrived in the second half of the year 1594 and brought along with him the plants he had in his private garden in Delft. These three men put plants in different parts of the garden, and with these plants, they also left their personal traces.

Most of the plant names in the catalogue concerning *area* 12 of *quadra* 3 feature the appendix *cret.* or *cretica*. Carolus Clusius sent seeds originating from Crete to the university before his arrival there.[14] The seeds were given to Pieter Pauw, who planted them in the garden, which was still then under construction. We therefore can assume that Pauw was in charge of this *quadra*. He planted the seeds in an interesting order: the plants are set alphabetically by name.

The reason for this chosen order was that Pieter Pauw was not yet truly a specialist in natural history and had just recently become an extraordinary professor in medicine. Furthermore, he probably did not know exactly what kind of plants he was dealing with, because these foreign specimens were mostly unknown in northern Europe. Moreover, he only had seeds, which meant he did not know what the actual plants would look like. Out of this lack of knowledge, he was not able to establish an order based on aspects like morphology or virtues. But since the seeds must have been labelled to make sure that all the plants could be identified after they flourished in the garden, the only thing he did know for certain were the names of the plants. Presumably, out of all these reasons, he simply decided to use the alphabet to order these seeds.

[13] For a biography of Clusius see Hunger (1927–1943).

[14] Letter and list transcribed in Molhuysen (1913–1924: Deel 1, Bijlage 213, 238* and Bijlage 214, 238*–242*).

However, ordering plants according to the alphabet was not an uncommon thing in the Middle Ages and the early modern period, when natural history was still a scientific field in the process of awakening and without an established system of classification.[15] For example, the natural historian Leonhard Fuchs ordered his herbal in this way. His Dutch colleague Rembertus Dodonaeus used the same order in his herbal, but only for the species he could not order into better categories, as he claimed (Dodonaeus 1608: 27). The known plants he ordered according to their growth, virtues, morphology and other criteria. The botanical garden in Montpellier—established at the same time as the one in Leiden—only classified its medical herbs according to the alphabet, which Guy de la Brosse criticized as follows: "[...] this manner of planting simples according to the order of the alphabet—as if they all required the same kind of cultivation. It is more ostentatious than rational or seemly" (Meier Reeds 1970: 87).[16]

Nevertheless, the alphabetical classification had an advantage: as long as the names were known, the individual species could easily be found, comparable to an alphabetical index in a printed herbal. However, what was lost was surplus information gained through the neighbouring plants, which could be used to compare and discuss the single species, putting their morphology or virtues into a bigger picture.

This alphabetical ordered *quadra* might be exactly the area of which Johan van Hoghelande gave critical report in a letter sent to his friend Clusius, who was still in Frankfurt on the Main. He wrote that Pauw had begun to construct a part of the garden, but that he was not sure whether "he does it seriously or as a joke".[17] After speaking to Pauw and giving him his criticism, Pauw answered that the other part of the garden should be made according to the ideas of Clusius (Hunger 1927–1943: vol. 1, 204).

Clusius was most probably in charge of the *quadra prima* and equipped it with plants from his private collection, which he brought to Leiden. This *quadra* features a different distribution of plants compared to the one Pauw equipped, which can already be seen in the catalogue on a typographical level. As an example, *area* 1 will be discussed (Fig. 3). Firstly, in this stripe, all different kinds of *Asphodelus* were set (*Hastula* was understood as another species of the same genus). We can therefore speak of a simple and scientific order in correspondence with the plants: one genus in one stripe.

Secondly, he also planted beautifully flowering plants, which he set in different rows in the long direction of the stripe. These were written vertically into the *Index stirpium*. Furthermore, the catalogue shows that three different flowers (*Pulsatilla*,

[15] On early modern plant classification, see Arber (1970: 163–184), for classifications according to the alphabet Arber (1970: 166).

[16] "cette maniere de planter les simples par l'ordre de leur alphabet, comme si toutes demandoient vne pareille culture, il y a plus d'ostentation en cette façon, que de raison & de bienseance" (de la Brosse 1636: 18).

[17] "sed ut vix sciam serio agat an ludat" (Hunger 1927–1943: vol. 1, 204).

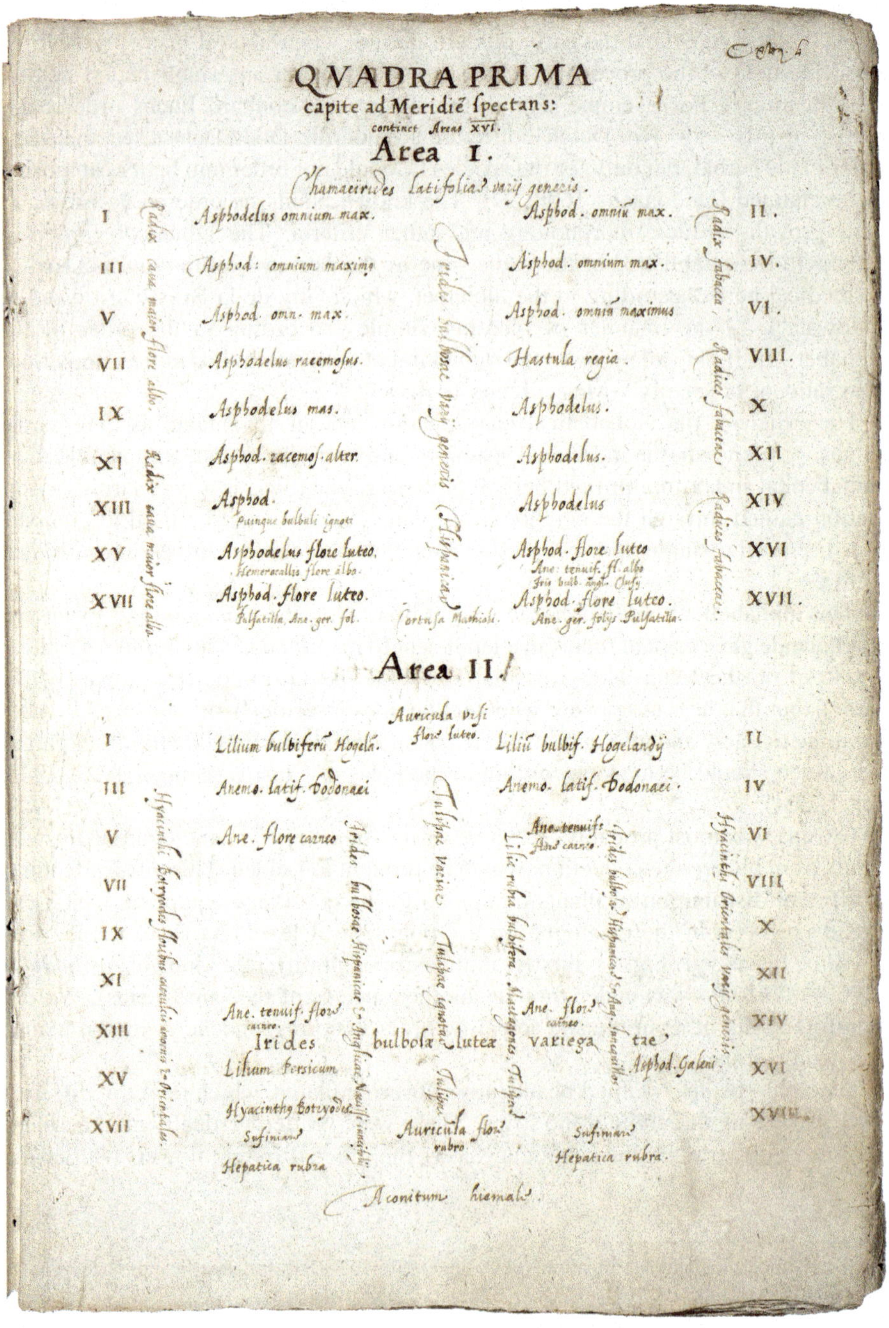

Fig. 3 Index stirpium, showing Area 1 and 2 of Quadra 1, 1594 (Leiden, University Library, Ms. AC1, Nr. 101, fol. 117Br)

Anemone and *Cortusa*) were planted on the small edge of the garden bed in a rhythm similar to A-B-C-B-A. They bloomed in March, April and May, and therefore showed blossoms during a quarter of the year. Similar patterns can also be found in adjoining stripes so that this colourful arrangement was repeated alongside the pathways. The stripes of the whole *quadra* were therefore linked together, establishing a unified appearance, which was aesthetically rather than scientifically motivated.

It comes as no surprise that Clusius was very fond of rare, exotic and beautiful plants. Before his time in Leiden, he was in charge of erecting different ornamental gardens in courts of the nobility (van Gelder 2011). For Clusius, the botanical garden had to be a device for research, as well as an aesthetic pleasure. The wonders of nature were not only scientifically interesting, but also pleasant to the eye.

The last two sections of the garden, *quadra secunda* and *quadra quarta,* were most probably assembled with plants of the private collection of our apothecary Cluyt. Here, the plants again follow a systematic order. The plants of the same genus were set next to each other. These two *quadrae* also feature more common plants and such with specific virtues, which could be practically used and were therefore sold by pharmacists.

As a first summary, we can say that, in the first years, the garden was not structured in a unified and stern system, but featured different ways to and ideas about ordering the plants. The consequent patchwork was the direct result not only of the different stocks of plants, but also of the three actors, their different knowledge of plants and their different interpretations of what the garden is, what should grow in it and what it should look like. The garden was, in its first year, not only a *hortus botanicus,* but also a medical as well as an ornamental garden.

The First Printed Catalogue

The first printed catalogue of the garden in Leiden was published in 1601[18] and was edited by Pieter Pauw, who became the prefect of the garden after the death of Cluyt in 1598. After some introductory notes and paratexts, an illustration of the garden was added (Fig. 4). This centerfold shows the structure of the garden, its plots, and their divisions. It also displays the already mentioned gallery, which was just newly established and could be used as a hothouse for exotic plants during winter and as a museum to display all sorts of curiosities. The drawing makes clear that some major adjustments of the garden beds were made. In short, the garden was simplified. *Quadra* A (1) and D (4) featured 4 by 4 stripes with 16 beds each. *Quadra* B (2) and C (3) instead featured 4 by 3 stripes with 24 plots each, which is clearly indicated in the engraving of the garden, showing the divisions in a similar way to the *Index*

[18] Pauw (1601) Catalogues in different editions filled out for the years 1600, 1602, 1603, 1608, 1614 and 1628 are now stored in the National Herbarium in Leiden, shelfmark HERBAR Folio room.

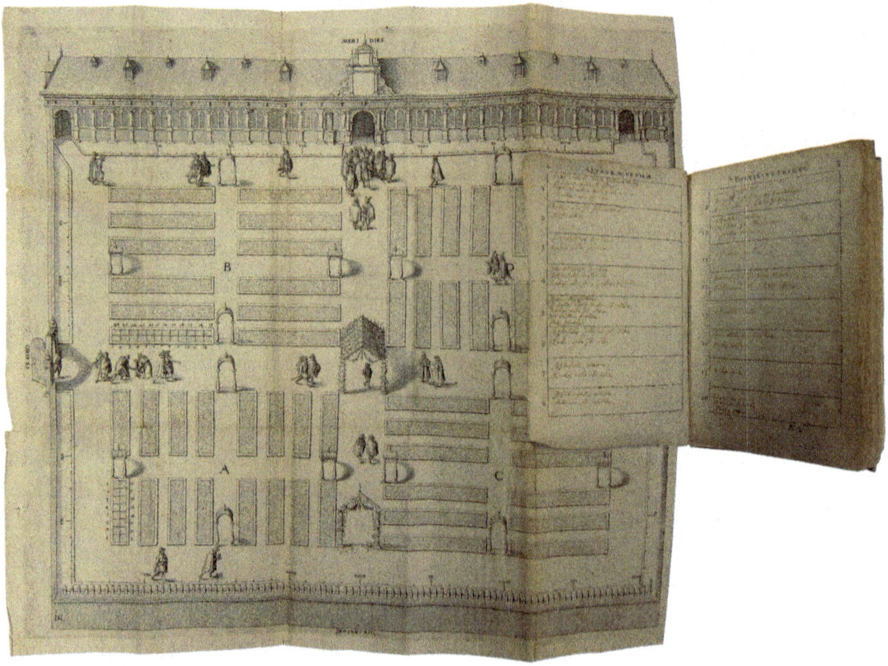

Fig. 4 Two pages with grids filled out by Pieter Pauws hand and the inbound view of the botanical garden in Leiden (etching by Jacques de Gheyn II), Pieter Pauw, Hortus Publicus Academiæ Lugduno-Batavæ [. . .], Leiden: Officina Plantiniana 1601, pp. 2–3 and view of the garden (Leiden, National Herbarium, Folio room)

stirpium. This adjustment in stripes and bed numbers was probably made because the garden was not an exact square but rather a stretched rectangle. The resulting plots became all more or less similar in size.

The most interesting feature of the catalogue is the following: on most of the pages, an empty grid was printed (Fig. 4), in which the name of the plants could be inscribed by hand (Swan 2005: 51–65 and 108–120). This had one big advantage: the printed catalogue could be used for different years and still feature a current inventory of the garden, an easy way to take care of the ever-changing project. As Pauw mentions in the foreword of the catalogue, it should not only be used by the prefect to make an up-to-date annual inventory, but also by the students during the classes held in the garden.

On each page, there is a grid with 8 boxes, geometrically relating to the plots of the garden, which had 16 or 24 beds each. Therefore, the stripes of *quadra* A (1) and D (4) can be shown on a double page, the stripes of *quadra* B (2) and C (3) on a double and a single page. Even the surrounding garden beds are structured in steps of eight plots. The correspondence of the number of garden beds with those of the grid in the catalogue is most certainly not a coincidence, but rather a direct result of the typographical needs of the catalogue. To use all the space the pages provide, it was necessary to unify the layout. The garden and the catalogue built a closely related system, depending on each other. The catalogue was certainly further linked to the actual garden by numbers, which enumerated the actual garden beds. It is known that such numbers written on stones were used in Padua and Montpellier (Rioux 1994: 29).

The idea of a flexible catalogue was, indeed, not a new invention. Exactly ten years earlier, a similar catalogue was printed at the University of Padua, where the most famous botanical garden of that time could be found.[19] The catalogue of Padua was the first printed catalogue of a botanical garden. It is not a big surprise that it was the role model for the one published in Leiden. Padua was the most famous school of medicine in the world and attracted many foreign students, a lot of them coming from protestant countries like the Netherlands. One of them was Pieter Pauw.

The garden in Padua had a more complex geometrical layout compared to the one in Leiden. It featured an aesthetic and ornamental structure, constructed on geometric patterns primarily based on squares and circles, which established spatial hierarchies: plants could be in the geometrical middle of a bed or at its border, and therefore also be the centre of attraction or not. In the Paduan catalogue, the grid is only a typographical necessity and has no direct similarity with the formal structure of the garden. The grid is only an aid to fill the pages in a more or less practical and efficient way. The catalogue features 144 empty grids for each *spaldo*—or quadrant. However, for example, *spaldo* 3 had only 121 numbered garden beds. Therefore, on three pages, grids were printed without corresponding garden beds, staying empty. Garden and catalogue were not as closely linked as in Leiden, where the botanical garden itself featured a non-hierarchical grid-like structure, similar to the catalogue. It seems that this structure of Leiden's garden and catalogue established a more objective and more scientific distribution of plants (Swan 2005: 108–120). Nevertheless—as we have seen in the catalogue from 1594—this grid could also be broken up and the stripes could be connected with each other.

The Arrangement of Plants

Pieter Pauw mentioned in the preface of the catalogue that in each lecture, one bed or stripe should be discussed. We can therefore assume that each stripe formed a set of plants belonging together. It is not in the framework of this paper to identify the mentioned plants and to set their pre-Linnaean names into modern nomenclature,[20] but to analyse the approaches by which the prefects acted in the garden according to their understanding of the plants. To recognize all mentioned plants, their

[19] The author of this catalogue is unknown, but it is assumed to be the engraver and publisher Girolamo Porro (1591); for the catalogue, see Helm (1968) and Swan (2005: 51–65 and 108–120).

[20] The garden was reconstructed twice according to the catalogue of 1594. For the newest reconstruction and the problems in identifying the plants, see Tjon Sie Fat (1991).

characteristics and their virtues, it is necessary to work with an accepted herbal of that time. The most promising book to use is the Cruydt-Boeck written by Rembertus Dodonaeus and first published in 1554.[21] It was reprinted several times and in different languages—a French translation by Clusius was published in 1557—and did not only have a huge influence in the field of natural history in the Netherlands, but was used all over Europe. It was reissued in 1608 and annotations by Clusius provided the newest scientific insights (Dodonaeus 1608). Both authors were close friends. Thus, it is no surprise that the names used in this herbal are similar to the ones used in the catalogues, making it possible to look up the plants and to understand how Clusius and Pauw perceived the plants.

Which considerations were used to distribute the plants? The clearest order in the stripes was to set all plants belonging to the same genus in one stripe, as we have already seen in the catalogue of 1594. However, there were many other ways to distribute the plants.

As the early botanical gardens were primarily understood as medical gardens—standing in the tradition of monastic herb gardens of the medieval ages—and botany itself not fully established, it is only logical that plants with the same virtues or healing powers were set next to each other. They served to collect and display the *materia medica*. An example of this arrangement is *area* 7 of *quadra* 4. Surprisingly, this stripe did not actually deal with medical herbs, but with deadly ones. All plants in it were very poisonous or even fatal. The catalogue shows that different sorts of *Aconitum*, *Napellus*, *Anthora*, as well as some species of *Christiophoriana*, were planted next to each other. If we look them up in the Cruydt-Boeck, the first three of them are discussed in Book 14, which deals with all fatally poisonous plants (Dodonaeus 1608: 760–824). The *Christiophoriana* is instead discussed in the section of the herbal dealing with climbing plants. Dodonaeus states that it is as poisonous as the *Aconitum* and claims that therefore it can be compared to this plant (ibid.: 697–759 and 719). It came to a slight shift between the classification in the herbal and in the garden. In conclusion, in this section of the garden it was not the aspect of growth which was of main importance, but the aspect of toxicity.

In *area* 9 of *quadra* 2, a set of plants with similar healing powers was put together. This stripe shows clearly that the garden was—at least in parts—still perceived as a traditional medical herb garden. As mentioned before, the garden was closely linked to the medical school of the university.

Next to a few plants, which do not share any replicable similarities, a second group containing *Saxifragia*, *Nasturthium*, *Petasites*, *Fumaria*, *Meium*, *Lithospermum*, *Hepatorium* and *Anetum* were planted in close vicinity. All of them had the same healing power: they were used to promote urination. Most of them were believed to be of benefit to rinse the liver, kidneys, spleen or gall bladder. The *Saxifragia*, *Nasturthium*, *Meium* and *Lithospermum* had a further specific healing power: they were thought to be able to dissolve kidney and bladder

[21] For this and other early modern herbals used and printed in the Low Countries, see Arber (1970: 79–92).

stones, resulting in the name of the latter. In the Cruydt-Boeck, there is no chapter dedicated to this special virtue, but rather the plants are discussed in different sections of the book.[22]

It is no surprise that these plants were put in a single stripe, establishing an order based on this distinct health issue. The prefect of the garden, Pieter Pauw, was not only working in the *hortus botanicus*, but also as an anatomist. He was responsible for the construction of the theatre of anatomy, which was built more or less at the same time as the garden. In the early modern period, it was a common practice that a professor would teach in a botanical garden during the summer—when the plants were in bloom—and in a theatre of anatomy during the winter—when the corpses did not rot too quickly.

The anatomical theatre at Leiden was equipped with a collection of curiosities. One of the exhibits was the bladder of the famous scholar Isaac Casaubon, which was sent to Pauw by one of his former students living in London.[23] In later times, Pauw's successor, Otto Heurnius, equipped the collection with a huge kidney stone and "seven stones, each one as big as a flat walnut [cut] from the bladder of the excellent and famous D. Joannes Heurnius", his father (Huisman 2009: 54).[24]Johannes Heurnius was not the only Professor of the University who suffered from kidney stones: Carolus Clusius did as well (Hunger 1927–1943: vol. 2, 195). Pieter Pauw and his successors therefore were able to discuss both the syndromes of the diseases in the theatre of anatomy and the herbs for their treatments in the garden, linking the two facilities together. This shows that the garden was still understood as a medical herb garden, featuring plants with specific virtues.

Other methods of plant classification could be found as well. In the upper part of stripe 3 of *quadra* 4, different species of the same genus named *Lapathum* grew in 1600, the common *Lapathum* and the *Acetosa* (Dodonaeus 1608: 1096–1102). Furthermore, we can find a close related species set next to them, the so-called *Tota bona*, which was described by Dodonaeus in the subsequent chapter as a close relative to the genus *Lapathum* (ibid.: 1102f). The different catalogues, which were filled out each year, show that, between 1603 and 1608, a fourth species of the same group was added, the *Patientia* (ibid.: 1096–1102), which could not be found in the garden in 1600. So, it is true that "day-to-day new species arrive in the garden", as the text adjoining the copper engraving claims. With these newly arrived plants, gaps signifying missing ones could be filled and plant sets completed. All of the plants in the upper half of the stripe are discussed in Dodonaeus' book in the same section, which describes "Eatable plants and vegetables, which can be mashed",[25]

[22] Dodonaeus (1608: 545–550, 1202–1203, 1027–1028, 85–87, 527–528, 122–123, 40–41 and 516–517).

[23] Inventory of the theatre, Leiden, University Library, shelfmark Ms. AC1, Nr. 228, no pagination.

[24] "Item seven steenen, een ieder van de groote en figure van een Walsche platte noote, asgrau van coleur wegende altemae levens vaer [...]. Dese sijn uyt de Blase ghesneden van den voortreffeliken en wijtberoembden D. Joannes Heurnius eerste Professor Medicinae in de Universiteit alhier [...]." (Barge 1934: 41).

[25] "Van de eetbaere oft Moes-cruyden" (Dodonaeus 1608: 1045).

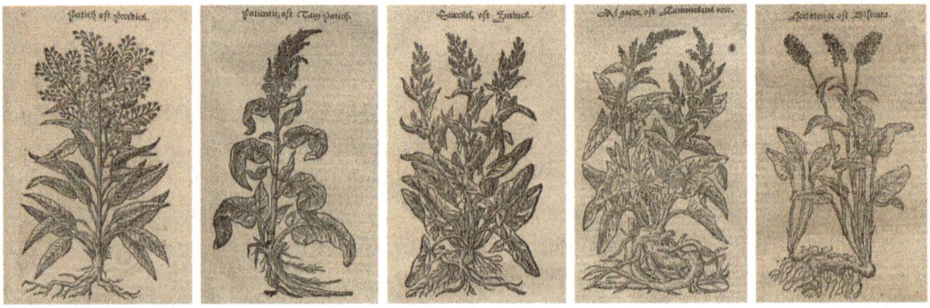

Fig. 5 *Lapathum, Patientia, Acetosa, Tota Bona* and *Bistorta* (Rembert Dodonaeus, Cruydt-Boeck [. . .], Leiden: Officina Plantiniana, 1608, pp. 1097, 1097, 1098, 1102 and 579) (ETH Zürich, Alte und Seltene Drucke)

focusing on the leaves of the plants (ibid.: Book 21, 1045–1121). On the lower part of this stripe, two different species of the same genus were set, the *Bistorta* and the *Britannica* (ibid.: 579–581). They are discussed in another section of the herbal, concerning "Plants whose roots can be used as medicine"[26] (ibid.: Book 11, 563–627).

So why were those two sets discussed in different chapters and with different virtues and uses put in the same stripe? The answer is given in the Cruydt-Boeck, in which it is mentioned by Dodonaeus that the *Bistorta* shows a lot of morphological similarities to the *Lapathum*. Reference is made to the classic authorities Dioscorides and Plinius, who already compared those two species with each other, stressing their comparable appearances (ibid.: 580), which can also be clearly seen in the accompanying illustrations of the plants (Fig. 5).

The plants did not have the same healing powers, but did have morphological similarities. Therefore, we can say that in the botanical garden of Leiden, empirically perceivable characteristics of plants played an important part in their classification and disposition. Plants with similar morphological aspects—with similar roots, stems or leaves, with similar smell or taste—were put together, as other stripes and examples also show. In doing so, the professors at Leiden participated in a fundamental change in the perception of natural history. No longer were the plants only a source for remedies or food. The plants themselves became interesting and an individual topic of research. With this new approach, the field of botany was slowly established (Ogilvie 2006: 1–24).

The garden was in constant change, as we can see while browsing through the catalogues of different years. Plants as living specimens could die. The exotic specimens especially had to be protected in heatable hothouses during winter (Tschira 1939). Nevertheless, lots of plants did not survive, leaving an empty space behind. On the other hand, new plants arrived on a regular basis, and they

[26] "Van de Wortelen die van de Medicijns gebruyct worden" (Dodonaeus 1608: 563).

were often just put wherever a space was free, blurring—together with the missing plants—the systematic order established in the beginning.

If we take a look at the stripe discussed before (*quadra* 4/*area* 3), a lot of new species were set in it, which is shown in the catalogue of 1608. Most of them did not in any way fit with the previously planted specimens: they did not feature similar virtues, looks or other perceivable similarities. Nevertheless, three of them established a suborder by themselves: they were all discussed in the subsequent chapters of Dodonaeus' herbal, in Chapters 20 (*Pyrola*), 21 (*Ophioglossum*) and 22 (*Lunaria*) of the fifth book (Dodonaeus 1608: 214–218). Their sequence in the garden was an exact counterpart to the one in the printed book, dealing with herbs which could not be put into a systematic order and were therefore mostly ordered according to the alphabet. They were transplanted from the adjoining stripe, which featured, in 1600, even more plants of the same section of the herbal, the plants discussed from Chapter 20 on up to Chapter 26, establishing an even closer relationship between the herbal and the garden.

Conclusion

The discussed examples clearly show that different systems for arranging plants in their beds could be found in the *hortus botanicus* in Leiden in its early years. They could be distributed according to their morphology, their virtues, or their qualities, but also according to their scent or flavour, or even the chapters of a plant book. In a lot of cases, plants were just put wherever there was a free space. Other orders were rejected: the plants were, in fact, not planted according to their natural habitat or their geographical origins or astro-botanical ideas and concepts.[27] Additionally, the alphabetical order was neglected in later years.

The divine creation could also arouse astonishment. Plants therefore were also placed according to their beauty, value and rarity. As we have seen, Clusius had already put effort into this aspect by 1594. He was very proud of the exotic, rare and beautiful plants he collected and had received. There was one plant of unparalleled interest for him and other admirers of plants in the Netherlands at the beginning of the seventeenth century: the tulip. It was a highly sought after, very rare and extremely expensive plant. The hype peaked in the so-called *Tulipmania* and ended in a burst speculation bubble in 1637 (Goldgar 2007).

Carolus Clusius—who played an important part in introducing the tulip to Europe (ibid.: 22–26)—was certainly proud of his beautiful and rare tulips and wanted them to be seen by all visitors to the garden. That is probably why they were the first sight of every incoming visitor. Like the reconstruction of the first printed catalogue shows, they were all planted alongside the main axis through the garden, putting them right into the centre of attraction. This exhibition must have been quite successful and resulted in the theft of tulips, which was not uncommon (ibid.: 57f).

[27] At least, such orders have not yet been found.

Soon, another kind of order had to be established to protect those valuable species: they had to be stored in a special compartment with a fence so as not to be stolen, clearly visible in the copper engraving of 1610 (Fig. 1).[28] We can therefore conclude that the public aspect of the garden led to another kind of order: a disposition of plants, which was not defined by aspects of the plants themselves, nor by their use or virtues, as previously discussed, but by aspects of their perception in society, that is to say, their value. Therefore, even sociocultural aspects led to a specific distribution of plants in the *hortus botanicus* of the University of Leiden.

Bibliography

Unpublished Sources

Cluyt, Dirk Outgaertsz. 1594. Index stirpium. Archief van Curatoren 1, shelfmark Ms. AC.1, Nr. 101, fol. 117A–T. Leiden: University Library.

Pauw, Pieter. 1601. Hortus Publicus Academiæ Lugduno-Batavæ, eius Ichnographia, descriptio, usus: addito quas habet stirpium numero et nominibus. Leiden: Officina Plantiniana. For the handwritten inscriptions, see Leiden, National Herbarium, shelfmark HERBAR Folio room, copies for 1600, 1602, 1603, 1608, 1614 and 1628.

Printed Sources

Brosse, Guy de la. 1636. Description dv iardin royal des plantes medecinales, estably par le roy Louis le Ivste, à Paris. Contenant le catalogue des plantes qui y sont de present cultiuées, ensemble le plan du jardin. Paris: s.n.

Dodonaeus, Rembertus. 1608. Cruydt-Boeck: met biivoegsels achter elck Capittel, wtverscheyden Cruydtbeschrijvers: item in't laetste een Beschrijvinge vande Indiaensche Gewassen meest getrocken wt de Schriften van Carolus Clusius. Leiden: Officina Plantiniana.

Porro, Girolamo. 1591. L'Horto dei Semplici di Padoua. Venice: Girolamo Porro.

Secondary Literature

Arber, Agnes. 1970. Herbals: Their Origin and Evolution. Darien CT.: Hafner.

Barge, J. A. J. 1934. De oudste inventaris der oudste academische anatomie in Nederland. Leiden: Stenfert Kroese's Uitgevers Mij.

Boon, K. G. 1984. Samuel de Swaef to Jan Thesing (Hollstein's Dutch and Flemish Etchings, Engravings and Woodcuts ca. 1450–1700, Vol. XXIX). Amsterdam: Van Gendt.

Findlen, Paula. 2006. Anatomy Theaters, Botanical Gardens, and Natural History Collections. In The Cambridge History of Science, Vol. III, Early Modern Science, eds. Katharine Park and Lorraine Daston, 272–289. Cambridge: Cambridge University Press.

[28] The plants put into this fenced compartment are listed in Pauw's catalogue for the year 1608 on an inbound leaflet, stored at Nationaal Herbarium Nederland in Leiden, shelfmark: HERBAR Folio room.

Gelder, Esther van. 2011. Tussen hof en keizerskroon: Carolus Clusius en de ontwikkeling van de botanie aan Midden-Europese hoven. Leiden: Leiden University Press.

Goldgar, Anne. 2007. Tulipmania: money, honor, and knowledge in the Dutch Golden Age. Chicago: University of Chicago Press.

Grämiger, Gregory. 2014. Verortungen von Wissen. Die Räume und Sammlungen der Universität Leiden 1575–1700. PhD-Dissertation ETH Zurich.

Helm, Johannes. 1968. Das erste 1591 im Druck erschienene Bestandsverzeichnis des Botanischen Gartens zu Padua. Ein Beitrag zur Geschichte der heutigen Indices seminum. Die Kulturpflanze 16: 75–90.

Huisman, Tim. 2009. The Finger of God: anatomical practice in 17th-century Leiden. Leiden: Primavera Pers.

Hunger, F.W.T. 1927–1943. Charles de L'Escluse (Carolus Clusius): Nederlandsch kruidkundige, 1526–1609. 2 vols. 's-Gravenhage: Martinus Nijhoff.

Jong, Erik de. 1991. Nature and Art: The Leiden Hortus as 'Musaeum'. In The authentic garden: A symposium on gardens, eds. Leslie Tjon Sie Fat and Erik de Jong, 37–60. Leiden: Clusius Foundation.

Jorink, Eric. 2010. Reading the Book of Nature in the Dutch Golden Age, 1575–1715. Trans. Peter Mason. Leiden and Boston: Brill.

Meier Reeds, Karen. 1970. Botany in Medieval and Renaissance Universities. New York and London: Garland.

Molhuysen, Philipp Christiaan (P.C.). 1913–1924. Bronnen tot de geschiedenis der Leidsche Universiteit. 7 vols. 's-Gravenhage: Martinus Nijhoff.

Müller-Wille, Staffan. 2004. Ein Anfang ohne Ende: Das Archiv der Naturgeschichte und die Geburt der Biologie. In Macht des Wissens, eds. Richard van Dülmen and Sina Rauschenbach, 587–605. Köln, Weimar, Wien: Böhlau Verlag.

Ogilvie, Brian W. 2006. The Science of Describing: Natural History in Renaissance Europe. Chicago: University of Chicago Press.

Rioux, Jean-Antoine (ed.). 1994. Le jardin des Plantes de Montpellier. Quatre siècles d'histoire. Graulhet: Odyssée.

Swan, Claudia. 2005. Art, Science, and Witchcraft in Early Modern Holland. Cambridge: Cambridge University Press.

Tjon Sie Fat, Leslie. 1991. Clusius' garden: a reconstruction. In The authentic garden: A symposium on gardens, eds. Leslie Tjon Sie Fat and Erik de Jong, 3–12. Leiden: Clusius Foundation.

Tongiorgi Tomasi, Lucia. 2005. Gardens of Knowlegde and the République des Gens de Sciences. In Barock Garden Cultures: Emulation, Sublimation, Subversion, ed. Michel Conan, 85–129. Washington D.C.: Dumbarton Oaks Research Library and Collection.

Tschira, Arnold. 1939. Orangerien und Gewächshäuser: Ihre geschichtliche Entwickung in Deutschland. Berlin: Deutscher Kunstverlag.

Veendorp, H., and L.G.M. Baas Becking. 1938. Hortus Academicus Lugduno-Batavus 1587–1937: The development of the gardens of Leyden University. Harlemi: Ex typographia Eschedaiana.

Garden Visits, Observations, Reading and Excerpts: Martin Fogel (1634–1675) and His Techniques of Acquiring Knowledge

Carola Piepenbring-Thomas

Abstract

The physician Martin Fogel developed special methods of accumulating knowledge in the seventeenth century. He collected more than 3600 books and he made notes on more than 30,000 little slips of paper, highlighting excerpts from his books or from received letters. He recorded information from oral accounts, from newspapers, experiences and travel adventures on his 4-year educational journey. He worked with each of his sources in a different methodical way. But he mainly consulted printed sources, and from this point of view, Fogel was more committed to the conservative historical approach than to modern experimental scientific work. But nevertheless, he was an innovative scholar. For example, he discovered the relationship between the Finno-Ugric languages. His literary estate was so important to Gottfried Wilhelm Leibniz that the latter acquired Fogel's books and borrowed his papers, ultimately never returning them to the inheritors.

Introduction

Knowledge has to be acquired, and this is always a complicated matter. This implies neither the simple accumulation of data, nor facts previously ordered, processed and invested with importance. Knowledge refers to an individual's complex, multiply-interlinked store of intellectual insights. One method of

C. Piepenbring-Thomas (✉)
Menzelstrasse 52, 30459 Hannover, Germany
e-mail: c.p-t@htp-tel.de

H. Fischer et al. (eds.), *Gardens, Knowledge and the Sciences in the Early Modern Period*, Trends in the History of Science, DOI 10.1007/978-3-319-26342-7_12

accumulating knowledge in the seventeenth century will be demonstrated here using the example of Martin Fogel.[1]

The Hamburg physician Martin Fogel exemplifies the polymath scholar of the seventeenth century.[2] Scholarship at that time did not simply mean completing one's studies and cultivating numerous contacts and relationships with scientists in different countries. Often, it also involved voluminous correspondence and multilingualism, and it implied a broad interest in various scientific disciplines. Introduced to the study of natural history by his teacher Joachim Jungius, Fogel devoted his special attention not only to historical, geographical, medical and other questions but also to botanical and horticultural issues. Fogel largely compiled and organized his knowledge in his collection of more than 30,000 paper slips.

This study is based on a manuscript collection in the Gottfried Wilhelm Leibniz Bibliothek in Hannover. It consists of three parts: Fogel's travel diary, many notes written on slips of paper, and his library. Currently, neither has the travel diary been published nor is it possible to work with the slips of paper systematically.[3] Only two great collections of slips of paper are known in Germany from the mid-seventeenth century. They belonged to Jungius and to his disciple Fogel. After Fogel died in 1675, at the age of 41 years, a catalogue of his library was published for an auction. Fortunately, Gottfried Wilhelm Leibniz was able to convince the Hanoverian Duke Johann Friedrich of the quality of the collection and commissioned the purchase of the entire library for 2000 *thaler*. In 1678, the library was transferred to Hannover. Today, the collection of books is no longer complete, because duplicates have been traded with other libraries, but the botanical books have only been slightly depleted.

When Leibniz travelled to Hamburg in 1678, he heard of the existence of Fogel's other manuscripts and his collection of slips of paper, similar to a modern card index system. He borrowed a part of the manuscript estate. The second husband of Fogel's widow asked Leibniz to return the papers, but Leibniz "forgot" to do so, and therefore these parts are still in Hannover. It is easy to recognize that Fogel worked with the three source genres (books, slips of paper and travel diary) differently, each used in a special way for the acquisition of knowledge. Only by analysing his different procedures as a whole do we obtain a complete picture of Fogel's practice of acquiring knowledge.

[1] Fogel's book and manuscript collection studied in Marten and Piepenbring-Thomas (2015). It examines the library, the slips of paper and further manuscripts like the travel diary appraisal from a source-critical point of view as sources for questions about the scientific and epistemological methods of acquisition of knowledge. I would like to thank Catherine Atkinson and James O'Hara for reviewing this text.

[2] The biographical remarks follow (Kangro 1969; Veenker 1986; Hoffmann 1855).

[3] See annotation no. 1 and annotation no. 21. See for the library: http: // www.leibnizcentral.de/. MenupünktDatenbanken / Leibniz-Arbeitsbiblithek; for he slips of paper: http: // www. leibnizcentral.de/. MenüpunktDatenbanken/ Handschriften/ Fogels Ordnungen.

Fogel's Biography

Martin Fogel was born in Hamburg in 1634, the son of the textile merchants Martin and Judith Fogel. Fogel graduated from grammar school having been taught by Joachim Jungius, Johann Adolph Tassius, Heinrich Lambeck and Aegidius Gutbier. Later, he studied theology at Giessen and Strasbourg. When his father died, Fogel had to return to Hamburg. To earn a livelihood, he taught various subjects, including oriental languages. Having been appointed one of the estate executors of his teacher Jungius, he organized the scholar's handwritten papers. After that, he worked on the edition of the "Doxoscopiae" (Jungius 1662) for 4 years. This is one of the most important works of the polymath Jungius. In the course of his 4-year educational tour (1662–1666), Fogel was accompanied by Martin Wevetzer the son of the businessman Valentin Wevetzer.[4] Fogel graduated as a doctor of medicine in Padua[5]and he established contacts with the scientists Francesco Redi, Ovidio Montalbani, Lorenzo Magalotti, Gottfried Welschius and Melchissedech Thevenot.

Back in Hamburg, he settled down as a physician. In 1671, he was admitted to the Hamburger Collegium Medicum.[6] In his early years, Fogel was mainly interested in linguistic studies (in addition to his professional activities). Grand Duke Cosimo III asked him to produce a Finnish dictionary.[7] In doing so, Fogel discovered the relationship between the Finnish and Hungarian languages.[8] He planned and worked on numerous publications and corresponded with scholars, including Henry Oldenburg, the secretary of the Royal Society. At the beginning of 1675, he was appointed professor of logic at the Academic Gymnasium in Hamburg.[9] In October 1675, he died of a fever, leaving his wife with two young children.

[4] Valentin Wevetzer owned a trading company in Hamburg that was significantly involved in Iberian trade in the 1630s. In 1645, we find Martin Vogel, Sr.with a turnover of 27,060 Marks and Valentin Wevetzer with one of 16,082 Marks. Reißmann (1975: 375)—Martin Wevetzer Jr., the son of Valentin Wevetzer, was baptized in 1642 in the parish of St. Catherine, married Margaret Hachtmann in 1682 and died in 1687.

[5] Wilckens (1770: 504), records the date as Jan. 29, 1663.—Moller names Hieronymus Frizimelica (1611–1683) as his professor in this context (Moller 1744: 260).

[6] The Hamburg Collegium Medicum was probably founded in 1644, following the suggestion of Paul Marquard Schlegel (1605–1653), who was appointed to the position of Hamburg Stadtphysicus in 1642 (Schrader 1840: 45ff).

[7] Cosimo III. de'Medici (1642–1723) Grand Duke of Florence, beginning in 1670.

[8] The transcription for Cosimo III.: Ms. Bibliotheca Nazionale Centrale Firenze—Cl. III.103 "Nomenclator Latino Finnicus". —Fogel's original manuscript is located in the GWLB: Ms IV, 574a. Mart. Fogelii de Finnicae linguae indole observationes, 29 Bl.

[9] On Feb. 25, 1675, he gave his inaugural address "De melioris logicae". He planned to open a college called *Collegium Analyticum et Disputatorum* in his house next to St. Catherine's Church. The requirement for enrolment was knowledge of Jungius' "Logica Hamburgensis" (Kangro 1969: 26).

Fogel was greatly interested in gardens and horticulture. He went to see both the garden of the Anckelmann family[10] in Hamburg[11] and the Gottorf garden in Schleswig,[12] and also visited not only famous libraries, art galleries and buildings, but also many famous gardens on his grand tour of such places as Italy, France and Spain. In his diary, he recorded impressions, descriptions and assessments of these gardens and vivaria, including observations on the vegetation of the local countryside and the cultivation and use of various plants. Later, in Hamburg, Fogel found enough time, alongside his professional life, for experiments and observations [e.g., relating to the colour changes in Larkspur (Rittersporn)]. He noted the results on his slips of paper and, above all, he managed to compile an extraordinary library (consisting of 3600 volumes). Fogel classified about 120 works as botanical, while other books were classified as medical reference books.

The Book Collection

Even a small selection of authors shows the breadth of Fogel's collection of books: Cesalpinus, Ursinus, Camerarius, Bauhin, Morison, Parkinson, Ray, Royer, etc. Examination of the books and their contents, and of the way they were subdivided into groups, shows that Fogel was less a bibliophile than a very systematic collector.

Fogel owned books by Theophrastus, herbal books, works about vernacular plant names, locations and living conditions, books about the arrangement and design of gardens and those containing planting instructions for the gardener with illustrations of useful gardening tools. He also had many works on local flora, plant lists from botanical gardens, commercial catalogues, books with microscopic images and, finally, the first publication on a single plant family, the Compositae.

The botanical books, comprising about 3.5 % of the total quantity, can be divided into 14 groups, the division being made from a historian's point of view[13]:

1. Authorities (Theophrastus)
2. Early herbals (Brunfels, Bock)

[10] Eberhard I. Anckelmann was a Hamburger councillor, who had already begun installation of the Hortus Anckelmannianus on the city fortifications before the middle of the century. His son Caspar (1634–1698) built an ornamental Baroquegarden on this foundation. It is well known today on account of the images (engravings) in the flora of the Hamburg flower painter Hans Simon Holtzbecker.—Cf. (Heilmeyer 2003; Roth 2003).

[11] For the horticultural life in Hamburg, see Schubert (2003).

[12] Holtzbecker also painted a Flora for the Gottdorfer Garden. Prints from the "Gottorfer Codex" can be found in the Kupferstichkabinett des Statens Museum for Kunst in Kopenhagen (Krausch 2003). Slips of paper in GWLB: Ms XLII 1923, phi 148 (fol. 11r, 12r). See also, on Wandsbeck: GWLB: Ms XLII 1923, phi 148 (fol. 11r). Hirschhörner, we(nn) sie abgeworffen, u. wied (er) gewaksen, haben auch Hahr. Das hab ich zu Wanbähkgesehn im Tiergarten.

[13] Based interalia on Jahn (2000) and Mägdefrau (1973).

3. The second generation of herbals (Lonitzer, Mattioli, Lobelius, Dodoneus, Tabernaemontanus, Daléchamps, Cordus, Clusius, Lauremberg, Gesner, Camerarius, Dorsten)
4. Astrological and magical plant books (Della Porta, Spigelius
5. Methodological and systematic works (Bauhin, Clusius)
6. Descriptions of botanical gardens and private gardens (Schenck, Vorstius, Montalbani, Castelli, Joncquet, Aldini, Paulli, Besler)
7. Local and regional flora (Chemnitz, Jungermann, Hoffmann, Pilleterius, Oelhafen, Pona, Ray, Royer)
8. Exotic plants (Alpinus, Cornut, Vesling)
9. Manuals for cultivation (Parkinson, Munting, Major, Soderini)
10. Religious writings (Mylius, Meursius, Schwenckfeld, Camerarius)
11. Herbal monographs (Pfautzius, Major, Morison, Ursinus, Alpinus, Horstius, Clavena)
12. On trees (Court, Della Porta, Rendella)
13. Works on microscopy and physiology (Grew, Digby)
14. Bibliographies (Bumaldi, Trichet du Fresne)

The selection of books reveals a methodical understanding of the subject. It also provides an insight into the spectrum of botanical literature available to the interested reader in the second half of the seventeenth century. Fogel owned English, Italian, French and Dutch works, and he used his books for intensive studies, annotating and excerpting relevant information. He placed special emphasis on the use of a unique terminology. Having been trained by his work on the edition of his mentor's "Doxoscopiae", he took particular care to make logical and systematic distinctions between significant and accidental characteristics in describing plants.

These books can be used in two ways. The inventory permits us to investigate the origin of the books as purchases or gifts, Fogel's motivation for collecting a book or not, and the library's monetary value. The individual books also allow an investigation of his method of work: Fogel worked hard on his books and his work can be traced by means of his underlinings, marginal notes and corrections (Fig. 1). In the

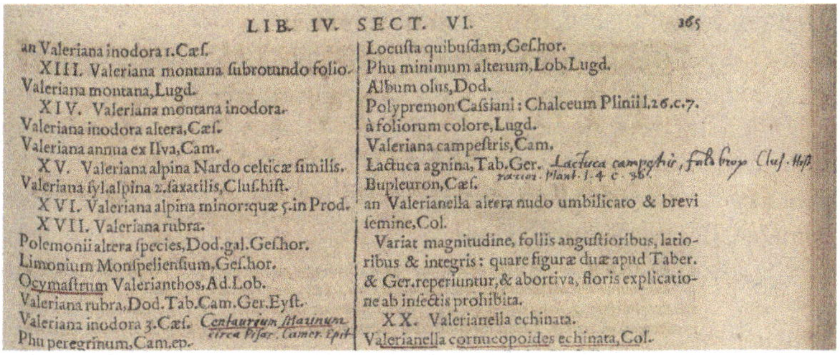
LIB. IV. SECT. VI. 165

an Valeriana inodora 1.Cæſ.
XIII. Valeriana montana ſubrotundo folio.
Valeriana montana,Lugd.
XIV. Valeriana montana inodora.
Valeriana inodora altera,Cæſ.
Valeriana annua ex Ilva,Cam.
XV. Valeriana alpina Nardo celticæ ſimilis.
Valeriana ſyl.alpina 2.ſaxatilis,Cluſ.hiſt.
XVI. Valeriana alpina minor:quæ 5.in Prod.
XVII. Valeriana rubra.
Polemonii altera ſpecies,Dod.gal.Geſ.hor.
Limonium Monſpelienſium,Geſ.hor.
Ocymaſtrum Valerianthos,Ad.Lob.
Valeriana rubra,Dod.Tab.Cam.Ger.Eyſt.
Valeriana inodora 3.Cæſ.
Phu peregrinum,Cam.ep.

Locuſta quibuſdam,Geſ.hor.
Phu minimum alterum,Lob.Lugd.
Album olus,Dod.
Polypremon Caſsiani: Chalceum Plinii l.26.c.7.
à foliorum colore,Lugd.
Valeriana campeſtris,Cam.
Lactuca agnina,Tab.Ger.
Bupleuron,Cæſ.
an Valerianella altera nudo umbilicato & brevi ſemine,Col.
Variat magnitudine, foliis anguſtioribus, latioribus & integris: quare figuræ duæ apud Taber. & Ger.reperiuntur,& abortiva, floris explicatione ab infectis prohibita.
XX. Valerianella echinata.
Valerianella cornucopoides echinata,Col.

Fig. 1 Detail from Caspar Bauhin: Pinax, Theatri botanici, p. 165 (Scan GWLB: N-A 7012)

margins, he often makes cross-references to similar passages in other works. The abbreviations he used can be resolved with the help of the auction catalogue.

The Collection of Slips of Paper

Fogel's collection of slips of paper also demonstrates his interest in botanical questions. Over the course of more than 20 years, he collected approximately 32,500 octavo sheets. Of course, the collection does not only include excerpts and notes on botany, but also countless further notes on his other special fields of interest. Paper sheets on his botanical interests can be found in about 75 sections (with an average of 20–30 sheets per section). There are copies of a large portion of the notes made by order of Leibniz. The language used on the slips of paper varies, depending on Fogel's main sources: Latin, Italian, Spanish, French, rarely English; German is often used for oral reports.

About 800 botanical sheets contain notes about plant structure, plant cultivation, and etymological observations on plant names, as well as notes on authors and their works and notes on the use and preservation of individual plants and their fruit (Fig. 2). They also include questions concerning changing colors, toxic plants, aromatic plants, herbaria, agriculture, horticulture, notes on prices (especially of wood), on the geographical location of plants (from published travelogues) and on fossilized plants. But most of the papers concern individual observations of plants (with mixed Latin and German names: Delphinium, Euphorbium, Tanacetum, horsetail (Schachtelhalm)), their fruits (apples, grapes, limes), roots and seeds.

Gardens are mentioned only sporadically, for example, on account of their remarkable plants (lilies, tulips). Fogel used numerous sources for his notes: oral communications, letters and newspapers, but mostly he worked with his library. He made excerpts and listed the authors and their works with abbreviations to identify them. This allows us to refer back to the originals and reconstruct Fogel's work. For example, "Park. Parad." means John Parkinson's: *Paradisi In Sole Paradisus Terrestris*.[14]

Fogel organized his notes in a similar way to his library, but he used different categories. He developed his index only with reference to subjects of interest to him; he did not aspire to systematic completeness.

What does a paper slip actually look like? Fogel frequently used scrap paper, paper still being a valuable resource; in the case shown below (Fig. 3), he drew right up to the border, but drawings appear very rarely on the slips of paper. This practice is similar to that of the Jungius card index system.

Although Fogel excerpted extensively from his botanic library (Bauhin, Della Porta, Lonitzer, Besler, Mattioli, Dodoneus, Gesner, Chemnitz. . .), it is remarkable how often he also made excerpts from non-botanical works. As an example, one

[14] Parkinson (1656). Fogel possessed the1656 edition. (GWLB: N-A 7079).

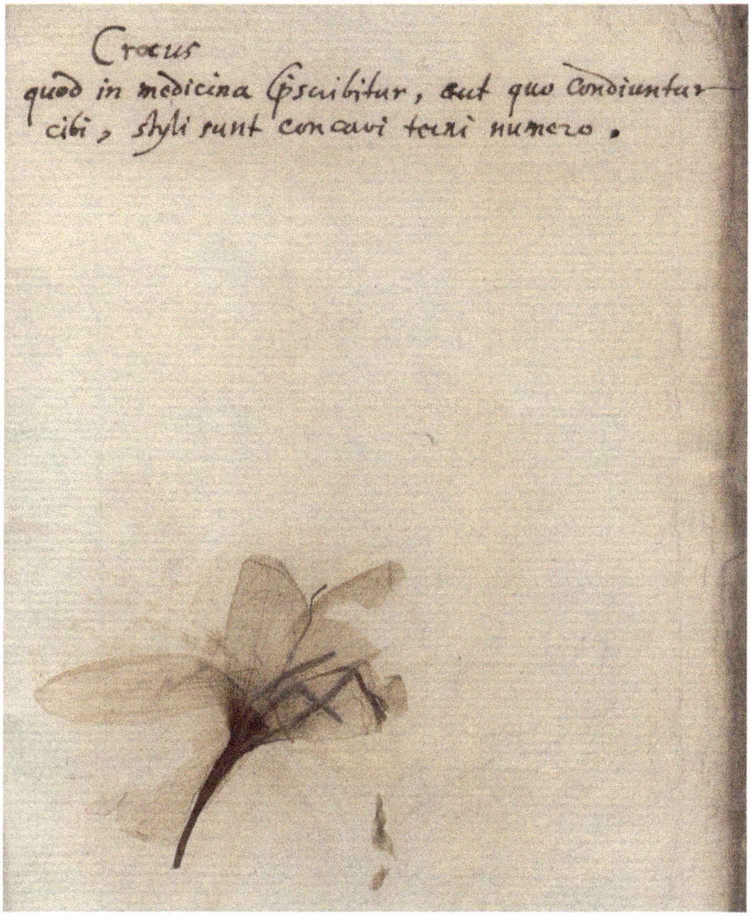

Fig. 2 Detail from a note on the medical use of Crocus. Scan GWLB: Ms XLII 1923, sigma 15 (fol. 3r): "Crocus quod in medicina proscribitur, aut quod condiuntur cibi, styli sunt concavi terni numero."

might quote an excerpt from the itinerary of John Wild's journey to Jerusalem.[15] In Chapter 28 (Beschreibung der Stadt Medinet en-Nebi, ihrer Beschaffenheit und Nahrung), one finds:

> Und vor der Statt ist ein grosser Garten/darinn stehen viel Dattelbaum/Wie ich denn auch nichts mehrers allda gesehen/als Datteln/die seyn wohlfeyl/und viel besser denn zu Al-cairo/es kauffens auch die Moren/und verkauffens zu Alcairo wider.

[15] Wild (1623) today: (GWLB: E-A 444).

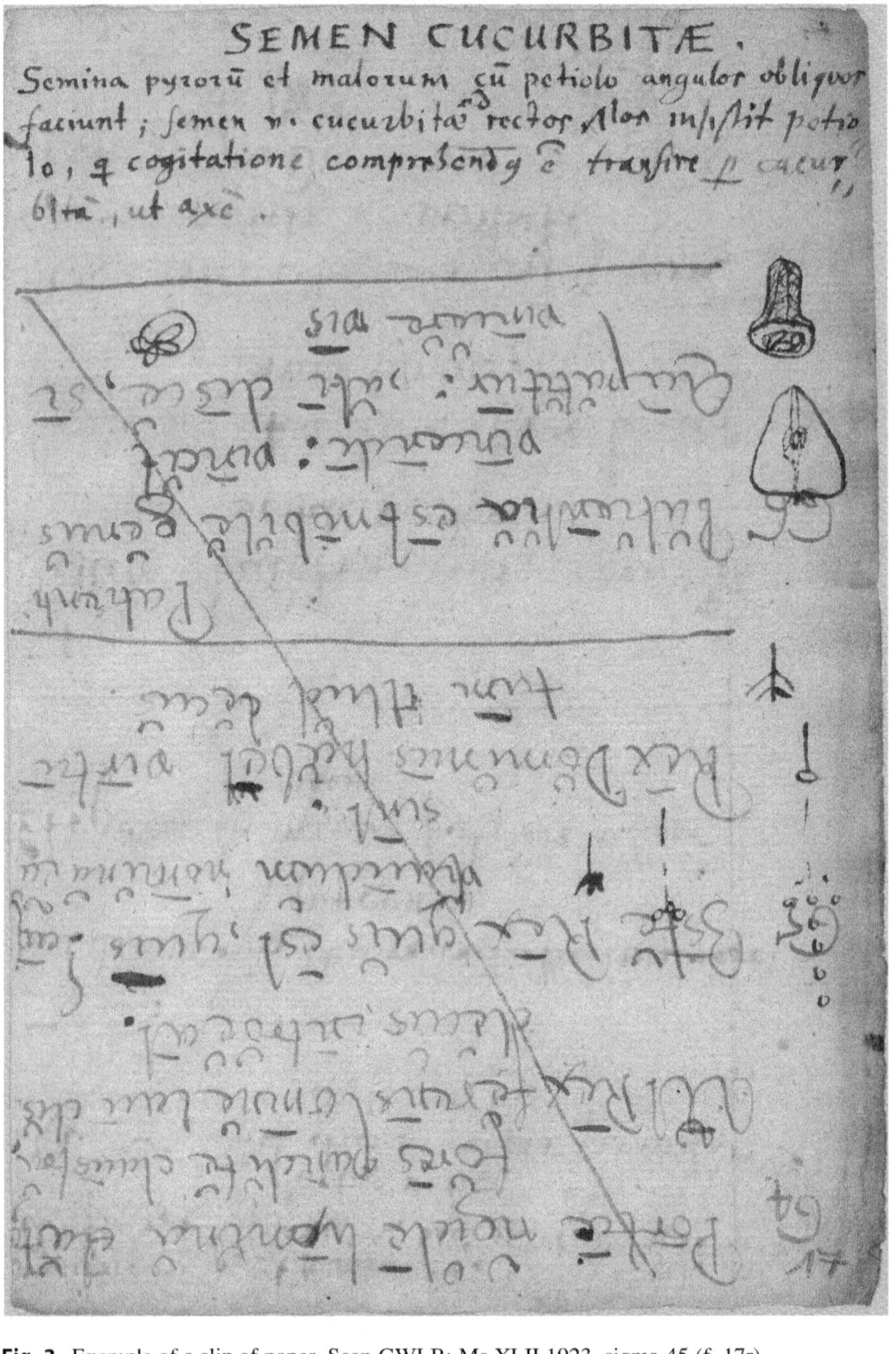

Fig. 3 Example of a slip of paper. Scan GWLB: Ms XLII 1923, sigma 45 (f. 17r)

On the paper slip, Fogel notes: "Datteln. Die zu Medina al nabi sind beßer als zu Alkairo, es kauffens auch die Morn u. ferkauens zu Alkairo wieder. Wild Reiseb.2 b. 28k."[16]

In order to generate knowledge, a scholar such as Fogel had to take into account a wide range of traditional writings and also needed to work with sources hitherto ignored, such as travelogues and newspapers from Hamburg and Altona, as well as from Haarleem in the Netherlands.[17] Fogel worked with numerous travelogues and evaluated the texts systematically for new and unrecognized plants and their use.

The Travel Diary

Finally, the travel diary gives an insight into Fogel's educational grand tour of 1662–1666.[18] The travelogue, comprising roughly 800 sheets of paper, documents the journey he undertook with the above-mentioned Martin Wevetzer. They travelled through Germany, Austria, Italy, France, Spain and the Netherlands, and visited well-known scholars, who occasionally functioned as their "guide". In his diary, Fogel noted details about the famous gardens, inscriptions, descriptions of landscapes, monasteries and villages, as well as records of the factories, libraries and universities visited.

The observations are written on slips of paper in quarto format, which are provided with dates (month and day) and distances (miles). But most of the slips are in a state of disorder (like his other slips of paper), due to the flooding in Hannover in 1946 when the River Leine broke its banks. Thus, it is quite difficult to group together all slips that relate to specific places or gardens. The predominant language is German; inscriptions are written in Latin.

The slips have a formal structure: the places seen and visited were written one under the other. Later additions are rare. The places, buildings and events that he had seen to the left of his route, Fogel noted on the left-hand side of the paper, and those seen to the right on the right-hand side (Fig. 4). The symbol "z.L." means the left-hand side (zur Linken), "z.R." the right-hand side.

Fogel had a very special way of writing the names of rivers and places (Fig. 4): "loär" means the river "Loire". Fogel underlined large and/or significant towns using a red pencil.

[16] GWLB: Ms XLII 1923 sigma 45 (f. 2r).

[17] Haerlemsche Courant, La Gazette d'Amsterdam (translated from French), Altonaische Relation, Hamburger Relations-Courier and Nordischer Mercurius.

[18] A publication of the diary is currently being prepared by the author.

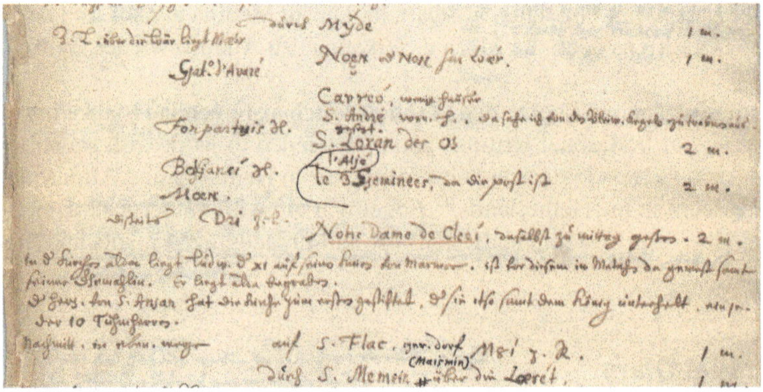

Fig. 4 Villages on the *left*- and the *right-hand*. Scan GWLB: Ms VI 618a, Mappe a (fol. 98r)

Another example is the report about Mantua, dated April 20, 1664. Fogel visited the Ducal Palace and the gardens, directed by the son of Petro de Castro.

There the son of Petro de Castro, to whom I brought a letter from Mr. Karl Offerd, showed us much friendship. On Sunday, 20 April he led us into a garden in the palace, which is very extensive, but not built in a particular architectural style. After this we went to the gallery upstairs and then below to the cabinet of curiosities called la Calleria del Paßino [Passerino], *fon Cerut*, which was arranged at the time of Ferdinand. In the first order a human heart is displayed which is supposed to have turned into stone. In the second there are then two beautiful, large sponges [spondiae] in the fourth a hippopotamus, on it Passarino's body.[19] There are also two beautiful *Lust*gardens at in the palace. In the afternoon we saw the stables and close to them three camels. Later we walked out to the *Te*, so-called because the trees were planted in the front of the palace in the shape of the letter T.[20] Therein two beautiful rooms painted by *Jul.* of Rome.[21] In the room where the room of the Giants a sky storm (*Himmelssturm*) is painted, you can talk quietly in one corner so that someone in the other opposite corner can hear it as at Caprarola[22] and

[19] This is clearly a reference to the room named after Rinaldo Bonacolsi, called "il Passerino" (Passerino Bonacolsi), whose mummified body was displayed in the cabinet of curiosities in the ducal palace and was commented upon by Joseph Furttenbach in 1626. Thanks to Catherine Atkinson for this.

[20] This is the Palazzo Ducale and Palazzo Te with the paintings of the giants; it is remarkable that the name is derived from the shape of the tree planting. Today, one reads thatthe "peculiar name is derived from Teireto, as the area was called". GWLB: Ms IV, 618 a, Mappe b (fol. 36r).

[21] "Jul. of Rome" was the architect and painter Giulio Romano.

[22] This refers to the Villa Farnese in Carprarola. Even today, the acoustic effects in the Hall of Angels are mentioned in guidebooks.

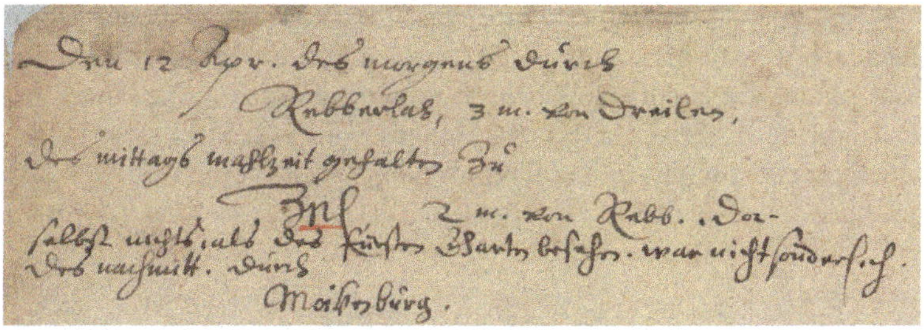

Fig. 5 Prince's garden. Scan GWLB: Ms VI 618a, Mappe a (fol. 30r)

Verona.[23] The room in which the bitter oranges are held in the winter is large and grand. The theatre was not deemed so worthy by our friends, that we should have seen it.[24]

Fogel's interest covered gardens, as well as nature cabinets, acoustic features, inscriptions, strange animals, collections of paintings, etc.

On Celle, he comments (Fig. 5): "2 m(eilen) fon Rebb.(erlah) darselbst nichts, als des Fürsten Garten besehn. War nicht sonderlich." ("Two miles from Rebberlah, have seen nothing there except the Prince's Garden, nothing special").[25] About a place near Marmoutier (near Tours), he writes: "Der Garten ist groß u. Lustig" ("The garden is large and pleasant").[26] The judgements of the 30-year-old scientist probably reflected the perceptions and opinions of an educated traveller: it was only possible to make superficial comments on so many diverse, strange and remarkable experiences in foreign countries.

The scientific and scholarly evaluation of the library, the slips of paper and the travel diary as sources has only just begun. Fogel wrote down his travel experiences relating to gardens and the vegetation of a particular landscape in abbreviated form.

[23] This refers to the Arena of Verona, a well-preserved Roman amphitheatre. Today, it provides for 22,000 opera spectators.

[24] "Daselbst hat der Sohn fon Pet. fon Kastro, an welchen ich schreiben gehabt fonH. Karl Offerd, uns alle freundschaft erzeiget. Sonntag, den 20. Apr. [1664N.S.] hat er uns in einen garten geführet im Pallast, so ein weitleuffiges gebäu ist, aber wenig nach der Bau kunst gebauet. Hernach in die Gallerie oben u. bald unten zu der Natuhrkammer, la Galleria del Paß—ino genannt, so fon Cerut. Zu ferdinands Zeiten angeordet. Darin in der ersten ordn. Ein Menschenherz gezeiget, so zu Stein sol geworden sein. Im 2 den 2 schöne große Swämme (spondiae) im 4 ein Märpferd (Hippopotam.) auf welchem Passarins körper steht. Es sind auch 2 gärten in der Lust am Pallast. Zu nachm. haben wir den Stal gesehen, u. in d nähe 3 kamele. Sind hernach heraußgangen, zum Te, also genant, weil die bäume, wie d buchstab t gesezt, for dem Pallast. Darin 2 schöne Zimmer fon Jul. fon Rohm gemahlet. In dem da der Riesen Himmelssturm abgemahlet, kan einer in einem Ek leise reden, das derjenige, so im Ek gegen über stehet, denselben wolfernimt, wie zu Caprarola und Verona. Das Zimmer, da die Pomeranzen des Winters gehalten werden, ist groß und prächtig. Das Theatrum ist nicht für wert gehalten worden fon unsern freunden, das wirs hetten besehen sollen."

[25] Rebberlahis a part of Eschede, 20 km north of Celle. Eleonore d'Olbreuse and the garden architects Henry Perronet and René Dahuron, who redesigned the French-style garden, only arrived in Celle after 1665.

[26] GWLB: Ms 618a, Mappe a (fol. 109r).

He worked intensively with the help of his library and used books from several disciplines in line with his botanical interests. Additionally, he collected information about plants, especially, being a physician, about narcotic drugs from both traditional and new sources. His new sources were weekly newspapers, printed travelogues, received letters and oral reports. This information was used by Fogel to build his knowledge about plants, gardens, gardening, healing powers, and the colours of flowers, but also in the new systematic thinking of a visionary like Robert Morison.

Fogel mainly consulted printed sources. His own experimental observations did not play a major role in his knowledge of biota. It seems that Fogel was more committed to the conservative historical approach than to modern experimental scientific work. Nevertheless, Fogel was a very innovative scholar. His literary estate, with all his papers, was so important to Gottfried Wilhelm Leibniz that he never returned the manuscripts. On the contrary, he ordered his amanuensis to copy many of them. Furthermore, Leibniz and Johann Georg von Eckhardt exploited Fogel's notes for their own studies.

Bibliography

Sources

Jungius, Joachim. 1662. Doxoscopiae physicae Minores, sive Isagoge physica Doxoscopia. Ex Recensione et Distinctione M.F.H. cujus Annotationes quaedam accedunt. Hamburg.

Moller, Johann. 1744. Cimbria Literata Sive Scriptorum Ducatus Utriusqve Slesvicensis EtHolsatici, Qvibus Et Alii Vicini Qvidam Accensentur, Historia Literaria Tripartita. Stockholm: Kisel.

Parkinson, John. 1656. Paradisi In Sole Paradisus Terrestris. London: R. N. (Printer) and Thrale.

Wilckens, Nicolaus. 1770. Hamburgischer Ehren-Tempel. Hamburg.

Wild, Johannes. 1623. Neue Reysbeschreibung eines gefangenen Christen. Nürnberg: Lochner.

Secondary Literature

Heilmeyer, Marina. 2003. Ein Blumengarten des Barock. Hortus Anckelmannianus. Mit einem Essay von Marina Heilmeyer. München et. al.: Prestel.

Hoffmann, Friedrich Lorenz. 1855. Martin Fogel und seine seit 1678 in der Königl. öffentlichen Bibliothek zu Hannover befindliche Büchersammlung. Nebst Briefen von Leibniz und Andern, die sich auf den Ankauf derselben beziehen. Serapeum 16: 97-110.

Jahn, Ilse (ed.). 2000. Geschichte der Biologie. Theorien, Methoden, Institutionen, Kurzbiographien. Heidelberg et al.: Spektrum, Akad. Verl.

Kangro, Hans. 1969. Martin Fogel aus Hamburg als Gelehrter des 17. Jahrhunderts. Ural-Altaische Jahrbücher 41: 14-31.

Krausch, Heinz-Dieter. 2003. Die Gottorfer Pflanzeninventare von 1655 und 1681. Abhandlungen des Naturwissenschaftlichen Vereines in Hamburg. N. F. 36: 269-282.

Mägdefrau, Karl. 1973. Geschichte der Botanik. Stuttgart: Fischer.

Marten, Maria and Carola Piepenbring-Thomas.2015. Fogels Ordnungen. Aus der Werkstatt des Hamburger Mediziners Martin Fogel (1634-1675). Frankfurt a.M.: Klostermann.

Reißmann, Martin. 1975. Die hamburgische Kaufmannschaft des 17. Jahrhunderts in sozialgeschichtlicher Sicht. Hamburg: Christians.

Roth, Dietrich. 2003. Die Blumenbücher des Hans Simon Holtzbecker und Hamburgs Lustgärten. Keltern-Weiler: Goecke und Evers.

Schrader, Friedrich Nicolaus. 1840. Das Hamburgische Collegium Medicum und der ärztliche Verein in Hamburg. Hamburg: Perthes-Besser & Mauke.

Schubert, Ingrid Anette. 2003. Von Bürgern und Botanikern, Geistlichen und Gärtnern. Die Entwicklung früher Hamburger Lustgartenkultur zwischen Späthumanismus und Frühbarock. Abhandlungen des Naturwissenschaftlichen Vereines in Hamburg. N. F. 36: 15-74.

Veenker, Wolfgang (ed.). 1986. Festvortrag zum Leben und zum Werk Martin Fogels. Memoriae Martini Fogelii Hamburgensis (1634-1675). Beiträge zur Gedenkfeier in Hamburg am 17. April 1984: 25-66.

Part IV

Botanical Knowledge and Horticulture

Watering the Renaissance Garden: Horticultural Theory and Irrigation Practice in Sixteenth-Century Tuscany

Anatole Tchikine

Abstract

Although essential for the life of plants, garden irrigation remains a little known aspect of Italian Renaissance horticulture. By drawing mainly on sixteenth-century Tuscan sources, this chapter attempts to reconstruct both the methods and theoretical underpinnings of this practice. It examines the modes of bringing water into Italian gardens; the specific uses of watering—for example, for fertilizing soil or medicating plants; the tools and techniques of garden irrigation; and various other factors, such as the origin and temperature of water, that were considered relevant by Renaissance garden theorists and practitioners. Special attention is paid to the methods of utilizing running water in the horticultural context. This chapter argues that the conservative nature of garden irrigation in the Renaissance, which in great measure continued the ancient tradition of plant cultivation, was primarily a response to the multiplicity of objectives that watering had to meet. It therefore reflected the time-honoured approach to the stewardship of land and its resources, in which manual work and diligent attention to the individual needs of plants could not easily be replaced by technological innovations. In conclusion, this chapter explains how the principles of garden irrigation influenced the design and appearance of sixteenth-century Italian gardens.

Introduction

Historians of Early Modern landscape architecture tend to see water primarily as part of garden aesthetics. They think of it as a fluid and malleable medium that animated various designed features—fountains, grottos, cascades, water-chains,

A. Tchikine (✉)
Dumbarton Oaks Research Library and Collection, 1703 32nd Street NW, Washington, DC, USA
e-mail: TchikineA@doaks.org

H. Fischer et al. (eds.), *Gardens, Knowledge and the Sciences in the Early Modern Period*, Trends in the History of Science, DOI 10.1007/978-3-319-26342-7_13

and fishponds—bringing them into narrative or thematic sequences and helping them convey specific subject matter and meaning (MacDougall 1978: 3). As a result, the waterworks of Renaissance villas are usually interpreted in terms of these discrete elements rather than as complex systems of interconnected channels, collecting and distribution pools, hydraulic mechanisms, and drainage outlets.[1] The questions of how water was given to plants, what dictated its patterns of circulation within a particular area, and what happened to it after it had left the garden, generally receive little attention in this context.

This approach derives from the old-fashioned view of Italian Renaissance villas, developed in nineteenth-century art history, as countryside or suburban retreats primarily dedicated to recreation, aesthetic pleasures, and intellectual pursuits. Later scholarship has corrected this largely distorted picture (see Beneš 2011). In particular, it emphasized the role of villas as centres of agricultural production surrounded by smaller holdings, usually cultivated by sharecroppers, as well as by hunting and fishing reserves. The satisfactory management of water resources was an essential aspect of the economy and the ecology of these estates. Accordingly, the importance of properly utilizing water in the horticultural context was reflected in the minute attention that Italian Renaissance authors paid to the theory and practice of garden irrigation. Given the growing emphasis on the environmental themes in contemporary landscape history, the lack of an adequate study of the principles and techniques of watering that these theorists advocated remains a conspicuous omission. The aim of this chapter, which is based primarily on the material of sixteenth-century Tuscany, is to provide a brief introduction to this subject as the first step towards filling this scholarly gap.[2]

Creating a Water Supply

The very choice of a garden's location often depended on the availability of water (Alamanni 1804: 138–140). The optimal solution was to divert a nearby river or stream, conducting its water to one's property through closed or open channels. Such conduits were usually lined with brick or stone walls, in which case they could be described as aqueducts. They carried water through narrow masonry channels or terracotta pipes composed of short interconnected conical tubes, the joints of which

[1] "It is regrettable that the gardens are almost never studied in relation to the development of aqueducts, pipelines, and other water features" (Battisti 1972: 4, n. 2). An important recent exception is Lamberini and Tamantini (2013).

[2] Although this study draws primarily on the works of Florentine theorists—Giovan Vettorio Soderini (1526–1597) and Agostino del Riccio (1541–1598), as well as Giovanni Rucellai (1475–1525), Luigi Alamanni (1495–1556), Giovambatista Tedaldi (1495–1575), and Bernardo Davanzati (1529–1606)—its conclusions are not exclusive to Tuscany. Other authors that have been consulted include the Sicilian Antonino Venuto (fl. 1516), the Brescian Agostino Gallo (1499–1570), the Milanese Bartolomeo Taegio (c. 1520–1573), and the Ravennese Marco Bussato (fl. 1592). The agricultural treatise by the early fourteenth-century Bolognese theorist Pietro Crescenzi, which remained an important reference source throughout the sixteenth century, is cited in its 1561 vernacular Venetian edition.

were sealed by oil-slaked lime (Fig. 1). Lead pipes were used, too, despite some authors' warnings against their potential health risks [Crescenzi (1561: 9r); for the contrary opinion, see Soderini (1902: 120)]. Open channels were occasionally preferred to underground ones, since they prevented the build-up of malodorous fumes. Their main disadvantage was that they could easily be polluted—for example, by laundresses who illicitly used them for washing linen—and were also likely to cause flooding.[3]

In the absence of nearby sources of running water, the most common alternative was digging a well. This work had to take place during late summer or early autumn—that is, between July and October, but preferably before September 22, the autumnal equinox—when temporary deposits of rainwater could not be mistaken for the proximity of an aquifer (Crescenzi 1561: 8r; Soderini 1903: 7). The site was determined by looking for specific signs, such as humidity and exhalations or the presence of moisture-loving plants—reeds, willows, poplars, or ivy—as well as mosquitoes and frogs. Another clue was the nature of the terrain—whether it was composed of soil, clay, sand, or rock—which usually revealed how much water one could expect to find in that particular location (Crescenzi 1561: 8r, with reference to Palladius). The next step involved making a trial hole that, as a rule, was no more than 2 m deep. A bronze or lead bowl (which was placed face downwards), a pot of unbaked clay, a piece of thick woollen fabric, or a burning oil lamp were left there overnight to test for the presence of groundwater. If the next morning the inside of the bowl was covered in condensation, the pot fell apart from absorbing moisture, the wool was soaked with water, or the lamp went out without consuming much oil, the digging of the well could begin (Crescenzi 1561: 8r–v; Soderini 1902: 125–126). All of these methods derived from the ancient tradition conveyed by the writings of Vitruvius, Pliny the Elder, and Palladius.

The well's shaft was typically round in plan and sustained by masonry walls, which could occasionally be plastered. Its depth dictated the choice of water-lifting mechanisms. These ranged from simple levers, pulleys, and windlasses to norias with multiple scoops attached to long ropes that were rotated on two wooden spools. Usually set in motion by a horse, they could reach relatively deeply.[4] Water was generally emptied into an adjacent stone trough. This arrangement is shown, for example, in the late sixteenth-century view of the Medici villa at Colle Salvetti by Justus Utens, where a circular well with a pulley stands next to a large rectangular trough in the center of a rustic courtyard between the villa and the service buildings (Fig. 2). The lower oblong basin connected to this receptacle must have been used for watering domestic animals.

[3] Soderini 1903: 15. Cf. ASF, MP, 527, 89r (Benedetto Uguccioni to Cosimo I, 11 March 1567); a copy of this document with an appended drawing is in ASF, CPG, neri, 1465: 73r–v. In the case of the Giardino dei Semplici in Florence, complaints about open channels causing flooding persisted through later, particularly eighteenth-century, documentation. See BBUF, MS 98, 37r; BBUF, MS 97, 47v, 156r, 256r, 408r, 760r–761v. For laundresses using this garden's water, also during the eighteenth century, see BBUF, MS 97, 394r, 409r.

[4] For a more detailed account of the nature and operation of water-lifting devices used in the gardens of sixteenth-century Italy, see Tchikine (2014: 142–147).

Fig. 1 Recently excavated water pipes below the demolished Medici villa, probably late sixteenth century. Pratolino (near Florence), Villa Demidoff. Photo: Anatole Tchikine. Even if these pipes are replacements that date from a later period, they represent the type that was ubiquitous in Tuscany during the Renaissance

Fig. 2 Justus Utens, Villa Colle Salvetti, c. 1599 (detail of the villa with service buildings). Oil on canvas. Florence, Villa La Petraia (formerly Museo Firenze com'era). Photo: by permission of Servizio Musei Comunali Fiorentini

Another option was to create a rainwater cistern that collected runoff from the villa's courtyard and gutters. Such sealed masonry tanks, although reliant on the quantity and frequency of rainfall, could retain surprisingly large amounts of water. To prevent leaks, their bottom had to be coated with an impermeable material, usually glazed terracotta, and regularly checked for fissures and cracks. The main problem with cisterns, however, was that they kept water motionless and still, thereby making it stale and fetid. One common solution was to introduce fish, especially freshwater eels, whose movement was supposed to create some degree of circulation (Crescenzi 1561: 9r). Open cisterns for collecting rainwater could also be built outside in the upper part of the garden (Fig. 3).[5] Another more basic yet very common method was to retain water in ditches, which could hold it for sufficiently long periods of time.[6]

Fig. 3 Garden with citrus trees and a cistern for their irrigation, late sixteenth–early seventeenth centuries. Pen and ink drawing. Florence, Archivio di Stato, Urbino, class I, division A, 4, 330r. Photo: by permission of the Ministero per i Beni e le Attività Culturali e del Turismo. This drawing dates to the time of Francesco Maria II della Rovere (r. 1574–1631), Duke of Urbino, and represents part of one of his properties, presumably the Villa Imperiale at Pesaro. The cistern is a rectangular enclosure in the central part of the garden attached to the back wall

[5] Cf. Soderini (1903: 7): "… facciasi nella più alta parte [del giardino] un gran truogolo che raccoglia l'acque del verno, murato, o nella stessa terra rilevata."

[6] Crescenzi (1561: 115v): "[L'orto] desidera parimente d'esser adacquato, & di haver di sopra un rio, per lo quale si possa per acconci solchi, quando sia bisogno, innaffiare, & se non potesse havere rio, habbia in se o piscina, o fonte, o pozzo in luogo di rio, & se non potrà haver nulla delle predette cose, si faccino delle fosse per lo horto, & molto picciole, nelle quali si ritenga per qualche tempo l'acqua piovana." Cf. Soderini (1903: 8), referring to "[l'acqua] delle fosse che vien da cielo."

The Uses of Watering

Water that came from these various sources was clearly not the same in terms of its quality and properties. It could, for example, be hard or soft and vary in colour, taste, and smell. Renaissance theorists were aware of these differences. An important part of their background knowledge was the ancient hierarchy of the types of water based on origin. In its most common version, it was known from a compilation of medical texts authored by the first-century Roman writer Aulus Cornelius Celsus [Celsus 1935–1938: I, 2.18; cf. the discussion of water in Crescenzi (1561: 225r–v)]. The best, softest and healthiest water, according to this source, fell in the form of rain. Then, in descending order, Celsus listed water from mountain or artesian springs, rivers, wells, melted snow, lakes, and swamps. At the bottom of this scheme, he placed seawater.

This hierarchy was formulated primarily with regard to the potability of water, reflecting Celsus' essentially medical perspective. The situation with plants was somewhat different. On the one hand, they were expected to be given the best water that was available, since this is what humans eventually consumed in the form of garden produce.[7] Eating juicy fruit, vegetables, and greens was supposed to help balance the cardinal humours and conduct vital nutrients to different parts of the body. On the other hand, what was good for humans did not always have the same effect on the plants, while inapt watering could impede their growth, weaken their stalks and root systems, and also impair their taste.

A crucial factor in this regard was the use of water not only for irrigation, but also for fertilizing soil. This was achieved by mixing it with manure that had to be dissolved into thick lumpy paste (*sugo*). Cold running water was not entirely suitable for this purpose, since it did not soak in easily and therefore was not readily absorbed by the roots of plants (Crescenzi 1561: 24r; Soderini 1903: 9; Del Riccio: I, 70r). Water that was optimal for conditioning manure had to be tepid and still, as was that of ditches and swamps. Alternatively, it could reach the desired temperature by being left in an open receptacle, such as a trough, to warm up in the sun for 1 or 2 days.[8] Muddy river water could likewise improve the quality of the soil, since it carried silt, the powers of which as a fertilizer were demonstrated by the classic example of the Nile in Egypt (Leon Battista Alberti 1565: 371; Del Riccio: I, 70r–v).

Furthermore, different kinds of plants required specific types of water for their growth. River water, for example, was the best for carrots, making them bigger,

[7] Cf. Del Riccio: I, 69v: "... perché quanto sia [l'acqua] più dolce, e di miglior' sapore, tanto sian' migliori gli erbaggi, che con essa s'annacqueranno... Et sieno assai migliori gl'erbaggi, alberi, e frutti, con l'acqua che è molta buona da bere che con quella che non ha buon' sapore."

[8] Soderini (1903: 25), referring to "l'acqua tiepida ... preparata così dal sole o dal fuoco, o ... acqua attinta d'uno o due dì addietro."

tenderer, and easy to cook (Soderini 1903: 104). It was also excellent for lettuce and other greens, helping them grow fast and develop stronger root systems (Soderini 1903: 179 for lettuce). River water was especially recommended for seedlings, as it was believed to make them germinate better than did water of springs and wells (Soderini 1903: 7 and 9). Olive trees, by contrast, preferred rainwater to that of rivers and springs (Crescenzi 1561: 98v). Palms needed water from wells to make sure that it was sufficiently cold; rainwater was considered the least favourable to their nature (Soderini 1904: 560). Seawater was supposed to be particularly good for radicchios, since it ostensibly took away some of their bitterness; it was also recommended for radish, rue, Swiss chard, and summer savory (*cunila*) (Soderini 1903: 8, 294 and 298). Cabbages liked seawater too, especially if they grew near the shore (Soderini 1903: 117). Water obtained by melting snow was considered bad for plants in general, but was especially detrimental to citron trees.[9]

Watering was also used to medicate plants, whether as a means of making them more resilient, improving their appearance and taste, killing parasites, or curing specific diseases. Citrons and oranges, for example, liked water to be sweetened with honey, a combination that was known as "acquamelata" (Soderini 1904: 126 and 352; Venuto 1550: 6v–7r). Lettuce, which was reputed to be a strong aphrodisiac, was sometimes watered with milk to become especially delicate, tender, and white (Soderini 1903: 182–183). When planting basil, its seeds were supposed to be sprinkled with boiled water (Soderini 1903: 59). Roses, when grown in vases, were sometimes given coloured water to assume a particular hue; dried mulberries, if they were ripe, served as the red pigment, while the unripe ones turned rosebuds yellow. These different colourings could also be alternated over a period time to produce a variegated effect (Soderini 1903: 335).

Plants could also be medicated by mixing water with a wide range of organic substances: wine, vinegar, wine spirit and lees, human or animal urine, different types of manure (mainly bovine, ovine, and porcine), ash, lime, and the oily sludge (*morchia*) that remained after straining olive oil. The use of these cures formed a whole chapter of Italian Renaissance viticulture. Grapevines were especially delicate trees that were susceptible to various diseases. If sterile, they needed to be watered with strong vinegar mixed with ash. During the heat, if growing in the open, they had to be given human, ovine, or bovine urine, making sure that it reached their roots. In this way, they could be expected to become more robust and produce tastier and more attractive fruit (Soderini 1902: 407, 1904: 225).

Other plants had their preferences too. Myrtles and quinces, for example, liked human or ovine urine (Soderini 1903: 217, 1904: 124–125, 127 and 376). Mulberries, savin juniper, peaches, and plums benefitted from wine lees put over

[9] Cf. Gallo (1572: 149) emphasizing that "quell'acqua è più contraria d'ogni altra a queste piante [di cedri]."

their roots (Soderini 1904: 126, 127, 228 and 594). Chestnuts and plane-trees preferred pure wine (Soderini 1904: 285 and 570). Water with added saltpeter was thought to be good for cabbages, making them leafier and bigger (Soderini 1903: 117–118). Palms liked water with added salt (although, with this single exception, it was generally considered harmful to trees) (Soderini 1904: 124–126). Olive oil sludge was recommended for killing various parasites; mixed with water, it was supposed to be poured over the roots of olive, plum, pomegranate, pear, and apple trees, especially during autumn and spring (Soderini 1904: 127, 224–225, 285 and 594). Porcine droppings soaked in wine spirit, an infusion of lupins, old human urine that was usually matured for at least 6 months, bovine manure diluted in water, and bovine urine provided common remedies against worms (Soderini 1904: 226 and 500; Venuto 1550: 18r, with reference to curing pomegranate trees).

Some of these recommendations clearly drew on classical precedents, such as the anecdotes about Emperor Nero watering his favourite holm-oak with excellent wine or his stepfather Passienus Crispus doing the same to a mulberry tree (Soderini 1904: 18).[10] Yet others must have derived from contemporary horticulture, since the authors of late antiquity, although already familiar with peaches, melons, and possibly lemons, never had to deal with such plants as oranges, spinach, aubergines, pumpkins, and prickly pears.[11] The pertinence of these cures therefore was not merely a bookish tribute to the ancient tradition revived through humanist scholarship. Rather, it seemed to point to an unbroken horticultural practice, in which old Roman methods were continually adapted to the changing realities of plant cultivation. In the case of water, this age-old experience underscored the multi-purpose nature of garden irrigation suggesting that no single solution could provide answers to the wide range of challenges that watering had to meet.

Watering Rules and Techniques

There were, however, some broadly interpreted general guidelines. Plants, for example, were not supposed to be watered when in bloom (Soderini 1903: 10). During the summer, they had to be given water before sunrise or after sunset, preferably at midnight; watering them in the afternoon could apparently scald their roots with excessive heat retained by the soil (Soderini 1903: 9; Soderini 1904: 227 and 229; Venuto 1550: 5v). The exception was basil, which needed to be watered at midday (Soderini 1903: 9 and 59). The opinions as to which water was the best for garden irrigation remained split. Some authors advocated the use of cold running water that supposedly rejuvenated plants; others maintained that it had to

[10] The anecdote concerning Passienus was reported by Pliny the Elder (1938–1963: IV, 16.91.242). Even though sixteenth-century authors, including Soderini, usually interpreted instances of pouring wine over the roots of trees in the context of garden irrigation, it is more likely that such acts had ritualistic significance.

[11] For the fruits and vegetables known in the ancient world, see Dalby (2003).

be tepid and still, as that of ditches and troughs.[12] Both schools, however, agreed that its temperature had to vary depending on the season and the nature of specific plants. During winter, for example, water was supposed to be heated to make saplings grow faster; this method, however, was not suitable for mature trees except citrons (Soderini 1903: 10, 1904: 125–126, 340, 348 and 352). By contrast, artichokes liked water to be warm at all times, as did taro (*colocasia*), a tropical plant of South Asian origin that started to appear in Italian gardens during the sixteenth century.[13] Scorching hot water was used to exterminate weeds (Crescenzi 1561: 159v).

The frequency of irrigation varied too, since either extreme could produce equally negative effects. Most plants needed water only during the summer, usually two or three times a week. This rule applied to the majority of orchard trees, with the exception of citrons that had to be watered throughout the year (Soderini 1904: 33, 125–126 and 330). By contrast, cypresses, pines, and firs that grew in Renaissance bosquets needed water only as seedlings (Soderini 1904: 126, 284–285 and 369). Frequent watering made oranges juicier, enabled myrrh and pistachio to produce more abundant or bigger fruit (which was also the case with quinces), and improved the flavour of pears and sorb-apples (Crescenzi 1561: 105r; Soderini 1904: 126–127 and 285; Venuto 1550: 13r). It also allowed almonds to develop softer, more easily breakable shells (Soderini 1904: 465). At the same time, too much water was considered harmful to mulberries and could turn pomegranates hard and sour (Soderini 1904: 126 and 285). Overwatered figs grew larger in size, but tended to rot more quickly (Soderini 1904: 285). No consensus was reached with regard to apricots; some authors believed that they needed watering, while others maintained that this made them taste watery and become more likely to be infested with worms (Soderini 1904: 285; for the opposite opinion, see Venuto 1550: 8v). The general assumption was that, when given less water, trees produced smaller fruit, which, however, was tastier and lasted longer (Bussato 1599: 2r); conversely, the fruit of well-watered plants was usually copious, but of inferior quality (Gallo 1572: 129).

Not all plants, however, could be irrigated in the same way. Palms, for example, needed to be watered gradually, in small drops, rather than by the bucketful (Soderini 1904: 126). Artichokes were supposed to have water poured far from the stem so that it did not cover their roots (Soderini 1903: 90; Davanzati 1853: II, 523; Tedaldi 1776: 81). Certain plants, rather than being watered at all, apparently preferred to grow in the proximity of water retained in ditches and troughs. This was the case, for example, with watermelons, melons, pumpkins, and cucumbers,

[12] See Soderini (1903: 9, 1904: 126), with references to Theophrastus and Pliny the Elder, who advocated the use of cold water. The opposite opinion was held by Crescenzi. As a compromise solution, Del Riccio recommended that water should be tepid—"mediocremente fredda" (Del Riccio: I, 69v).

[13] Davanzati (1853: 524) for artichokes; Soderini (1903: 94) for taro. Taegio, however, seemed to believe that artichokes needed warm water only in winter (Taegio 1559: 158; note inconsistent page numbering in this edition).

which became tastier when they were given little water, but benefitted from growing in the proximity of its sources (Soderini 1903: 281–282, 401 and 404; Alamanni 1804: 194 for cucumbers). Alternatively, one could insert bull or ram horns among planting beds to collect and hold small amounts of rainwater (Soderini 1904: 127–128). This method was particularly advantageous for growing capers, which suffered both from too much and too little watering (Soderini 1903: 100 and 101). A related solution involved the use of large porous stones buried under the roots of trees, which were supposed to absorb and retain moisture (Soderini 1904: 126; Venuto 1550: 28v with reference to the cultivation of plum trees).

One crucial factor that dictated the choice of watering techniques concerned a specific phase in the planting or cultivation process. Seeds, for example, had to be watered with particular care, making sure that the thin layer of soil that covered them remained undisturbed; otherwise, they became easy prey for the birds (Bussato 1599: 5r). In this case, gardeners employed a small broom or a sponge; having dipped it into a bucket or a trough filled with water, they then vigorously shook it over the newly planted area (Del Riccio: I, 71v for the use of brooms; Soderini 1904: 197 mentioning sponges). Brooms were likewise used for sprinkling cabbages with water that had added saltpeter; this method was supposed to help in killing caterpillars and moths (Soderini 1903: 117–18). Once the seeds had sprouted, they could be watered from bronze or terracotta watering cans with a perforated knob at the end of the spout, which produced a gentle shower of minute drops (Del Riccio: I, 71v–72r) (Fig. 4).[14] A cheaper alternative was to use dried and hollowed out pumpkins pierced a few times with an awl or a thick needle. They could either be carried around the garden like ordinary watering cans or suspended over potted plants, especially citrus trees, with water trickling through a long straw (Rucellai 1804: 263–264; Soderini 1904: 334–335 and 340). Another watering implement that was particularly favoured in Tuscany was the so-called "padella da ortolani" ("gardeners' pan"), which must have looked rather like a conventional frying pan with a long wooden handle (Del Riccio: I, 72v; Soderini 1903: 9–10 and 15). Its use required some experience and skill. Raised with a brisk hand motion as if flipping a pancake, water was shattered against the pan's rim, falling in a myriad of small drops.

Whether sprinkled with a broom, poured from the nozzle of a watering can, dripped from the opening in a hollow pumpkin, or splashed with a "gardeners' pan", water came down in a semblance of rain— the analogy made explicit in at least one contemporary text (Rucellai 1804: 263).[15] This similarity was not coincidental. Rainfall was considered the best method of watering plants, which allowed water

[14] Cf. the description of a typical watering can by Raffaello Borghini (1584): "... un grande inaffiatojo di terra, di quelli, che sono pieni di piccioli pertugi, con cui si inaffiano i giardini" (Borghini 1730: 31).

[15] The effect of rainfall was also frequently imitated in the design of Italian Renaissance fountains (Tchikine 2010: 57, 61 and 64).

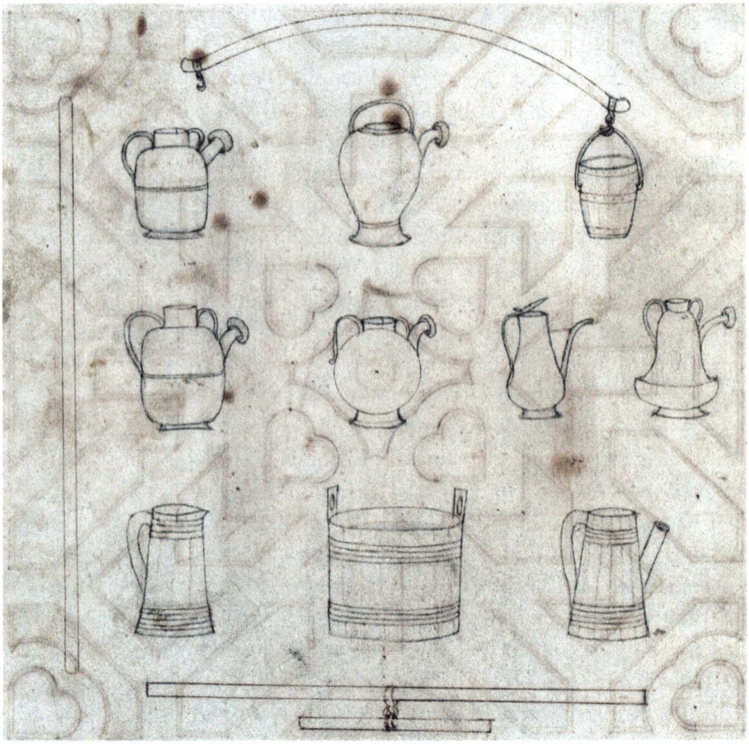

Fig. 4 Watering cans (*clepsidre*) and other watering devices, c. 1588. From Giuseppe Casabona (att.), "Libro di compartimenti di giardini". Pen and ink drawing. Pisa, Biblioteca Universitaria, MS 464, 14v. Photo: by permission of the Biblioteca Universitaria di Pisa

gradually to penetrate the soil without washing it away. It also supposedly killed insect larvae (Soderini 1903: 8 with reference to Pliny the Elder; Del Riccio: I, 70r). Nature, in other words, clearly knew best how to take care of itself, making the art of Renaissance gardening a faithful yet creative student of its ingenious solutions.

Utilizing Running Water

Unlike gentle rainfall, running water of rivers or streams tended to erode the soil, thereby exposing the roots of plants and leaving the land barren (Crescenzi: 24r; Soderini 1903: 9; Del Riccio: I, 70r). It was largely the need to prevent this erosion that made garden irrigation such a slow, painstaking process. As a result, watering, for the most part, had to be done by hand, with the aid of only some very basic devices. The availability of cheap manual labour in the Renaissance was clearly fundamental to enabling and sustaining this conservative practice. Some contemporary authors, in fact, even saw the creation of gardens as a means of providing employment for the poor.[16]

This situation had two main consequences. Firstly, it made the development—or rather mechanization—of irrigation technology if not entirely superfluous, then at least purely discretionary. No mechanical devices could follow the complicated watering routine or adequately substitute for human experience, dedication, and skill. The slight efficiency gains that they offered were generally offset by their elevated maintenance, repair, and replacement costs. For this reason, in the garden context, even such advanced water-lifting mechanisms as pneumatic displacement pumps could not always successfully compete with primitive pulleys (Tchikine 2014: 147).

Secondly, this approach meant that, unlike their Islamic counterparts typified by the presence of irrigation brooks, Italian Renaissance gardens did not depend on the use of running water. Its introduction, in fact, often generated additional problems, such as humidity, flooding, and the need to manage the runoff. This paradoxical situation helps to explain why many Renaissance gardens, including a number of Medici properties—such as Trebbio, La Magia, and the already-mentioned Colle Salvetti—were not provided with aqueducts. Unlike the better known grand ducal gardens like Pratolino, these villas were centres of productive estates, where the exigencies of princely display did not have to take precedence over the basic demands of horticulture.

Yet, whenever it was available, running water was always put to opportune use. Large agricultural properties were normally irrigated by means of furrows or ditches, which allowed water to be gradually absorbed by the soil. This was also the traditional practice of fruit and vegetable growers and vendors (*ortolani*) operating in the area of Prato north-west of Florence, who diverted water from the river Bisenzio for this purpose (Del Riccio: I, 70r, 73r–v). In sixteenth-century Lombardy, this method was practiced on the river Po and its tributaries (Leandro Alberti: 361r). Many kitchen gardens were irrigated in the same way, as, for example, the late sixteenth-century view of the Medici villa L'Ambrogiana by Justus Utens demonstrates (Fig. 5).[17] Water ran there across geometric patterns of planting beds in a series of narrow brooks that widened into miniature circular pools; it then collected in small rectangular fountains at the head of the four main compartments.[18] The use of such brooks, however, was more common in southern

[16] Del Riccio: I, 27r: "... et poi fate gran carità a dare aiuto a pov[e]ri huomini in dar' che far loro poi che si vanno sostendando et vivendo allegramente con le loro famiglie."

[17] For the use of irrigation channels in sixteenth-century kitchen gardens, see Alamanni (1804: 171), Soderini (1903: 10–11 and 14–15); for the water supply of the Villa L'Ambrogiana, see Del Riccio: I, 43r.

[18] Channels cut through Renaissance gardens were mentioned by many Italian authors, including Boccaccio; it is not clear, however, whether they were used for irrigation or merely for conducting water from one part of the layout to another (Boccaccio 1573: 139–40).

Fig. 5 Justus Utens, Villa L'Ambrogiana, c. 1599 (detail of the kitchen garden showing irrigation channels). Oil on canvas. Florence, Villa La Petraia (formerly Museo Firenze com'era). Photo: by permission of Servizio Musei Comunali Fiorentini

Italy, where they might have survived as vestiges of Islamic influence. Sixteenth-century Bolognese traveller Leandro Alberti, for example, noted them in the gardens of Campania and particularly Calabria, where, with a gentle and pleasant murmur, they irrigated groves and espaliers of citrus trees.[19]

Ditches and brooks needed to be replenished only intermittently, usually overnight and no more than a few times a week. This practice meant that the flow of water had to be regulated. This was done by collecting it in closed or open reservoirs, most commonly fishponds, from where it was channelled to different parts of the garden. Within a terraced layout, water moved by gravity, descending from the higher to the lower levels. For this reason, fishponds were generally built above the areas that they were supposed to irrigate. During the sixteenth century, they were still largely utilitarian structures usually built on the periphery of pleasure gardens—as, for example, was the case in the Villa Lante at Bagnaia—or concealed in groves or bosquets (as in the Medici villa at Castello). They also typically contained fish, which, apart from attracting visitors and supplying the owners' kitchens,[20] were seen as a means of keeping water in circulation (Del Riccio: I, 69r–v); this idea must have originated in antiquity, since it was mentioned by Palladius. A more advanced method of creating motion involved passing water through a system of interconnected fishponds. This arrangement was exemplified, for example, by the two cascades in the Medici villa at Pratolino, where water descended a sloping hill by flowing through multiple reservoirs of different size and configuration (Del Riccio: I, 32v).

The most creative application of running water, however, concerned its use for irrigating espaliers of citrus trees, which were among the most distinctive features of Italian Renaissance gardens. The traditional method for watering these plants involved the use of medium-sized pots (*mezzine*) of baked unglazed clay; buried near the roots—usually two or three per tree—they had to be replenished manually overnight so that they would gradually empty themselves through a hole in the bottom (Bussato 1599: 13r; Del Riccio: I, 78r associated this method with Rome). A more elegant solution involved planting espaliers on low earthen walls above which lay a stone border with a thin channel carved along its inner edge (Fig. 6a). At regular intervals, this duct opened into a series of shallow circular basins, the rims of which were slightly slanted in the opposite direction (Fig. 6b). When water ran

[19] Alberti 1550: 174r, 189v, 199v (referring to "chiar' acque [che trascorrono] per i ruscelletti imbriacando le radici de i Cedroni, Aranci, & altri simili alberi") and 200r. The connection between the use of irrigation brooks and the perseverance of Islamic traditions in southern Italy was pointed out to me by Clemens Wimmer.

[20] For the pleasure derived from watching fish in garden fishponds, see Leandro Alberti: 301r.

Fig. 6 (**a**) Espalier of orange trees with an irrigation channel near Buontalenti's Grotto. Florence, Boboli Gardens. Photo: Anatole Tchikine. Although presumably carved in the eighteenth century, this irrigation channel reproduced the type that had been commonly used in Florence during the Renaissance. (**b**) Irrigation channel (detail of a basin). Florence, Boboli Gardens. Photo: Anatole Tchikine

through this channel—which was supposed to happen three times a week—it gently overflowed the adjoining receptacles, infiltrating the soil (Del Riccio: I, 74r).[21] This elegant arrangement, which can still be seen today in two different locations in the

[21] Cf. Soderini (1904: 262): "... tavoloni di pietra sì grossi che vi si facci dentro un canale in mezzo che vi si corra, e a ogni tante braccia in un vaso a nicchi a uso di rinfrescatoio, al medesimo piano della tavola, che vi sia fatto un po' cupo, risalti e v'entri dentro riempiendolo, e scorrendo poi in un raccetto che cali et abbi più capi da scolare l'acqua in un luogo, di donde la si cavi poi a adacquare, senza spandersi o perdersi; oltre alla bella vista."

Boboli Gardens,[22] must have been relatively common in sixteenth-century Tuscany, since it was recorded in a number of contemporary documents. Such channels, for example, also existed in the Giardino dei Semplici, north of the Piazza San Marco, and the garden of Antonio Salviati in Borgo Pinti.[23]

Irrigation and Design

The sheer multiplicity of tasks that Renaissance garden irrigation was supposed to meet inevitably turned it into a feat of micromanagement. This approach was clearly a reflection of an economic system that favoured long-term sustainability over momentary financial or efficiency gains. Besides, it relied on the availability of skilled yet inexpensive practitioners, whose specialist expertise gave the profession of gardener the artisan status that it enjoyed throughout the Renaissance.[24] Finally, the theoretical principles that underlay this practice were rooted in the humanist outlook of the time, in which the revived interest in Galenic medicine and Vitruvian hydraulics went hand in hand with the survival and enduring influence of the ancient methods of plant cultivation.

Yet, this approach to irrigation also had an undeniable impact on Italian Renaissance garden design. One of the crucial implications of its principles was that trees and shrubs could not be watered in the same way as herbs, vegetables, and greens. For this reason, areas for growing these different kinds of plants had to be functionally and aesthetically separated. In practical terms, this situation resulted in a fundamental division of the overall layout of Italian Renaissance gardens into orchards and bosquets on the one hand and kitchen, medicinal, and flower gardens on the other.[25] The discrete nature of these areas is fully evident, for example, in the

[22] The surviving irrigation channels near Buontalenti's Grotto, shown in Fig. 6, and in the former "Giardino d'agrumi" at the top of the Viottolone must be of eighteenth-century origin (see Galletti 1995: 214; Figs. 5 and 6). In the sixteenth century, the espalier of orange trees planted alongside Vasari's Corridor to the left of the grotto had a series of niches behind it, which were somehow linked into its water supply. An interesting document (April 13, 1586) indicates that this water was tapped by the owner of a neighbouring property, who had drilled through one of these niches; the ensuing investigation, however, confirmed that he had done that with the permission of Grand Duchess Bianca Cappello (ASF, CPG, neri, 1469: 4r–6r, 42r). A similar irrigation method was presumably used in the Medici garden of Castello, as described by Giorgio Vasari: "... dove sono fra nicchia e nicchia fonti, e fra le fonti spalliere di melaranci e melagrani" (Vasari 1878–1885: VI, 78).

[23] ASF, CPG, neri, 1467: 342r (reference to "pietre concie" for "panche per le spalliere" for the Giardino dei Semplici, 1587–1589); ASNSP, AS, II, 136, 17r, 78r, 79r (payments for "panche e pile di pietra" for Salviati's garden, 1573–1575).

[24] The métier of gardener was professionalized in the Renaissance; see Garzoni (1588: 894–900), especially 899.

[25] Cf. Soderini (1903: 10): "... dove sono arbori fruttiferi, sotto non vi hanno a essere erbaggi; e dove sono erbaggi non v'hanno a essere sopra i frutti; e così l'erbe da insalata e l'erbe da cuocere, come i frutici et erbe medicinali, le quali deono essere di per se spartite dalle altre, senz'uggia o occupazione d'altro." One should note, however, that this approach is at odds with the current vernacular practice in Tuscany, where vegetables are often planted in between rows of fruit trees.

late sixteenth-century bird's-eye view of the Boboli Gardens by Justus Utens, where the large parterre to the left of the Pitti Palace, surrounded by walls with masonry niches, is sharply separated from the nearby orchard by the geometrical blocks of the bird-trapping grounds (*ragnaie*) (Fig. 7).

Moreover, the laborious nature of garden irrigation was, to a large extent, responsible for the distinctive character of sixteenth-century Italian gardens. Although impressed with their statuary, waterworks, and pergolas, contemporary visitors were especially captivated by the abundance of flowers and fruit trees, noting, in particular, the citrus plants.[26] Fragrant and colourful, these gardens had little to do with the somber geometry of gravel paths and tall evergreen hedges imposed upon them during the subsequent centuries. What lay behind their sensory appeal, however, were precise horticultural methods, meticulous routine, and special care given to the needs of individual plants. The rejection of this philosophy in favour of a standardized approach was one of the reasons why both the appear-

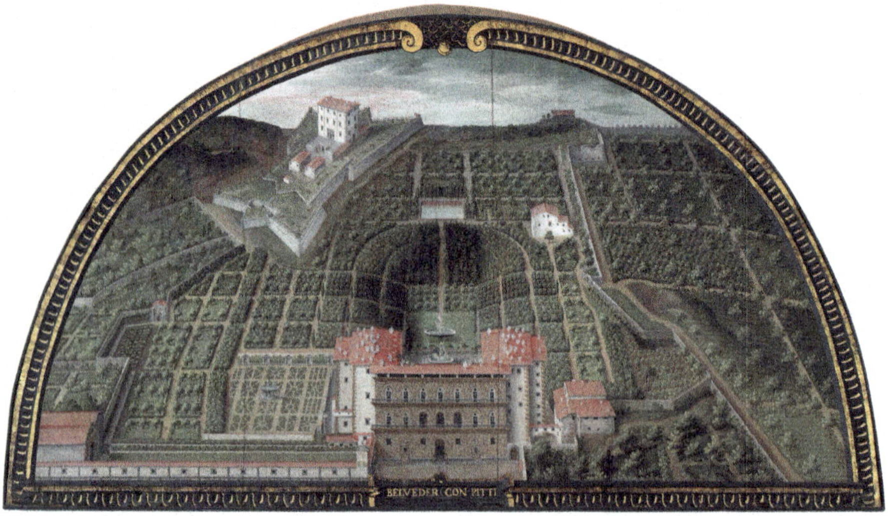

Fig. 7 Justus Utens, The Pitti Palace and the Boboli Gardens, c. 1599. Oil on canvas. Florence, Villa La Petraia (formerly Museo Firenze com'era). Photo: by permission of Servizio Musei Comunali Fiorentini. The kitchen garden defined by geometrical patterns of flower and vegetable beds, with Stoldo Lorenzi's Fountain of Neptune in the middle, is to the left of the grand ducal palace; Buontalenti's Grotto is in the bottom left corner, with Vasari's Corridor running in front of it

[26] The fundamental role of citrus trees in Italian gardens was noted by many authors from the sixteenth through the eighteenth centuries (Leandro Alberti, *passim*; Lastri 1821: 106). For the role of flowers, see Masson (1972).

ance and the experience of so many of these gardens are now irredeemably altered.[27]

Acknowledgements My main thanks are owed to Joachim Wolschke-Bulmahn, Volker R. Remmert, and Hubertus Fischer, who invited me to participate in the "Gardening and knowledge" symposium in Hanover in September 2012 and encouraged me to contribute this chapter to the resulting volume. I would also like to express my gratitude to John Beardsley, Mirka Beneš, John Dixon Hunt, Clemens Wimmer, and especially Raffaella Fabiani Giannetto, for their suggestions and invaluable critical feedback.

Bibliography

Manuscript Sources

Del Riccio, Agostino. Agricoltura sperimentale. 3 vols. Florence, Biblioteca Nazionale Centrale (BNCF), MS Targioni Tozzetti, 56 I–III.

Florence, Archivio di Stato, Capitani di Parte Guelfa, numeri neri (ASF, CPG, neri), 1465.

Florence, Archivio di Stato, Capitani di Parte Guelfa, numeri neri (ASF, CPG, neri), 1467.

Florence, Archivio di Stato, Capitani di Parte Guelfa, numeri neri (ASF, CPG, neri), 1469.

Florence, Archivio di Stato, Mediceo del Principato (ASF, MP), 527.

Florence, Biblioteca Botanica dell'Università degli Studi (BBUF), MS 97 ("Documenti, dissertazioni, lettere, bilanci, etc. etc. della Società Botanica fiorentina dall'anno 1724 al 1774").

Florence, Biblioteca Botanica dell'Università degli Studi (BBUF), MS 98 ("Diari della Società Botanica fiorentina dal 1718 al 1774").

Pisa, Archivio della Scuola Nazionale Superiore, Archivio Salviati (ASNSP, AS), series II, 136.

Printed Sources

Alamanni, Luigi. 1804. La coltivazione. In La coltivazione di Luigi Alamanni e Le api di Giovanni Rucellai. Milan: Società Tipografica de' Classici Italiani. Ed. Giuseppe Bianchini da Prato, 1–231.

Alberti, Leandro. 1550. Descrittione di tutta Italia... Bologna: Giaccarelli.

Alberti, Leon Battista. 1565. Dell'architettura... Trans. Cosimo Bartoli. Florence: Franceschi.

Boccaccio, Giovanni. 1573. Il Decameron. Florence: Giunti.

Borghini, Raffaello. 1730. Il Riposo. Florence: Nestenus e Moücke.

Bussato, Marco. 1599. Giardino di agricoltura. Venice: Combi.

Celsus, Aulus Cornelius. 1935–1938. De medicina. 3 vols. Trans. W. G. Spencer. London: Heinemann; Cambridge, Mass.: Harvard University Press (Loeb classical library).

Crescenzi, Pietro. 1561. Pietro Crescentio bolognese tradotto nuovamente... Trans. Francesco Sansovino. Venice: Sansovino.

Davanzati, Bernardo. 1853. Toscana coltivazione. In Le opere, ed. Enrico Bindi, vol. 2, 487–532. Florence: Le Monnier.

Gallo, Agostino. 1572. Le vinti giornate dell'agricoltura... Venice: Borgomineri.

[27] Cf. Claudia Lazzaro's important observation on adapting the maintenance of twentieth-century Italian gardens to the use of electric tools (Lazzaro 2005: 159).

Garzoni, Thomaso. 1588. La piazza universale di tutte le professioni del mondo... Venice: Somasco.
Lastri, Marco. 1821. Orti Oricellarj, ora degli Stiozzi, e giuochi di magia. In Osservatore Fiorentino 3: 92–108.
Pliny the Elder. 1938-1963. Natural history.10 vols. Trans. H. Rackham, W. H. S. Jones, and D. E. Eichholz. Cambridge, Mass.: Harvard University Press (Loeb classical library).
Rucellai, Giovanni. 1804. Le api. In La coltivazione di Luigi Alamanni e Le api di Giovanni Rucellai. Milan: Società Tipografica de' Classici Italiani. Ed. Roberto Titi, 239–301.
Soderini, Giovan Vettorio. 1902. I due trattati dell'agricoltura e Della coltivazione delle viti. In Le opere, ed. Alberto Bacchi Della Lega, vol. 1. Bologna: Romagnoli Dall'Acqua.
Soderini, Giovan Vettorio. 1903. Il trattato della cultura degli orti e giardini. In Le opere, ed. Alberto Bacchi Della Lega, vol. 2. Bologna: Romagnoli Dall'Acqua.
Soderini, Giovan Vettorio. 1904. Il trattato degli arbori. In Le opere, ed. Alberto Bacchi Della Lega, vol. 3. Bologna: Romagnoli Dall'Acqua.
Taegio, Bartolomeo. 1559. La villa. Milan: Moscheni.
Tedaldi, Giovambatista. 1776. Discorso dell'agricoltura. Ed. Marco Lastri. Florence: Allegrini.
Vasari, Giorgio. 1878-1885. Le opere. Ed. Gaetano Milanesi. 9 vols. Florence: Sansoni.
Venuto, Antonino. 1550. D'agricoltura... Venice: Rampazetto.

Secondary Literature

Battisti, Eugenio. 1972. "Natura artificiosa" to "natura artificialis." In The Italian garden, ed. David Coffin, 1–36. Washington, DC: Dumbarton Oaks.
Beneš, Mirka. 2011. Methodological changes in the study of Italian gardens from the 1970s to the 1990s: a personal itinerary. In Clio in the Italian garden: twenty-first-century studies in historical methods and theoretical perspectives, eds. Mirka Beneš and Michael G. Lee, 17–54. Washington, DC: Dumbarton Oaks.
Dalby, Andrew. 2003. Food in the ancient world from A to Z. London, New York: Routledge.
Galletti, Giorgio. 1995. Agrumi in casa Medici. In Il giardino delle Esperidi: gli agrumi nella storia, nella letteratura e nell'arte, eds. Alessandro Tagliolini and Margherita Azzi Visentini, 197–215. Florence: Edifir.
Lamberini, Daniela, and Maura Tamantini. 2013. Le acque del Giardino di Boboli. Florence: Sillabe.
Lazzaro, Claudia. 2005. Politicizing a national garden tradition: the Italianness of the Italian garden. In Donatello among the Blackshirts: history and modernity in the visual culture of Fascist Italy, ed. Claudia Lazzaro and Roger J. Crum. Ithaca, NY: Cornell University Press, 157–69.
MacDougall, Elizabeth B. 1978. Introduction. In Fons sapientiae: Renaissance garden Fountains, ed. Elizabeth B. MacDougall, 3–14. Washington, DC: Dumbarton Oaks.
Masson, Georgina. 1972. Italian flower collectors' gardens in seventeenth century Italy. In The Italian garden, ed. David Coffin, 63–80. Washington, DC: Dumbarton Oaks.
Tchikine, Anatole. 2010. "Giochi d'acqua": water effects in Renaissance and Baroque Italy. Studies in the history of gardens and designed landscapes 30: 57–76.
Tchikine, Anatole. 2014. "L'anima del giardino": water, gardens, and hydraulics in sixteenth-century Florence and Naples. In Technology and the garden, eds. Michael G. Lee and Kenneth I. Helphand, 129–153. Washington. DC: Dumbarton Oaks.

Gardening Nature, Gardening Knowledge: The Parallel Activities of Stabilizing Knowledge and Gardens in the Early Modern Period

Alette Fleischer

> *The links between different places in time and space are completely modified by this fantastic acceleration of immutable mobiles [books] which circulate everywhere and in all directions in Europe.* (Latour 1986: 11)
>
> *"[..] a garden extends not only in space but in time. The garden demonstrates not only power to control a part of the world but a peculiar sort of confidence because it indicates an expected continuation of that power in the future."* (Miller 1993: 56)

Abstract

This chapter discusses how nature and knowledge were domesticated by the parallel activities of constructing early modern gardens and publishing botanical books. In both activities, students of nature collected botanical knowledge and plants to make them solid, mobile, reproducible, and combinable, in order to understand nature's workings. Naturalists made exotic flora transportable by labeling and wrapping seeds and bulbs, plants and sapling so that they might safely arrive in the Low Countries. In Dutch gardens, the specimens were planted and domesticated, so unfamiliar flora could be admired and examined. Alternatively, plants were made immutable by being turned into pressed and dried specimens in herbaria or by being depicted and described in publications. As images in publications, exotic plants could circulate and be examined by botanists and amateurs elsewhere. Knowledge and plants, both made solid and mobile in books, helped in the formation of agreement about nomenclature and

I thank the participants of the conference for their remarks, and in particular, Professors Lissa Roberts and Glyn Parry for their generous help in the final stages of this paper.

A. Fleischer (✉)
Lodewijk Tripstraat 4L, 1052 ES Amsterdam, Netherlands
e-mail: alette.fleischer@gmail.com

H. Fischer et al. (eds.), *Gardens, Knowledge and the Sciences in the Early Modern Period*, Trends in the History of Science, DOI 10.1007/978-3-319-26342-7_14

the early modern idea of nature's workings. The stabilization and domestication of nature enabled producers and consumers to take power over local and exotic plants and make nature combinable and controllable.

Hieronymus van Beverningh (1614–1690) owned a country retreat called Oud-Teylingen, just outside Leiden, where he resided from 1663 onwards. He was a diplomat and also curator of Leiden University's Hortus Botanicus. Beverningh's estate consisted of "gardens, orchards, planted trees, canals and walls [. . .] measuring five *morgen* (ten acres)."[1] The statesman was a *liefhebber*, an amateur-inquirer into natural history and a cultivator of exotic plants. Fellow enthusiasts, gardeners and botanists regularly visited his garden and inspected Beverningh's collection of plants and his botanical books and manuscripts (Veendorp and Baas Becking 1938: 82) (Fig. 1). After his demise, part of his collection went to the Leiden University Library and Herbarium (today, the collection of Naturalis).

A large number of plants and dried specimens in Beverningh's collection came from the botanist and physician Paulus Hermann. In 1670, Beverningh urged Hermann to apply for the position of medical officer to the Dutch East India Company (also known by its Dutch acronym VOC) at Ceylon. Immediately after his return from the East in 1680, Hermann became professor of botany at Leiden University. A brief eastbound stopover at the Cape of Good Hope in South Africa allowed Hermann to collect Cape flora. He dried and pressed the plants and kept these exsiccates in his herbarium. Each specimen was accompanied by a descriptive name, sometimes its original site, references to other publications and/or indigenous names and locally gathered (oral) sources regarding its use. In Ceylon, the physician regularly collected local plants, bulbs, seeds, drawings, letters and herbaria which he shipped to his contacts in the Low Countries. Besides Beverningh, Hermann supplied various private collectors with his materials, as well as the botanical gardens of Leiden and Amsterdam.[2]

Since plants and (local) botanical knowledge are difficult to transport, a botanist needs to stabilize both in order for them to survive time and space. One step in the stabilization of botanical knowledge and materials was the transformation of flora into mobile but immutable parcels and papers. Another step towards stabilizing nature and knowledge was the production of botanical books. The Dutch amateur-botanist Jacob Breyne (1637–1697) produced several books on exotic plants from around the world. As an acquaintance of Beverningh, Breyne had easy access to the vast collection of plants and prints at Oud-Teylingen. Breyne's work enabled

[1] Regionaal Archief Leiden: Archiefnr. 512, inventarisnummer 687 ". . .de thuijnen, boomgaerden, plantagie, gragten ende cingelen daer aen ende toebehorende [. . .] groot gevonden vijff morgen. . ."

[2] For more on Hermann's contacts, see: Gunn and Codd (1981), Molhuysen (1920), Wijnands (1988: 73) and den Hartog and Teune (2002: 195).

Fig. 1 Jan de Baen, *Hieronymus van Beverningh*, 1670. With kind permission of the Rijksmuseum Amsterdam, SK-963

further stabilization. As printed knowledge, Beverningh's botanical collection could be enjoyed and examined anywhere.

A garden demonstrates, according to the philosopher Mara Miller, "not only power to control a part of the world but a peculiar sort of confidence because it indicates an expected continuation of that power in the future" (Miller 1993: 56). This means that, as an enclosed space, a garden is a site where local conditions are controlled as much as possible so as to achieve the same (botanical) results repeatedly over time and regardless of the geographical setting. Every visitor should encounter the same plants during the same time of year, every year. A garden thus was a site of demonstration, where nature was made immutable over time and space.

This paper explores the parallel and intertwined processes of making gardens and botanical knowledge stable, combinable and mobile. Both processes allowed nature and knowledge to be enjoyed and examined repeatedly, throughout time, and at different places (Roberts 2011). Or, to put it in Latourian terms, cultivators of nature and knowledge made flora immutable and immobile in gardens, and they transformed natural knowledge into immutable and mobile materials, such as publications and prints (Latour 1986). Cultivators of nature visited public and private gardens, discussed, collected and exchanged specimens and learnt about ways to define and combine nature. Seeds, plants, and saplings, as well as letters, garden designs, illustrations and herbaria, circulated between sites and over time. Examining and experimenting with nature taught students of nature how to grow vegetation out of season or outside of its natural region. Furthermore, it enabled cultivators to identify, name and classify specimens, and it allowed them to experiment on unfamiliar flora for their medicinal, edible, aesthetical, or commercial characteristics. Nature domesticated and demonstrated, either in gardens or in books, helped in the construction of botanical knowledge.

In order to map the activities which stabilized nature and knowledge so that they might survive over time and space, this paper first discusses the efforts made to transport plants and botanical knowledge from distant countries to Holland and Europe. Secondly, it introduces the botanical books produced by the amateur-botanist Jacob Breyne. The third section shows how plants and names are both mobile and mutable. To conclude, this paper shows how cultivators of nature, in their quest to stabilize nature and knowledge, were actually performing transformative acts.

Shipping Plants and Seeds

Transporting natural specimens involved making them stable enough so that they could be replanted in European gardens. For plant material to survive a long sea voyage, it had to be well packed and carefully stored. In a letter of December 23, 1686, Daniel Desmarets, supervisor of the gardens of the Dutch stadholder Prince William III of Orange, advised the English physician Hans Sloane on how to ship seeds, bulbs, trees and shrubs. Sloane was preparing to go to Jamaica in 1687, where one of his goals would be to collect local flora and send them to England. Desmarets advised Sloane that shrubs had to be placed in wooden caskets with their roots covered in moist soil and then wrapped in cloth, to protect them against seawater. Seeds should be dried and then kept in glass jars covered with a pig's bladder, or in flasks closed with a cork. Bulbs had to be dried in a hot breeze and placed in wooden boxes.[3]

Plant parts such as seeds, bulbs, roots and saplings were fairly easy to ship, whereas whole trees and large shrubs needed constant care and fresh water. When the Dutch sailor Cornelis Segewaart was preparing a botanical expedition to the Canary Islands, Desmarets sent him a letter of advice. Desmarets informed him that the precious plants could best be protected against rats and mice by placing breadcrumbs and bacon soaked in arsenic around the caskets.[4] Other perils such as stormy seas, incoming salt water and seamen thirsting for fresh water also threatened the fragile flora. Piracy and shipwreck equally hindered the successful transportation of precious cargo. The immutability and mobility of exotic plants also depended on further variants, such as caring sailors, reliable ships, good weather conditions and proper navigation home.

In order to propagate their findings, cultivators of nature like gardeners and botanists had to find ways to make nature immutable. As mobile and solid matter, plants could be dispersed to distant sites of accumulation where they could be "seen" by others, as Latour argues in "Visualization and Cognition." By circulating botanical specimens, other naturalists could (re-)use, adopt and adapt them. This led to centuries of transportation, discussion and transformation of botanical matter by naturalists from many different places around the globe.

Making nature stable and transportable therefore went hand in hand with prudent packing and shipping. However, on site, gatherers of nature had to number each specimen and list its name, so that the distant receiver knew what was kept in the various containers, parcels, flasks, bags, and caskets. Without proper labels and registration, names could not travel together with the materials, and thus remained unstable and immobile. Since names are part of botanical knowledge, it was eminently necessary that both plants and names were made mobile. Even upon safe delivery, it was still uncertain if a planted seed, bulb or shrub would take root. Cultivators had to find ways to domesticate these unfamiliar plants, by growing them in hothouses, for instance. After a plant took hold on Europe's soil, botanists and amateurs could then begin to examine, describe, compare and categorize a particular specimen.

In general, in the late seventeenth century, European botanists were interested in the ornamental qualities of plants. Whether an indigenous plant was medicinally or economically interesting depended on domestic necessities (Jarvis 1973; Fleischer

[3] British Library London, Sloane MS 4036, fol. 21-22, letter Daniel Desmarets to Hans Sloane 23 December 1686. "Si à la Jamaique on trouve des Aloe Euphorbia & des autres plantes lactescentes pour les envoijer il foudra les mettre avec un peu de sable sec autour de leur racine dans de la mousse sechepareillement dans une simple caisse de planches de la longueur et grosseur des dittes plantes la moindres humidité quelles rencontreraijent les ferayt tot pourrir.

Pour les graines il n'ija point da une facon que de les mettre bien secher dans une boëte bien fermee, mais pour les oignons de fleurs ou bulbes ils faut les laisser secher au vent hors du soleil jusques au que leur verdure fait toute fanée et en cest estat les mettre parmis de la mous se seche dans une caisse de bois carée." For more on ways of shipping plants: (Meister 1692: 155–156).

[4] British Library London, Egerton MS 1717, fol. 98 verso, memoire to Cornelis Segewaart "En opdat de ratten en muijsse op't schip de plante niet beschadigen, [...] dat men broodt meel speck of diergelijcke spijse met arsenicum of rattenkruijdt bestroijt, om op de kistjes legge."

2011). Why eat a bulb from an unfamiliar plant (potato) if a known European plant, such as grits and beans, provided enough food? Furthermore, although botanical materials were gathered, packed and shipped, indigenous knowledge about those materials and their uses by and large remained behind at the site of origin. Therefore, the immutability and mobility of plant materials did not necessarily lead to a stabilization of botanical findings to go with the flora.

In Beverningh's garden near Leiden, exotic plants from Asia, South Africa, North and South America, and the Mediterranean area were domesticated with great effort. The collection of plants growing there enabled naturalists to examine every sample at close range. They had made foreign flora immobile and mutable again, albeit by controlling the local surroundings and simulating warmer habitats. The recreation of a plant's original climate entailed the creation of a warm and light environment. Around circa 1680, Dutch gardeners had invented hothouses to enable the growth of prized flora and fruits such as the pineapple and citruses (Sikkens-de Zwaan 2002; Mukerji 1999). This type of hothouse soon conquered Europe, showing that gardens could become sites of demonstration. The garden experiment could be repeated somewhere else and with the same outcome, every year (Shapin and Simon Schaffer 1989).

The task of transforming nature into a mobile immutable involved assembling and collecting herbaria, prints and books. Garden owners like Beverningh not only had vast plant collections, they also owned libraries filled with botanical publications and herbaria. An herbarium filled with dried plants was a step towards making nature and knowledge stable and reproducible. It contained pressed specimens, usually with their names in one or more languages, their place of origin and a reference to well-known botanical publications. Sometimes, collectors compiled an herbarium for personal use, as a form of diary that a botanist could use as a basis for a publication. However, botanists also compiled herbaria to give as presents to other botanists and important patrons. A document titled *MS. Apparatus Botanicus, qui olim studiis inseaviit visi D. Hieron. Van Beverning, XVII Voll.*, gives an idea of the size and shape of the extensive herbaria collection that Beverningh received from his botanist friends. This document was created when the Leiden University Library purchased twenty-seven herbaria in 1741.[5] Some of the herbaria in Beverningh's collection originally came from the amateur-botanist Jacob Breyne, who compiled multiple herbaria and produced several books on botany, as the next section will show.

Producing Botanical Books

Breyne was a regular visitor to Oud-Teylingen. A merchant of Dutch descent who lived in the harbor town of Danzig (today Gdansk, Poland), Breyne traded in dyes, including Polish cochineal, a red dye made from insects. His trade brought him

[5] This manuscript is kept in the Naturalis Biodiversity Center, Leiden. Not all of the listed herbaria have survived.

frequently to Leiden, where a number of of his relatives lived, and which, besides its famous university, was known for its textile industry. While in Leiden, Breyne would visit Beverningh, as both shared a passion for botany, which led to their growing collection of plants, exsiccates, flower drawings and botanical books. Between 1659 and 1690, Jacob Breyne compiled several herbaria and produced multiple publications on foreign and local flora. He received many of the exotic specimens in his collection from Beverningh, via his patron's wide connections, or through his own trading network (Fig. 2).

While visiting gardens in Holland and journeying around Poland, Breyne collected local and exotic plants that he dried, pressed and placed in herbaria. His first

Fig. 2 Danzig plants. From Breyne: *Herbarius vivus* (1659), f. 5. With kind permission of the Naturalis Biodiversity Center, Leiden

herbarium was the *Herbarius vivus* (1659).[6] It consisted of specimens named mostly in Dutch, and sometimes additional information about the plant's natural environment supplied by Breyne. Later, possibly in the late eighteenth century, another botanist renamed the plants according to the binominal system introduced by the Swedish botanist Carl Linnaeus. A hundred years after that, the *Herbarius vivus* was still functioning as a repository of botanical knowledge, where the nomenclature was kept up-to-date. Breyne's next herbarium, *Plantae rariores Borussicae et Cassubicae*, consisted of four volumes of exsiccates collected in the Prussian-Polish region around Danzig in 1673.[7] In 1697, he wrote to his English friend, the apothecary James Petiver, that he planned to publish a book on Polish plants. However, he felt the need to undertake one more expedition to collect further plants, providing his health improved.[8] Breyne died that same year.

Compiling herbaria formed, for Breyne, an integral part of the production of his botanical books. He first transformed a living plant into a dried specimen, after which he was able to make it reproducible in print. This process of making plants stable, mobile and combinable resulted in several books. In 1678, Breyne published his *Exoticarum aliarumque minus cognitarum Plantarum centuria prima* (*Centuria* for short) with pictures and descriptions. In 1680 appeared the first part of *Prodromus Fasciculi Rariorum Plantarum*, filled with names, some descriptions and very few depictions. Volume two was published in 1689. A third book based on the earlier two volumes was published posthumously by Jacob's son Johann Philipp Breyne as *Prodromus Fasciculi Rariorum Plantarum, Primus et Secundus* (Danzig 1739). This version was enlarged with prints and included a biography of Jacob Breyne (Fig. 3).

Books with illustrations and descriptions, such as the *Centuria*, were meant for wealthy amateurs and garden owners, as they were costly to produce. To a collector, the *Centuria* was a type of display case filled with exotic plants, which could be acquired from the cultivator or obtained from afar. At the same time, the *Centuria* showed off the wealthy patron Beverningh's vast and prized collection of exotica to other amateurs and estate owners. A curious collector could learn from this book in which Dutch garden the actual specimen could be admired and studied.

Through his publications, Breyne demonstrated how he acquired and transformed esoteric nature and knowledge. He informed his readers about two precious Japanese plants, tea and ginger, that his contacts in the East had sent him.

[6] This is an incomplete manuscript with various plants missing. Kept at the Naturalis Biodiversity Center, Leiden.

[7] Kept at the Naturalis Biodiversity Center, Leiden.

[8] British Library London, Sl. MS 4066, f. 271, Jacob Breyne in a long (undated) letter in which he asks Petiver not to publish them "because I do design and figure them myself for the publique". [...] "I am now fully resolved to publish my Viridarium of Prussia & Cassubia, as soon as ever I [...] take a journey one summer more through Prussia &Cassubia to take a view and reckonize those plants which I had determined to have this year and had made some beginning too with with [sic] wonderfull success as to the observations tho' my health was not answerable to succeed for it...."

Fig. 3 Frontispiece. From Breyne: *Exoticarium aliarumque minus cognitarum Plantarum centuria prima* (Danzig 1678). With kind permission of the Naturalis Biodiversity Center, Leiden

Breyne also introduced the flora acquired by his patron van Beverningh, as well as local plants he had collected during his Polish expeditions. Breyne received *naturalia* either as living or dried samples, in the form of manuscripts and written descriptions, or as images. By domesticating nature into garden plants, exsiccates or printed matter, Breyne made nature and knowledge demonstrable, combinable, reproducible and mobile. At the same time, through his books, Breyne added to previously constructed knowledge, either through introducing new plants or through suggesting new nomenclature.

For his own publications, Breyne examined the publications of physicians and botanists from other eras and other places. These reference books revealed a search for a common nomenclature, clear classification and detailed descriptions. The goal was to enable other botanists to use these books to make identification possible and to create a communal language (Ogilvie 2003b). Through his findings, Breyne contributed to a growing body of botanical knowledge while simultaneously adopting and adapting established plant knowledge and names. These descriptions and depictions reflected an agreement between earlier generations of botanists and Breyne on the stabilization of a botanical language, both in words and pictures.

The process of establishing a shared plant idiom went hand in hand with the acts of gardening and botanizing. In gardens, cultivators of nature could discuss and exchange plants and knowledge. By sharing the same language regarding the names and nature of plants, botanists and amateurs could shape a garden and adopt nomenclature. It so happened that new ideas regarding this idiom formed at the same time that a new garden ideal became fashionable, as we will discuss in the next section.

Renamed Plants and Replanted Nature

In the late eighteenth century, Beverningh's garden disappeared, but Jacob Breyne's manuscripts give us some idea as to which plants grew in the gardens of Oud-Teylingen. From 1675, the garden's layout was kept in the geometrical 'baroque' style, until Beverningh's descendants decided to sell it a century later in 1770 (Rijken 2005: 235). Beverningh acquired plants from the Mediterranean region, Turkey, Japan, Ceylon, the Cape of Good Hope and South America (Peru and Surinam). These were sent to him by a wide variety of contacts, like the physician Paulus Hermann. The prized plants were diligently cared for, safeguarded from the cold in the winter, their seeds and fruits gathered and cultivated, bulbs and roots packed and stored, and in the summer and fall, fruits were gathered, eaten or presented as gifts. However, ultimately, all plants perish. The sole remembrances of Beverningh's plant collection can be found recorded in the various letters, manuscripts and books written by its visitors.

In the 1680s, the English gardener George London toured Holland, visiting public and private gardens on behalf of his employer Henry Compton, the Bishop of London. His aim was to collect plants from private and public botanical gardens in The Hague, Amsterdam, Haarlem and Leiden for the Bishop's garden and

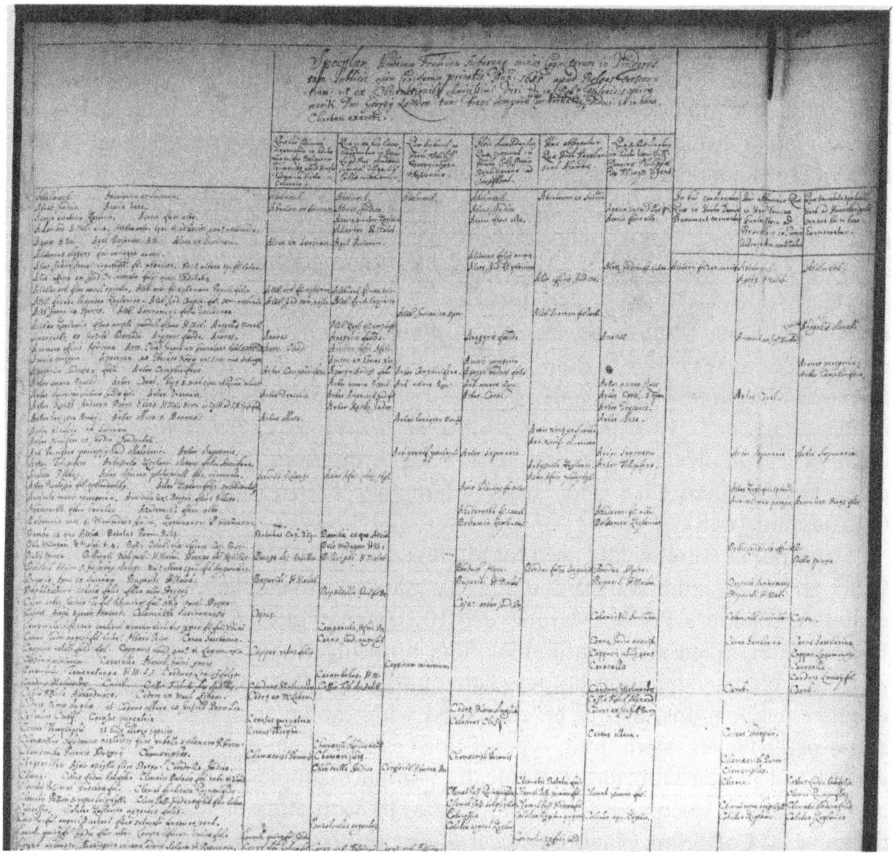

Fig. 4 Leonard Plukenet, *Speculum Herbarum*.

botanical collection at Lambeth Palace. London listed all the plants he had gathered. Later in England, the botanist and physician Leonard Plukenet transcribed London's notes into a schematic overview. On one large sheet of paper, we find named the gardens of the Prince of Orange, the Hortus of Leiden and Amsterdam, the gardens of the diplomat Gaspar Fagel, the merchant Philips Des Flines and his cousin the rich widow Agnes Block, also known as Mrs. Des Flines, and the estate of Hieronymus van Beverningh (Fig. 4).[9]

Like their contemporaries, Breyne, Plukenet and London used descriptive names for the unfamiliar flora they encountered. This type of nomenclature was largely based on a plant's morphology. It often meant that the same plant had more than one name, or variants thereof. For instance, Plukenet mentioned the Alth. Javanica

[9] Leonard Plukenet's manuscript *Speculum Herbarum* is in the British Museum of Natural History, London, Sloane Herbarium 91.

species, which in Breyne's *Centuria* is named as Alcea Javanica arboresens, flore pleno (p. 121); Plukenet's Arbor Camphorifera was called Arbor Camphorifera japonica in the *Centuria* (p. 11); Plukenet's Chrysantimum Africana Breynij is represented in the *Centuria* by three types of Chrysanthemum Aizooides Africanum, of which the second type, the S. Teretifolium (p. 161), and a third type (p. 163) happened to grow in Beverningh's garden.

When applicable, Breyne cited the works of well-known botanists in his plant descriptions. For example, he referred to names used in Jean and Gaspar Bauhin's *Pinax theatri botanici* (Basel 1623). The *Pinax* listed about 6,000 plant names. Amongst others, Breyne also mentioned the works of the Italian botanist Fabius Columna *Minus cognitarums tirpium aliquot ac etiam rariorum nostro coelo orientium ekphrasis* (Rome 1616), the Scotsman Robert Morison, *Praeludia botanica* (*London* 1669), the Polish-Scottish botanist Jan (John) Jonston, *Dendrographias, sive historiae naturalis de arboribus et fructicibus, tam nostri quam peregrini orbis, libri decem, figuris aeneis adornati* (Frankfurt 1662) and the physicians Willem Piso and Georg Marcgrave's *Historia Naturalis Brasiliae* (Amsterdam 1648).

These books were widely used for reference and comparison throughout Europe. Other amateurs and botanists could verify plant specimens from their copy of, for example, Bauhin's *Pinax* (Margocsy 2010; Bouman et al. 2007; Ogilvie 2003a). Breyne categorized some unfamiliar flora by using Robert Morison's method of arranging plants using morphology. Other unfamiliar flora, not named in the *Pinax* or other publications, had to be compared with known samples. By studying the shape of its leaves, stem, root, type of flower or fruit, an unknown plant could be named and paired with morphologically similar examples. This way of working allowed Breyne to compare different species to each other, noting similarities and differences. Collecting plants, exsiccates, descriptions and depictions allowed him to study, identify and combine different varieties. By referring to other books, he could also discuss and introduce exotic flora. This helped to establish a communal and stable idiom. Breyne thus used and adapted elements of previously constructed and published knowledge to produce botanical knowledge, which he propagated through his publications.

To communicate their findings, botanists needed to establish a common language in both writing and drawing, which would enable them to discuss plant characteristics, use and nomenclature. Their communal idiom helped botanists to come to a mutual agreement about plants, and allowed for the stabilization of botanical knowledge. In order to overcome the problem that a plant could carry multiple names, botanists tended to refer to other botanical books to underscore their findings (Margocsy 2010). This allowed for clearer communication and transportation of botanical materials and knowledge. Today, the plants mentioned above are known under their Linnaean names: Hibiscus rosa-sinensis L.; Cinnamomum Camphora (L.); and Cylindrophyllum calamiforme (L.) Schwantes (or its synonym Mesembryanthemum calamiforme L.).[10] After botanical confidence in the Linnaean system became established, this became the dominant way of classifying nature. The stabilization of the binomial system allowed botanical knowledge to become mobile and communicable amongst naturalists.

This process of domestication and communication occurred on planes both two-dimensional (books) and three-dimensional (gardens). For not only did nomenclature change in the eighteenth century, the style in which nature was ordered in gardens changed as well. A garden such as Oud-Teylingen had been designed according to geometrical rules. The seventeenth century considered regularity and symmetry an antidote against the chaos and disorder that occurred after the expulsion from paradise. By recreating God's Garden of Eden, humans could obtain a deeper understanding of nature's divine workings (Prest 1988).

In the eighteenth century, a new ideal of nature emerged. Garden designers took inspiration from landscape paintings, and therefore came to believe that a garden should resemble a 'natural' landscape, painted with trees and water instead of oil on canvas. This so-called 'picturesque' or English landscape garden consisted of curved lines (of paths and water features), unclipped shrubs, and irregularly placed groups of trees. The grounds were adorned with classical temples, Roman ruins and Gothic bridges, and encompassed broad vistas without visible fences, so as to visually incorporate the surrounding landscape into the garden (Dixon Hunt 1992).

This garden style took root in the Low Countries, where many estate owners adopted the new fashion. In 1774, the new owners of Oud-Teylingen decided to redesign their garden. In order to prepare the terrain, their first step was to auction off all the existing plants. The auction catalog listed circa three thousand trees (beech, lime, poplar, ash, alder, holly, palm), two thousand fruit trees (apple, pear, plum, cherry, and peach) and two thousand herbs, flowers, seeds and bulbs.[11] The sale and subsequent transformation of the garden went hand in hand with the changed garden style. After the auction, the plants from Oud-Teylingen were moved to other gardens. In their new surroundings, their new owners replanted the trees, shrubs and flowers in new groups, placed in a new setting according to the latest notions of 'nature'. Thus, living plant specimens could be transported and transplanted in order to illustrate the latest idealizations and explanations of nature.

Domesticating and demonstrating nature in gardens and books allowed plant materials and knowledge to survive over time and space. However, over time,

[10] I thank Gerard Thijsse of the Naturalis Biodiversity Center, Leiden for this information.

[11] Regionaal Archief Leiden, Inventaris van het huisarchief van de heeren van Warmond, 1347–1900 (1911). Archive no. 512. Inventory no. 164: "Men is van meninge in het openbaar te doen veijlen, ende te verkopen:1000 stuks, Elst, Esse, Italiaanse, France, en Inlandse Abele, [onleesbaar], groene, en bonte Hulst; 2000 stuks, Linde, en Beuke, van 7, 7, a 8 duijm dik, twaalf en veertien voeten hoog. 2000 stuks vrugtboomen soo ongevend als gevend, Peeren, Appelen, Pruijmen, Kersen, en Persiken etc. Alle de bovenstaande in eene koop. Nog 2000 stuks Aardvrugten, planten, kruijden, bloemen waaronder verscheijde angelieren, alle uyt zaad soo van een, twee, alsmede jaaren oud, ook jonge bequaam om te verplanten. Verscheijde oude, en jonge Palm, en andere soorten van planten, alle in eene koop."

cultivators conceived other ideas regarding nomenclature and different conceptions of an ideal nature. Styles in designing and categorizing nature are prone to change, but at the same time, humans seek to stabilize plants and knowledge in order to communicate their findings, as well as incorporate new interpretations of nature.

Conclusion

Early modern naturalists looked for ways to make materials and knowledge stable, mobile, combinable and reproducible in order to communicate their findings either in gardens or in books. Gardens functioned as immobile sites of demonstration and domestication of nature, where flora was accessible for passing visitors. At the same time, nature itself was mobile, as when shrubs, bulbs and seeds were packed, labeled and shipped, or when botanists produced drawings, prints and exsiccates. However, the length of this immutability is dictated both by nature itself and by humans who seek to transplant, domesticate and categorize plants to fit the latest insights.

For nature and knowledge to move over time and space, it had to be made stable, while simultaneously establishing the reliability of descriptions and depictions and agreement on communication, nomenclature and classification. Botanists referred to reputable publications to validate their own writings while adding their findings to a growing body of natural knowledge. As plant materials and knowledge circulated from one site of demonstration to the next, both went through a process of translation and transformation. Botanists and amateurs adapted knowledge and plants to fit local requirements and/or combined these with other findings. In turn, plant materials and knowledge could again be transported to other gardens.

For cultivators of nature, both gardens and books formed fertile soil for botanical examination and combination. These parallel processes enabled botanists and naturalists to attribute a new meaning to, or explanation of, nature's workings. The formation of botanical knowledge in books and gardens thus supports the notion that humans strive to control nature, which is confined within a certain period and timeframe. The stabilization of nature and knowledge lasts only until the moment when new insights into nature and knowledge replace the former ideals.

Bibliography

Sources: Printed

Bauhin, Jean and Gaspar Bauhin. 1623. Pinax theatri botanici. Basel.

Breyne, Jacob. 1678. Exoticarum aliarumque minus cognitarum Plantarum centuria prima. Danzig.

Breyne, Jacob. 1680. Prodromus Fasciculi Rariorum Plantarum: 1. Danzig: David Fridericus Rhetius.

Breyne, Jacob. 1689. Prodromus Fasciculi Rariorum Plantarum: 2. Danzig: David Fridericus Rhetius.

Breyne, Jacob, and Johann Philipp Breyne.1739. Prodromus Fasciculi Rariorum Plantarum, Primus et Secundus. Danzig: Thom. Joh. Schreberi.
Columna, Fabius. 1616. Minus cognitarum stirpium aliquot ac etiam rariorum nostro coelo orientium ekphrasis. Rome.
Meister, George. 1692. Der Orientalisch-Indianische Kunst- und Lust-Gärtner. Dresden. Eds. Friedmann Berger and Wilfried Bonsack (reprint 1973). Weimar: Gustav Kiepenheuer Verlag.
Morison, Robert. 1669. Praeludia botanica. London: Jacobi Allestry.
Piso, Willem and Georg Marcgrave. 1648. Historia Naturalis Brasiliae.Leiden: Franciscus Hack; Amsterdam: Ludovicus Elzevier.

Manuscript

Breyne, Jacob. 1659. Herbarius vivus. Leiden: Naturalis Biodiversity Center.
Breyne, Jacob. 1673. Plantae rariores Borussicae et Cassubicae. Leiden: Naturalis Biodiversity Center.
Breyne, Jacob. Letter to James Petiver.Sl. MS 4066, f. 271. London: British Library London.
British Library London. Memoire for skipper Cornelis Segewaart. Egerton MS 1717, fol. 98. London: British Library London.
Desmarets, Daniel. Letter Hans Sloane. Sloane, MS 4036, fol. 21-22. London: British Library London.
MS. Apparatus Botanicus, qui olim studiis inseaviit visi D. Hieron. Van Beverning, XVII Voll. Leiden: Naturalis Biodiversity Center.
Plukenet, Leonard. Speculum Herbarum. Manuscript Sloane Herbarium 91. London: British Museum of Natural History.
Inventaris van het huisarchief van de heeren van Warmond, 1347-1900 (1911). Archive no. 512. Inventory no. 164 and 687. Leiden: Regionaal Archief Leiden.

Secondary Literature

Bouman, Ferry, Bob Baljet, and Erik Zevenhuizen. 2007. Kruidenier aan de Amstel, De Amsterdamse Hortus volgens Johannes Snippendaal (1646). Amsterdam: Amsterdam University Press.
Dandy, J.E. 1958. The Sloane Herbarium, an annotated list of the Horti Sicci composing it: with biographical accounts of the principal contributors. London: British Museum.
Dixon Hunt, John. 1992. Gardens and the Picturesque; Studies in the History of Landscape Architecture. Cambridge and London: MIT Press.
Fleischer, Alette. 2007. The Beemster Polder: conservative invention and Holland's great pleasure garden. In The Mindful Hand: inquiry and invention from the late Renaissance to early industrialization, eds. L.L. Roberts, Simon Schaffer and Peter Dear, 145-166. Amsterdam: Edita.
Fleischer, Alette. 2011. Trading Places: (ex)changing nature and knowledge at Cape of Good Hope, circa 1652-1700. In Centres and cycles of accumulation in and around the Netherlands during the early modern period, ed. L.L. Roberts, 110-127.Münster: LIT Verlag.
Gunn, Mary and L.E. Codd. 1981. Botanical Exploration of Southern Africa. Cape Town: A.A. Balkema.
Hartog, Elizabeth den, and Carla Teune. 2002. Gaspar Fagel (1633-88): His garden and plant collection at Leeuwenhorst. Garden History, Journal of the Garden History Society 30: 191-205.
Jarvis, P.J. 1973. North American Plants and Horticultural Innovation in England, 1550-1700. Geographical Review 63: 477-499.

Jonston, Jan (John). 1662. Dendrographias, sive historiae naturalis de arboribus et fructicibus, tam nostri quam peregrini orbis, libri decem, figuris aeneis adornati. Frankfurt: M. Meriani.

Latour, Bruno. 1986. Visualization and Cognition: Thinking with the eyes and hands. Knowledge and Society: studies in the sociology of culture past and present 6: 1-40.

Livingstone, David N. 2003. Putting Science in its Place, Geographies of Scientific Knowledge. Chicago/London: The University of Chicago Press.

Margocsy, Daniel. 2010. 'Refer to folio and number': Encyclopedias, the Exchange of Curiosities, and Practices of Identification before Linnaeus. Journal of the History of Ideas 71: 63-89.

Miller, Mara. 1993. The Garden as an Art. Albany: State University of New York Press.

Molhuysen, P.C. 1920. Bronnen tot de Geschiedenis der Leidsche Universiteit, 18 febr. 1682 – 8 febr. 1725. Den Haag: Martinus Nijhoff.

Mukerji, Chandra. 1999. Storehouses to Stoves: Built Environments and the Early Dutch Plant Trade. Unpublished paper.

Mukerji, Chandra. 2002. Material Practices of Dominion: Christian Humanism, the Built Environment and Techniques of Western Power. Theory and Society 31: 1-34.

Ogilvie, Brian W. 2003a. The Many Books of Nature: Renaissance Naturalists and Information Overload. Journal of the History of Ideas 64: 29-40.

Ogilvie, Brian W. 2003b. Image and text in natural history, 1500-1700. In The Power of Images in Early Modern Science, eds. Wolfgang Lefèvre, Jürgen Renn and Urs Schoepflin, 141-166. Basel, Boston, Berlin: Birkhäuser.

Prest, John. 1988 (1981). The Garden of Eden, the Botanic Garden and the Re-Creation of Paradise. New Haven and London: Yale University Press.

Rijken, Henk. 2005. Leidse Lustwarande Geschiedenis van de tuinkunst op kastelenen buitenplaatsen rond Leiden, 1600 – 1800. Leiden: Primavera Pers.

Roberts, Lissa L. (ed). 2011. Centres and Cycles of Accumulation in and Around the Netherlands during the Early Modern Period. Münster: LIT Verlag.

Shapin, Steven, and Simon Schaffer. 1989 (1985). Leviathan and the air-pump; Hobbes, Boyle, and the experimental life. Princeton: Princeton University Press.

Sikkens-de Zwaan, Marisca. 2002. Magdalena Poulle (1632-99): a Dutch lady in a circle of botanical collectors. Garden History, Journal of the Garden History Society 30: 206-220.

Veendorp, H., and L. Baas Becking. 1938. Hortus Academicus Lugduno Batavus, the development of the gardens of Leyden University 1587-1937. Haarlem: Joh. Enschedé.

Wijnands, D.O. 1988. Hortusauriaci: de tuinen van Oranje en hunplaats in de tuinbouw enplantkunde van de late zeventiendeeeuw/Hortus auriaci: the gardens of Orange and their place in late 17th-century botany and horticulture. Journal of Garden History 8: 61-86.

Gardening Knowledge Through the Circulation of Agricultural Treatises in Portugal From the Sixteenth to Eighteenth Centuries

Ana Duarte Rodrigues

Abstract

In this paper, I demonstrate that from the sixteenth to eighteenth centuries, gardening horticultural knowledge in Portugal was much more accessible through Spanish and French agricultural treatises than through any other kind of garden literature. So, focusing on agricultural treatises, the aim of this paper is to present the results achieved so far by investigating the circulation of these books, to assess which authors were the most popular and which editions and translations had the greatest success, and to evaluate their depth and scope as regards the character and key features of Portuguese gardens. As far as possible, I will do this by comparing these with horticultural literature sources used in other European countries.

Furthermore, I highlight the following texts as the works with the greatest circulation in Portugal: Gabriel Alonso de Herrera's *Agricultura General* (1513), Miquel Agustí's *Libro de los secretos de agricultura, casa de campo, y pastoral* (1626), Louis Liger's *Oeconomie générale* (1700) and João Garrido's *Agricultor Instruído* (1730). On this basis, I aim to establish a connection between the theoretical knowledge available during the said period and the Portuguese gardens themselves. Until recently, horticulture was understood as an art conveyed exclusively by *praxis*.

Introduction

Agricultural treatises were the best way to spread horticultural knowledge and theory in Portugal during the Early Modern period until the eighteenth century. This study provides insights into the theory of garden art accessible in Portugal

A.D. Rodrigues (✉)
Faculdade de Ciências, Universidade de Lisboa, 1749-016 Lisboa, Portugal
e-mail: amnrodrigues@fc.ul.pt

H. Fischer et al. (eds.), *Gardens, Knowledge and the Sciences in the Early Modern Period*, Trends in the History of Science, DOI 10.1007/978-3-319-26342-7_15

based on the results of research into the literary sources of interest for gardening, such as those found in Portuguese libraries or described in historical sources held in archives.

State of the Art

The study of the circulation of gardening knowledge in Portugal from the sixteenth to eighteenth centuries is taking its first steps into unexplored territory in Portuguese historiography. Only a few scholars have paid any attention to this subject. Helder Carita published many primary sources important for Portuguese garden history in *Tratado da Grandeza dos Jardins em Portugal ou originalidade e desaires desta arte* (1990). In her PhD thesis (1995) on the essence of Portuguese gardens, Aurora Carapinha pointed out the importance of Islamic and Spanish authors as the architects of the theoretical program of Portuguese gardens and identified many of the fundamental authors of this knowledge. In his *História da Teoria da Arquitectura no Ocidente* (2010), Simões Ferreira dedicated a chapter to western garden literature in general, but without specifying the Portuguese case.

On the way to my PhD, I embarked on this research by studying the literature of interest for garden art and the contents of the architects' and patrons' libraries (Rodrigues 2009; published in Rodrigues 2011a: 237–265). The Real Mesa Censória (1768) holdings include lists of books held by many libraries drawn up by order of the Marquis of Pombal, who was concerned with censorship. This source revealed some important libraries belonging to architects such as Manuel Caetano de Sousa, who created gardens (Rodrigues 2011a: 633–642), and to estate owners such as Paulo de Azevedo Lima, the owner of an estate in Almeirim with magnificent libraries (Rodrigues 2011a: 642–645). This proved to be the first step of the research project *Tratados de Arte em Portugal/Art Treatises in Portugal* (2010–2013). Results relating to the theoretical knowledge available to architects and gardeners in Portugal from the fifteenth to nineteenth centuries, with a focus on agricultural and garden art treatises, literature and botany, were published in 2011 (Rodrigues 2011b: 119–144). Recently, this study was enhanced by the exhibition on treatises and theoretical sources on garden art held at the Biblioteca Nacional de Portugal (BNP) between May 15 and July 31, 2014, the catalogue of which includes all copies of the books found at the National Library (Rodrigues 2015b). Further studies have been pursued on the impact specific treatises had in Portugal, such as those by Alonso de Herrera (Rodrigues 2014b: 11–28) and Olivier de Serres (Rodrigues 2015b: 99–114), as well as on the evaluation of certain libraries, such as the one owned by Eugénio dos Santos. He was the architect responsible for the reconstruction of Lisbon's town center after the 1755 earthquake and he had a unique library on garden art in a Portuguese context (Rodrigues 2015a: 127–150).

The research hitherto conducted into the Spanish agricultural treatises has been dealt with in various publications and helps us to understand the character and the cultural and ecological system of Portuguese gardens. However, it is clear that we are still taking first steps into this field of study, and further archival research will throw more light on the scope of this theoretical knowledge and determine whether or not it was transferred to the management of the *quintas* (estates).

Methodology

The studies by Helder Carita (1990), Aurora Carapinha (1995) and myself (2009) were an important guide to further research using quantitative methods. The main aim of the research project "Art Treatises in Portugal"[1] was to compile a database of the "artistic literature", found in the most important Portuguese libraries, on architecture, sculpture, painting, drawing, iconography, garden art and music. Each area was supervised by an expert, and the library research was carried out by experienced professionals with a list of treatises and books considered important to that art to guide the research.

The starting point for the research was the assumption that for garden art, garden art treatises were not the only books that contributed to horticultural theory and practice. In addition, all possible writings, whether in manuscript or printed form, relating to garden art as part of the greater subject of "artistic literature", such as agricultural treatises, architectural treatises, poems, romances, note-books, descriptions of monuments, travel-guides and even books on hydraulics, also needed to be considered. Information on all their various editions, number of copies, owners and places of provenance was gathered.

Thus far, information has been collected from the following libraries: Biblioteca Nacional de Portugal (National Library of Portugal), Biblioteca do Palácio Nacional da Ajuda (Library of the National Palace of Ajuda), Biblioteca do Palácio Nacional de Mafra (Library of the National Palace of Mafra), Biblioteca Pública de Évora (Public Library of Evora), Biblioteca da Universidade de Coimbra (Library of the University of Coimbra), Biblioteca da Academia de Ciências de Lisboa (Library of the Academy of Sciences in Lisbon) and Biblioteca da Academia Nacional de Belas Artes (Library of the National Academy of Fine Arts).

The result is a corpus of treatises and books of relevance to garden art in circulation in Portugal during the Early Modern period. Further research has been made independently, with a closer look at some treatises, cross-referencing this data with archival research and historical sources. Furthermore, I have conducted research at Dumbarton Oaks (2013) on literary sources relating to garden arts unknown in Portugal in order to become familiar with the contents of English language treatises and to be able to compare this theoretical framework with that in use in Portugal during the Early Modern period.

[1] Art Treatises in Portugal was a research project (with the reference PTDC/EAT-EAT/100496/2008) undertaken by the Faculdade de Ciências Sociais e Humanas (Faculty of Social Sciences and Humanities) of the Universidade Nova de Lisboa (New University of Lisbon), with funding from the Foundation of Science and Technology/Ministry of Science.

The Agricultural Treatises Circulating in Portugal from the Sixteenth to Eighteenth Centuries

I have now identified more than one hundred copies of 24 different agricultural treatises. In addition, there are more than two hundred botanical publications alongside the 20 copies of treatises on garden art. Treatises on garden art were extremely rare in Portugal: we found only a few examples, such as Dézallier d'Argenville's *La Theorie et la Pratique du Jardinage* (1709), Whately's *Observations on Modern Gardening* (1770), Duchesne's *Sur la formation des jardins* (1775), Morel's *Théorie des Jardins* (1776) and Hirschfeld's *Théorie de l'Art des Jardins* (1779) in the French translation, as well as some descriptions, such as Salomon de Caus' *Hortus Palatinus* (1620), Blondel's architectural treatise on villas, *De la distribution des maisons de plaisance* (1737), and Briseux's *L'Art de bâtir les maisons de champagne* (1743).

Therefore, it would seem legitimate to conclude that the theory of garden art was more accessible in agricultural treatises than in treatises on garden art. Although agricultural treatises were preponderant throughout the Early Modern and Modern periods, we should, however, comment on some noticeable differences: in the earlier part of this period, the difference between the numbers of copies of agricultural treatises and of garden art treatises was much more significant; in the latter part, books on botany became more relevant.

I would like further to point out some of the perceptible differences in the more than one hundred books on agriculture of relevance to garden art that have already been identified: Spanish was the most commonly used language in the early part of this period, giving way to French later on.

In the sixteenth century, above and beyond the books in Spanish, the most successful works were modern editions of ancient works written in Latin, such as those of Cato (234 B.C.–149 B.C.), Varro (116 B.C.–27 B.C.), Columella (first century) and Palladius Rutilius, whose texts entitled *De Re Rustica* appeared in several different editions, either separately or together. In the modern editions of authors of classical Antiquity, villas and country life were praised and contrasted with urban ambition, commerce and a social life packed with intrigue. In the case of the modern editions, the classical authorities on farming were placed more freely at the disposal of landowners, but in this case, owing to the revival of all classical arts and culture, it is difficult to recognize whether they were just editions and translations of ancient learning or, in actual fact, intended as guides to farming.[2]

Two important works of the intermediate period (between the Roman Empire and Renaissance)—the *Geoponika,* a collection of 20 texts on agricultural pursuits compiled in Constantinople during the tenth century for the emperor Constantine

[2] The first collection of the Latin authors Cato, Columella, Varro and Palladius was printed in Venice in 1472, in Reggio in 1482, 1496 and 1499, and in Bologna in 1554. Nine other editions of this collection were printed during the sixteenth century. All information on these sixteenth-century editions can be found in Fussell (1969).

VII and also used by the Moors (and which, through them, arrived on the Iberian peninsula) and especially the treatises of the *Geoponike* Al-Andaluz (Carapinha 1995: 148–164, 2015: 49–65)—are fundamental to our understanding of the essence of Portuguese gardens and were probably known in this territory (Carapinha 1995: 159–160). For example, the twelfth-century Islamic treatise by Abu-Zacaria already mentions orange trees, considered a key feature of Portuguese gardens (Carapinha 2014: 6–15).

Pietro de Crescenzi's *Liber Ruralium Commodorum* (ca. 1305), of which there are some sixteenth-century copies in Italian in Portuguese libraries, was also probably well-known in Portugal, because the tower in the middle of a field described as a pigeon house can be seen in the Paço das Alcáçovas in Alentejo, probably dating to the seventeenth century.

Gabriel Alonso de Herrera's *Agricultura General* (1513), Gregorio de los Rios'*Agricultura de Jardines* (1592), Miguel Agustí's *Libro de los secretos de agricultura, casa de campo, y pastoral* (1625), Francisco Gilabert's *Agricultura Pratica con la qual puede uno llegar a ser perfeto Agricultor* (1626), Diego de Carballo y Sampayo's *Elementos de Agricultura* (1790) and Fr. Theobaldode Jesu's *Agricultor Instruído* (c. 1730), all originally works in Spanish, stand out as the most successful agricultural treatises in circulation in Portugal.

Spanish and Latin books clearly prevailed in number in Portuguese libraries in the sixteenth century. In addition, French books such as Charles Estienne's *Maison Rustique* (first edition 1564; augmented by Jean Liébault (1572), Nicolas de Bonnefond's *Les delices de la champagne* (1665), Carlo d'Aquino's *Nomenclator Agriculturae* (1736), Louis Liger's *Oeconomie Générale* (1700), Agustin Roux's *Traité de la culture et de la plantation des arbres* (1750), La Salle de l'Etang's *Manuel d'agriculture pour le laboureur* (1764), *Catéchisme d'Agriculture ou Bibliothèque des gens de la Campagne* (1773), François Rozier's *Cours complet d'agriculture* (1791–1796), and Italian works such as Luigi Alamanni's *La Coltivazione* (1546), Giovanni Maria Bernardo's *Le Richezze dell'Agricultura* (sixteenth century), Giovanni Battista della Porta's *Villae* (1592), Giuseppe Falcone's *La nuova, vaga, e dilettevole villa. Opera d'agricoltura* (1612) and Giovanni Battista Ferrari's *Florum Cultura* (two editions: 1633 and 1638) and *Hesperides* (1646) also circulated in Portugal.

Giovanni Battista Ferrari's *Hesperides* is an essential book for understanding the Portuguese garden as the garden of Hesperides, where the orange tree from Lisbon, *Auratium Olisiponensis*, is included.

The first book of this type in the Portuguese language does not appear until the eighteenth century and is a translation: Fr. Theobaldo de Jesu Maria's *Agricultor Instruído com as prevençoens necesarias para annos futures, recuppilado de graves autores* (1730), translated by the Portuguese João António Garrido. This book was quite popular, with more than ten copies discovered so far of the 1730, 1749, 1764, 1790, 1789 and 1837 editions. In addition, also in Portuguese, there are some copies of Afonso Toar da Silveira's *A nobreza dos lavradores e a vida de S. Izidro* (Lisboa, 1741), which promoted farmers and gardeners, as well as nine copies of *Memórias de Agricultura* (1788). Nevertheless, *Agricultor Instruído* maintained the early tradition of agricultural treatises by including one part dedicated to gardening, mixed with tree cultivation, vineyards and animals.

Dictionaries on agriculture and gardening dedicated to gardeners and farmers also became much more common in the eighteenth century, and are an important complement to the knowledge divulged through agricultural treatises. They were mostly in French, such as the *Dictionnaire pratique du bon menager de Campagne et de ville* (1715), the *Dictionnaire universel d'Agriculture et de Jardinage* (1751), Philip Miller's *Dictionnaire des jardiniers* (1785) and Jean Jacques Fillassier's *Dictionnaire du jardinière françois* (1789).

The Preponderance of Spanish Agricultural Treatises

Spanish agricultural treatises played a prominent role up until the War of Restoration (1640–1668), especially during the Philippine dynasty between 1580 and 1640, when important gardeners, including Rodrigo Álvares, who had worked in Aranjuez and Casa de Campo, probably brought Spanish agricultural treatises to Portugal (Carapinha 1995: 121–128).

While defining the essence of Portuguese gardens, Aurora Carapinha (1995) was the first to draw attention to the role books such as Alonso de Herrera's *Agricultura General* (1513) or Gregorio de los Rios's *Agricultura de Jardines* (1592) played in the conception of Portuguese gardens and how they could help us to understand the essence of those gardens (Carapinha 1995: 116–121).

Gabriel Alonso de Herrera's treatise on agriculture is based on his long experience of farming in Talavera and Granada, mixed with classical sources and Islamic traditions, as well as on his own travels to Spain, France and Italy (Rodrigues 2014b: 11–26). It was published for the first time in 1513 and, by the nineteenth century, had gone through several editions and translations.

So far, I have found 25 copies of Alonso de Herrera's book in Portuguese libraries, with the oldest edition in Portugal dating to 1528.[3] This work continued to be revised down to the nineteenth century and it circulated in Portugal until it was translated into Portuguese with the title *Nova arte que ensina a criar, tartar, escolher e curar cabras*, published in Lisbon in 1841 and 1849 (all 25 copies are identified in Rodrigues 2014b: 17–20).

Although there are fewer copies of works by Gregorio de los Rios, two of them contained inside the 1620 Herrera edition, we consider it one of the most important texts for the Portuguese *horto*. In fact, the typical features of sixteenth-century Portuguese gardens can be found in de los Rios' text. The values of color, fragrance

[3] There is one methodological issue that needs pointing out here, besides all those relating to the database and already discussed in *Art Treatises in Portugal* (Rodrigues 2011c): this is the risk of associating a sixteenth century copy with its actual usage in that century, because sometimes libraries were only established much later and the book may have been bought at a time when it was no longer up-to-date. But in this case, it seems probable that Herrera was known and being used in the early sixteenth century.

and sound are present in de los Rios just as they are in Portuguese gardens: the importance of ornamental flowers that only had to be "agreeable to the eyes", which was not underlined in Herrera's book, with the identification of colors by species; the praise of the fragrance of jasmine; and finally, the pages dedicated to the sounds of nightingales as indispensable in every garden.

The other major interconnection between Portuguese gardens and Gregorio de los Rios' works is the orange trees, because both consider them ornamental. First, the author says that fruit trees should not be in a garden because they belong properly to orchards. Later, in a completely different chapter, he dedicates a great number of pages purely to orange trees, now ranked as properly belonging to gardens because of their beauty, color and fragrance. Orange and citrus trees, which were rare and exotic in the Roman Empire, became very common in the Iberian Peninsula after the Moorish conquest at the beginning of the eighth century, one of the best-known examples being the Naranjos courtyard in Córdoba; this has remained one of the main features of an Iberian garden up to the present day. As mentioned, Aurora Carapinha considers orange trees to be a key feature of the Portuguese garden, thereby remaining heir to the Mediterranean horticultural matrix (Carapinha 2014: 6–15).

The works by Herrera and Gregorio de los Rios are also understood as a legacy of the text written by the Moor Ibn Al-Awwam, who also exerted a great influence on Portuguese gardens, particularly in terms of water technology: for example, the distribution of water through small channels all over the terrain in the open air and the noria irrigation technique, so common in Portuguese villas, were a great advance for the Mediterranean world, where rainfall is light, and well replaced the older, passive system of trapping rainfall temporarily and releasing it to be spread by gravity. *Kitab al-Felahah* or *Book of Agriculture* by Ibn-Al-Awwam was written in the latter half of the twelfth century for the benefit of Spanish farmers and was known during the Medieval Ages throughout the Iberian peninsula under Moorish rule, and although it is not quoted by Alonso de Herrera,[4] it is true that the Islamic agricultural tradition was assimilated by the Spanish authors, alongside that of the classical authorities.

The books we are studying in this paper focus less on decorative elements, such as grottoes, sculptures, pergolas, benches, tiles, and paintings (the subject of treatises on garden art), but rather on fruit trees, vegetables and flowers, organized with the two aims of production and fruit-bearing. This close relationship between the art of gardening and agricultural traditions—one that was observed from Roman times until the age of Le Nôtre (Lambin 1975: 29)—is also confirmed by the literary sources in use, because when architectural and theatrical scenes became important

[4] *Kitab al-Felahah* by Ibn-al-Awwam was rediscovered in the Royal Spanish Library of San Lorenzo del Escorial in the middle of the eighteenth century and hailed as the greatest of all medieval treatises on agriculture, but was totally unknown in most of Europe. Due to the conflict between Moors and Christians, it was known in the southern part of the Iberian Peninsula up until the fifteenth century and condemned for a long time in the northern Christian area (Olson and Eddy 1943: 100–108).

features in gardens, one can assume that other treatises besides the agricultural ones were being used in landscape gardening.

Agricultural treatises, held in convent libraries and the libraries of villa owners, were probably of much greater importance for garden art in the sixteenth century than later. In the latter part of the period, they would have had to compete with treatises on garden art, the influence of which becomes easily recognizable in baroque gardens. It seems that the privileged owners of such books were doctors and botanists (Rodrigues 2011b: 136–142), in addition to members of the Royal household and villa owners.

Nevertheless, Miquel Agustí's *Libro de loss ecretos* (1617/1626) enjoyed a success similar to that of Alonso de Herrera: ten editions, ranging from the first edition in 1626 to the 1781 edition. In Portugal, copies exist of many different editions (1625, 1646, 1665, 1695, 1722 and 1762). Among the owners of this book were the writer D. Francisco Manuel de Melo, a foreign doctor, and Paulo de Azevedo Pita Lima, owner of a villa in Almeirim. It was also kept in the library of the Marquis of Castelo Melhor (who also had a copy of Herrera's treatise of 1598), the owner of a palace in central Lisbon; that palace was destroyed by the 1755 earthquake, but it must have been similar to Fronteira's garden, judging from certain written descriptions (Rodrigues 2011a: 76). Most of the owners of the books listed here were the libraries of convents and monasteries, but that was not the case with this particular work. This confirms what Luca Pablo Nunez says: [L]la parte más importante de sus lectores posiblemente fueron los medicos rurales o boticarios" ("[R]ural doctors and botanists probably figured most significantly among his readers") (Nunez 2007–2008: 204).

Agustí's *Libro de Agricultura* has contents similar to Herrera's *Agricultura general* and Olivier de Serres' *Le Théâtred'Agriculture et Ménage des Champs* (1600), and although he lays claim to originality, it is, in fact, quite closely aligned to Estienne's *L'agriculture et maison rustique* (1554).

Nevertheless, the Spanish book by Miquel Agustí had much more success in Portugal than the French texts, probably due to the language. The fact that certain ideas reached Portugal more easily through a Spanish version than in the original is not a unique case. If we consider mythographies, Baltazar de Vitoria's *Theatro de los Dioses* (1620) was much more popular than Cartari's *Imagini degli dei* (1556) (Rodrigues 2010: 48–67); and in architecture, the classical style finally arrived through Diego Sagredo's *Medidas del Romano* (1526) rather than via Alberti's *De re aedificatoria* (1485) (Rodrigues 2011c: 21–26). It would again appear that Spanish books proved much more accessible because of language and geographical proximity. However, in this case, three other aspects help explain the number of copies of the 1626 edition. First, in 1626, Portugal belonged to the Spanish crown. Second, Agustí's book was a smaller compendium than the French treatise published shortly before by Olivier de Serres (1600) and achieved greater success in Portugal (Rodrigues 2011b: 126–127). Third, the first edition, dated 1617, was published in Catalan and was only completed in 1626, with nine more parts written in Spanish and with a vocabulary in six languages—Spanish, Catalan, Latin,

Portuguese, Italian and French. This might also explain the success this compendium had among the Portuguese.

From the seventeenth century, another Spanish agricultural treatise stands out: Francisco Gilabert's *Agricultura Pratica* (1626). The Catalan nobleman Don Francisco Gilabert (1559–1638) was a disappointed courtier who returned to managing his estates after finding Philip III's court in Madrid full of vanity and empty promises (Rubies 1995: 7). He wrote *Agricultura Pratica*, published in 1626, 10 years after his most important text on Catalan society and politics: *Discourse* (1616). The Portuguese nobleman and historian Francisco Manuel de Melo used him as a source in his contemporary account of the Catalan revolt (*Historia de los movimentos y separación de Cataluna*, Lisbon, 1643). It starts with chapters defending the nobility of agriculture and the qualities of the perfect peasant, just as the classical authors had done earlier and just as these aspects had been defended by the French author Charles Estienne when eulogizing villas and country life.

The Growth in Importance of the French Treatises Throughout This Period

Charles Estienne's *Praedium rustic* was considered by R.E. Prothero to be the first methodical work on French agriculture. Published for the first time in 1554, it was augmented and later published by his son-in-law Jean Liébault as *L'agriculture et maison rustique* (1567). By 1600, this encyclopedia of rural life had gone into several editions and had been translated into German in 1579, Italian in 1581, English in 1600 and Dutch as late as the seventeenth century, but it was not translated into Spanish or Portuguese (Fussell 1969: 544). The calendar and the topics discussed (vineyards, olives, citrus fruit and so on) are typical of southern Europe, although Germans had been interested in this book ever since it was published. This French tradition of treatises on production and fruit-bearing continued in Olivier de Serres'*Théâtre d'agriculture* (1600), which also circulated in Portugal, although just two copies have been found. Pascal Julien discovered that Olivier de Serres' treatise had been in the library of the Marquis of Fronteira, though by then it had already disappeared (Julien 2011: 147). However, there are still copies of the 1623 and 1675 editions in the University of Coimbra's library. Olivier de Serres (1539–1619) is considered the father of French agronomy, and his compendium of more than a thousand pages approaches different issues, such as the villa owner's duties, land and vineyard cultivation, water distribution and food management. Among these subjects, there is one chapter of more than one hundred pages totally dedicated to gardens and gardening, identifying different typologies of French gardens created by that time, as well as technical and horticultural questions.

From the seventeenth century on, we continue to see a lot of agricultural treatises in Portuguese libraries. Nevertheless, the significance of these works differs considerably from that of the agricultural treatises like Herrera's editions from the sixteenth century or Olivier de Serres' *Théâtre d'Agriculture* in that the latter included chapters especially dedicated to gardens and gardening when there was little else on this topic. Treatises on garden art had taken a separate route from agriculture at the beginning of the seventeenth century, and garden literature had started to conquer a territory of its own. However, agricultural treatises from that point on dealt with similar cultivation problems, and were mostly to be found side by side in the same libraries, because both were of interest to villa owners.

There are two first editions of Joannes de la Quintinye's *Instruction pour les jardins fruitiers et potagers avec un traité des Orangers* (1690), but also later copies, for example, the 1739 and 1770 editions. Jean de La Quintinye (1626–1688) was the lawyer and gardener who created the Potager du Roi at Versailles to produce vegetables and fruits for the king. All his theoretical knowledge, tested by experience, which he had written in *Instruction pour les jardins fruitiers et potagers,* was published two years after his death. This treatise was translated by John Evelyn,[5] three years after the first edition, and entitled *The Compleat Gard'ner; or Directions for Cultivating and Right Ordering of Fruit-Gardens and Kitchen-Gardens; with Divers Reflections on several parts of Husbandry* (1693); it was subsequently translated into Italian seven years later. This book disseminated ideas throughout Europe on how to clip, crop, cut back and weed out fruit-trees, which had become such an important feature in gardens.

Nevertheless, the circulation of agricultural treatises intensified in the eighteenth century, albeit mostly in French. The treatises that reached Portugal were not only by French authors, they also derived from translations into French. For example, the agricultural treatise by the German botanist Georg Andreas Agricola, *L'agriculture parfaite*, arrived in a French translation in 1732, as did *L'agriculture réduite à ses vrais principes* by Johann Waller, which was translated into French from the Latin version edited in 1774. *Les Agrémens de la Campagne* (1750), written by the Dutchman Pieter la Court Van der Voort, also circulated in Portugal in French. Philip Miller's English dictionary of gardeners and cultivators was used too, but in a French translation (1768, 1785). Furthermore, *Le Jardinier solitaire, ou Dialogues entre un curieux et un jardinière solitaire* was a translation of Mr. Bradley's treatise in English. The copy I quote here belonged to the Countess of Vimieiro, in whose memory her husband ordered an obelisk for their garden at the Vimieiro palace.

Louis Liger's *Oeconomie Générale,* based on Estienne's text, became renowned as *La Nouvelle Maison Rustique,* which was expanded and reformulated by Jean-François Bastien in 1798 and was one of the most popular agricultural treatises of the eighteenth century. Copies have been found of *Oeconomie générale* (1701) and the 1749, 1768 and 1790 editions of *La Nouvelle Maison Rustique*. It was a book that had also circulated in Portugal in the Spanish version, translated by Francisco de la Torre y Ocón (1720). Louis Liger's treatise is a book in praise of country life and of instruction in how to farm, seeking to advise on the pleasures and profit to be derived from rural pursuits. Nevertheless, it became one of the best ways to divulge know-how about the cultivation of gardens.

[5] On John Evelyn see Michael Leslie's paper in this book.

This book was found in the libraries of the most famous monasteries and colleges, including the Monastery of São Vicente de Fora, the library of the Jesuit College in Campolide and the library of João Rebelo and Manuel Joaquim Rebelo. In the library collections mentioned above, there are eight copies in French, but also a Spanish translation (1720). Further works in French found in circulation in Portugal were Nicolas de Bonnefond's *Les Délices de la champagne* (1665), Carlo d'Aquino's *NomenclatorAgriculturae* (1736), Roux's *Traité de la culture et de la plantation des arbres* (1750), La Salle de l'Etang's *Manuel d'agriculture* (1764) and François Rozier's *Cours complet d'agriculture* (1791–1796), although on a smaller scale.

Comparing the Circulation of Agricultural Treatises in Portugal With That in Other Countries

What is remarkable about garden art in Portugal is the fact that, while Italian and French gardening treatises were divulging the Renaissance model of gardens, the essence of Portuguese gardens essentially remained quite different until 1650, probably due to different influences, but also simply because some of those treatises never circulated in Portugal. Gardening failed to be informed by this branch of knowledge. We have not been able to find some of the most important Italian, French and English Renaissance treatises on the development of garden art, such as the Italian work *La Villa* (1559) by Bartolomeo Taegio; or the French works *Traité du jardinage selon les raisons de la nature et de l'art* (1638) by Boyceau and *Thé âtre des plans et jardinages* (1652) by Mollet. Manuscript texts such as Agostino del Riccio's *Agricultura Teorica,* so important in the development of Italian Renaissance gardens and divulged through manuscript copies, never reached Portugal. English was not, at the time, an accessible language in Portugal, so it is not surprising that we have found no copies of Thomas Hill's *A Most Brief and Pleasant Treatyse* (1563), or William Lawson's *The Country House-Wife's Garden* (1618). Sixteenth-century German books on farming and gardening, such as Conrad Heresbach's *Rei rusticae* (1570), Martin Grosser's *Kurze und gar einfeltige Anleitung zu der Landwirtschaft* (1590) and Johann Coler's *Kalendar* (1591) and his *Oeconomia ruralis et domestica* (1594), also did not circulate in Portugal. The sources in southern Europe and in the north were different, but then, they had to be, because advice on farming and gardening useful in the south would not be useful in completely different geographical and biophysical environment. Even Heresbach has been criticized for divulging methods for cultivating figs and almonds, because what was fruitful in Italy and had been transmitted by classical authors could never be successful in Germany (Fussell 1969: 548).

While some treatises did circulate in Portugal, and probably in southern Europe too, many books on garden art from northern Europe did not reach Portugal. But it is also true that very common books on the Iberian Peninsula were almost unknown in northern countries (Rodrigues 2014a). For example, Alonso de Herrera's treatise was the most accessible book in Portugal and had probably been known there since

1528, as it represents the oldest copy found in Portuguese library collections, yet it was totally unknown in England until the middle of the seventeenth century. The oldest copy that exists there is a rare book from the British Library owned by John Evelyn, which was offered to him by the Earl of Sandwich, ambassador to Spain in 1665 (Grice-Hutchinson 1988: 123–129). The edition given to John Evelyn is that of 1645, which is of particular interest because it contains the treatises by Juan de Arrieta and Diego de Salinas on agriculture, Gonzalez de las Casas on growing silk, Luiz Mendez de Torres on bee-keeping, and Gregorio de los Rios on gardening, one of Evelyn's passions.

Conclusion

Early Modern Portuguese gardens appeal to our senses and maintain a strong relationship with the countryside. They were places of spatial unity and intimacy, full of colors, sounds, fresh odors, brilliance and different textures, combining aesthetic and recreational purposes with a traditional utilitarian and rural character. When we compare them with Italian and French gardens, these values stand to the fore. In comparison, the Renaissance Italian and French gardens appear as abstract and more erudite constructions, and in villas, the aesthetic and recreational elements dominate.

It is important to stress that my concern in this research was not to point out evidence that a theoretical knowledge of agricultural treatises was used directly and experimented with in Portuguese gardens of the Early Modern period. However, the Spanish agricultural treatises and the Hispano-Arabs texts that influenced them help us to understand the character of Portuguese gardens. This is not the case with other agricultural treatises such as that of Louis Liger, which has no connection with the Portuguese *quinta*. Furthermore, for a smaller group of gardens of the Early Modern and Modern periods, garden art treatises are essential for understanding the erudite sources of their artistic forms.

The Portuguese *horto* is clearly more closely related to the principles of garden art explained in the Spanish agricultural treatises of the sixteenth century than to the garden art propounded in Italian, French or English treatises. It was logical for owners of Portuguese "quintas" (estates), which were mainly devoted to cultivation, to have agricultural treatises in their libraries. Through these much more accessible books, a taste for recreational and aesthetic outdoor spaces was nurtured and developed.

Bibliography

Bourde, André. 1955. Agronomie et agronomes en France au XVIIIe siècle. Birmingham: Cambridge University Press.

Carapinha, Aurora. 1995. A essência do Jardim Português. PhD thesis on Landscape Architecture presented at the University of Évora.

Carapinha, Aurora. 2014. The Guardian of the Mediterranean Matrix – the Portuguese Garden. Gardens & Landscapes of Portugal 2: 6-15.

Carapinha, Aurora. 2015. Sucursal del Paraíso: o Libro del principio de la belleza y Fin de la sabiduría que trata de los fundamentos da arte de la agricultura, de Ibn Luyun. In Uma história de jardins. A arte dos jardins na tratadística e na literatura. ed. Ana Duarte Rodrigues, 49–65. Lisboa: Biblioteca Nacional de Portugal.

Carita, Helder. 1990. Tratado da Grandeza dos Jardins em Portugal. Lisboa: Círculo de Leitores.

Ferreira, José M. Simões. 2010. História da Teoria da Arquitectura no Ocidente. Lisboa: Vega.

Fussell, G. E. 1969. The Classical Tradition in West European Farming: The Sixteenth Century. The Economic History Review, New Series 22: 538-551.

Grice-Hutchinson, Marjorie. 1988. Some Spanish Contributions to the Early Activities of the Royal Society of London. Notes and Records of the Royal Society of London 42: 123-132.

Julien, Pascal. 2011. La bibliothèque du Marquis de Fronteira et la conception du jardin de son palais. In Tratados de Arte em Portugal/Art Treatises in Portugal, eds. Rafael Moreira and Ana Duarte Rodrigues, 145–152. Lisboa: Scribe.

Lambin, Denis. 1975. Pleasure with Profit: Some aspects of Landscape Gardening. Garden History 3: 29-36.

Le Dantec, Jean-Pierre. 1996. Jardins et Paysages: texts critiques de l'Antiquité à nos jours. Paris: Larousse.

Nunez, Luis Pablo. 2007-2008. Ediciones e historia textual del Libro de los secretos de agricultura de Miguel Agustín. Butlletí de la Reial Academia de Bones Lletres de Barcelona, 51: 199-223.

Olson, Lois, and Helen L. Eddy. 1943. Ibn-Al-Awam: A Soil Scientist of Moorish Spain. Geographical Review 33: 100-109.

Pérez, Joaquín Fernández, and Ignacio González Tascón (eds.). 1991. A propósito de la "Agricultura de jardines" de Gregorio de los Rios. Madrid: Real Jardín Botánico.

Rodrigues, Ana Duarte. 2010. The Most, and Less, Known Mythographers in Portugal. European Review of Artistic Studies 1: 48-67.

Rodrigues, Ana Duarte. 2011a. A Escultura de Jardim das Quintas e Palácios dos Séculos XVII e XVIII em Portugal. Textos Universitários de Ciências Sociais e Humanas. Lisboa: Fundação Calouste Gulbenkian and Fundação para a Ciência e Tecnologia/ Ministério da Ciência e Ensino Superior.

Rodrigues, Ana Duarte. 2011b. O conhecimento teórico ao alcance de arquitetos e jardineiros em Portugal durante a Idade Moderna. In Tratados de Arte em Portugal/Art Treatises in Portugal, eds. Rafael Moreira and Ana Duarte Rodrigues, 119-144. Lisboa: Scribe.

Rodrigues, Ana Duarte. 2011c. The circulation of Art Treatises in Portugal between the XV and the XVIII centuries: some methodological questions. In Tratados de Arte em Portugal/Art Treatises in Portugal, eds. Rafael Moreira and Ana Duarte Rodrigues, 21-42. Lisboa: Scribe.

Rodrigues, Ana Duarte. 2014a. O saber da arte dos jardins em circulação na Europa (1500-1900). In Mobilidade e Circulação: Perspectivas em História da Ciência e da Tecnologia, eds. Alice Santiago and Pedro Raposo, 87-93. Centro de História d'Aquém e d'Além-Mar and Centro Inter-Universitário de História da Ciência e Tecnologia.

Rodrigues, Ana Duarte. 2014b. Sustainability in Alonso de Herrera's treatise and the construction of the Mediterranean landscape. In The Garden as a Lab where cultural and ecological systems meet in the Mediterranean context, ed. Ana Duarte Rodrigues, 11-28. Évora: Centro de História da Arte e Investigação Artística/University of Évora.

Rodrigues, Ana Duarte. 2015a. The quality of Eugénio dos Santos Library on gardens in a Portuguese context. In Views on Eighteenth Century Culture: Design, Books and Ideas, eds. Leonor Ferrão and Luís Manuel A. V. Bernardo, 127–150. Cambridge: Cambridge Scholars Publishing.

Rodrigues, Ana Duarte. (ed.). 2015b. Uma história de jardins. A arte dos jardins na tratadística e na literatura. Ed. Ana Duarte Rodrigues, 49–65. Lisboa: Biblioteca Nacional de Portugal.

Rubiés, Joan-Pau. 1995. Reason of state and constitutional thought in the crown of Aragon, 1580–1640. The Historical Journal, 38, 1: 7.

Commerce and Erudition: Civic Self-Representation Through Botany and Horticulture in Germany, Sixteenth to Eighteenth Centuries

Iris Lauterbach

Abstract

Through several case studies, this paper examines the important role gardens played in civic self-representation and the creation of a civic "corporate identity" in the sixteenth to eighteenth centuries. Several German cities—and not residential cities—were crosspoints in the history of botany and horticulture. The progress of scientific botany depended on international networking, commercial and business contacts and communication among scholars. Autopsy, observation and description as basic scientific techniques for approaching nature and its phenomena were familiar to the physicians and apothecaries, as well as to the naturalists and botanists. The exchange of letters, drawings and written observations, about seeds and plants, were important steps in the development of botany as a science. The first garden books in the German language were written not by court gardeners, but by learned citizens: physicians, apothecaries, and Protestant pastors.

Horticulture as practised in the Free Cities of the Holy Roman Empire in Southern Germany was important for the introduction of rare plants into Central Europe. In many cities, in the seventeenth to eighteenth centuries, the urban elite had a particular interest in horticulture. Though private, the gardens are considered and shown as a part of the public urban space.

Garden history, when dealing with the early modern age, the sixteenth to eighteenth centuries, emphasizes the importance of gardens of the imperial, royal, and princely courts. The courts disposed of everything necessary for the creation and maintenance of large and important gardens: a property near the residence and estates near the residential city, together with the political authority to have these properties formed; the artistic potential of prominent architects and garden artists; and the manpower of

I. Lauterbach (✉)
Zentralinstitut für Kunstgeschichte, Munich, Germany
e-mail: i.lauterbach@zikg.eu

H. Fischer et al. (eds.), *Gardens, Knowledge and the Sciences in the Early Modern Period*, Trends in the History of Science, DOI 10.1007/978-3-319-26342-7_16

gardeners in the service of the court administration. This article, though, seeks to understand the development of horticulture as a science and an art in the sixteenth to eighteenth centuries, as an achievement of both the courts and the cities.

Certainly, the progress of scientific botany during the sixteenth century lead to the creation of botanical gardens, for example, at the imperial court in Vienna, under the direction of Carolus Clusius (1526–1609), or at the court of Wilhelm IV, landgrave of Hessen-Kassel, who asked the Nuremberg physician and botanist Joachim Camerarius the Younger (1534–1598) for his advice (Wickert 1993; Camerarius-Florilegium 2004). But scientific botany was mostly studied and practiced by academic institutions—a university or an academy of sciences—and displayed in specific spaces, the botanical gardens. This is the case for the French court, where botanical research was established roughly around 1600 at the Jardin du Roi, the botanical garden in Paris. At the Bavarian Court in Munich, a botanical garden was created only after the foundation of the Bavarian Academy of Sciences in 1759 by the Elector Max III Joseph. The botanical interest of Karoline Luise, landgravess of Baden, in the 1770s was an exception even in a century of garden passion such as the eighteenth century, and resulted from a personal concern with gardens and plants (Lauts 1979). All of the important botanical gardens, from the sixteenth century onward, belonged or still belong to universities, in Pisa, Padova, Bologna, Leiden, and Montpellier, as well as in Germany, be it Leipzig, Jena, Heidelberg or Altdorf (near Nuremberg) (Fig. 1).

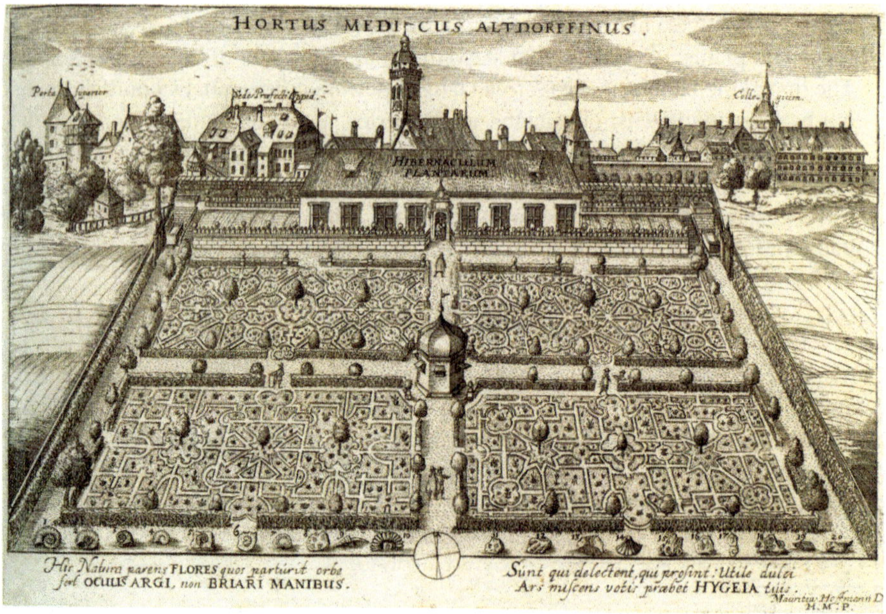

Fig. 1 Altdorf, Botanical garden, copperplate engraving by Mauritius Hoffmann, 1662. © After: Pavord, Anna. 1999. The tulip. London: Bloomsbury, p. 66

Several German cities—and not residential cities—were crosspoints in the history of botany and horticulture. The progress of scientific botany depended on international networking, commercial and business contacts, and communication among scholars undertaking their research at prestigious Italian and French universities (Nissen 1951; Lack 2001; Camerarius-Florilegium 2004). Autopsy, observation and description as basic scientific techniques to approach nature and its phenomena were familiar to the physicians and apothecaries, as well as to the naturalists and botanists. The exchange of letters, drawings and written observations about seeds and plants were important steps in the development of botany as a science. The first garden books in the German language were written not by court gardeners, but by learned citizens: physicians, apothecaries, and Protestant pastors (Wimmer/Lauterbach 2003; Peschel 1597/2000).

In his publication "De hortis Germaniae liber" (Gessner 1561), among famous gardens in Germany, the Swiss naturalist Konrad Gessner (1516–1565) lists only a few residential gardens, for example, the Munich "Hofgarten", but numerous private gardens: in cities along the Rhine river, starting with Basel and Strasbourg, in Lindau and Konstanz, in Augsburg, Esslingen, Nuremberg, Breslau, Leipzig, Torgau and many others. Gessner had himself visited the gardens he mentions, which mostly belonged to physicians and apothecaries.

The great German cities of some commercial and political importance were trade cities such as Leipzig and Breslau, the Free Cities of the Holy Roman Empire such as Frankfurt, Augsburg, Nuremberg, and the Hanse Cities such as Königsberg and Hamburg. Local pride and patriotism were at the origin of horticulture and botany as practised by many of the citizens and patricians of these cities, which, although not being residential cities of princely courts, had a prosperous garden culture. Gardens often played an important role in civic self-representation and the creation of a civic "corporate identity", in the sixteenth and early seventeenth centuries, and starting again after the 30 Years War. Descriptions of the vegetation and flora and of horticulture in general were highly suited to effusions of local patriotism. In city chronicles from the sixteenth to eighteenth centuries, the vegetation and flora are often used on a metaphorical level in order to describe the prosperity of these communities. Since gardens show how citizens identified with their prosperous cities, they were interpreted as civic and collective achievements. The blooming vegetation of private gardens is shown as a result of a successful city government. In his garden book from 1682, the Nuremberg patrician Wolf Albrecht Stromer von Reichenbach only mentions a few private German gardens, those of Free Cities of the Holy Roman Empire in or near his native Franconian area: "[T]he magnificent and beautiful gardens surrounding Ulm, Augsburg, Regensburg and Nuremberg, our beloved home country, in which city the art of gardening is flourishing, as prove the well-ornamented and widely spread flower gardens"[1]

[1] "[...] die herrlichen und schönen Gärten um Ulm, Augspurg, Regenspurg und Nürnberg, in welchem letztern Ort, unserem geliebten Vatterland, sonderlich die Gärtnerey in vollem Schwange geht, wie die schön-ausgezierte Blumen-Gärten daselbsten es erweisen, mercklich und weitläuffig vergrössern."

(Stromer von Reichenbach 1673: vol. 2, 276). A catalogue of shrubs and perennial flowers adapted to the Franconian region is included in Stromer's book (Stromer von Reichenbach 1671: vol. 1, 199–286).[2]

From the seventeenth century, flower books and plant lists more often referred to a certain city or a delimited region. One further example is the physician and botanist Johann Lösel's publication on the Prussian flora: "Plantae in borussia sponte nascentes e manuscriptis parentis" (Königsberg 1654). Set against the city silhouette of Königsberg, the frontispiece shows a man climbing up a cherry tree and a young woman adorned with and holding flowers.

Collecting and cultivating rare, Mediterranean or exotic plants in the sixteenth century was a challenge, not only in the residential culture but also as a point in princely self-representation. A garden with extravagant species was a sort of botanical "kunstkammer". Academic qualification and scientific achievement in general had a large potential in strategies of civic self-representation. In this context, since the late sixteenth century, scientific botany played an important role. Publications on horticulture and garden design, as well as on botany in general, gained importance (Nissen 1951).

Horticulture as practised in the Free Cities of the Holy Roman Empire in Southern Germany was important for the introduction of rare plants, such as tulips and citrus and others, into Central Europe. In the forefront were Augsburg, which, in the sixteenth century, maintained close trade links with Italy and the rest of Europe and even the overseas territories through the Fugger family, and Nuremberg, where patricians and other citizens maintained intensive commercial and academic contacts with Italy. Italian Renaissance gardens, adorned with works of sculpture, fountains and flora such as citrus trees, received an overwhelmingly enthusiastic response from scholars, poets and merchants.

Communication among scholars studying and undertaking their research at prestigious French and Italian universities was crucial to the cultural transference between Italy and Germany that took place from the sixteenth century onwards in the field of horticulture. The academic relations between German cities such as Nuremberg or Augsburg and, for example, the city and the university of Bologna, had a long history since the Middle Ages. The importance of academic networks for horticulture is obvious in the letters and publications of researchers and botanists of the early Modern period, in Gessner's publication "Hortis Germaniae liber" (1561) as well as in his manuscript "Historia plantarum" (before 1565). Gessner's famous manuscript with notes and drawings was bought by Camerarius in 1580 and today is conserved in the Erlangen university library (Conradi Gesneri Historia Plantarum 1972–1980).

[2] "Catalogus der Bäume und Stauden so bey uns in Francken wachsen und in den Gärten gezielet werden."

Augsburg

Rare plants were often first introduced and cultivated in the Free Cities of the Holy Roman Empire in Southern Germany. Through the Fugger family, in the sixteenth century, Augsburg maintained close trade links with Italy and the rest of Europe, and the overseas territories.

In April 1559, Gessner admired a rare red tulip in the Augsburg citizen Johann Heinrich Hörwarth's garden (Fig. 2). A woodcut after his drawing, one of the earliest tulip pictures, illustrates Gessner's later publication, as a botanical portrait (Pavord 1999; Goes 2004).

From the early sixteenth century, the gardens of the immensely wealthy Fugger family in Augsburg met with an enthusiastic response by the visitors (Heerwagen 1915; Trauchburg 2001): Jakob Fugger's garden "extra muros", Raymund Fugger's famous garden, adorned with beautiful fountains, statues and pleasure houses, among other things. A portrait by Martin Schaffner shows Raymund Fugger

Fig. 2 Augsburg, Tulip in Johann Heinrich Hörwarth's garden, April 1559, drawing by Conrad Gessner, Universitätsbibliothek Erlangen. © After: Pavord, Anna. 1999. The tulip. London: Bloomsbury, p. 189

holding a citrus fruit, the precious fruit being a symbol both of Fugger's international commercial network and of his personal passion for horticulture and botany (Fig. 3) (Teget-Welz 2008). Gardeners often had themselves portrayed holding a

Fig. 3 Martin Schaffner, Portrait Raymund Fugger, holding a citrus fruit, oil on canvas, about 1530, private collection. © Teget-Welz, Manuel. 2008. Martin Schaffner. Leben und Werk eines Ulmer Malers zwischen Spätmittelalter und Renaissance. ed. Haus der Stadtgeschichte - Stadtarchiv Ulm (Forschungen zur Geschichte der Stadt Ulm, 32). Ulm/ Stuttgart: Kommissionsverlag W. Kohlhammer, pl. 90

piece of fruit or a bunch of flowers, a touching image of a life's passion, and a commitment to the fragility and fugacity of life and beauty.

Not only the wealthy Fugger family but other Augsburg merchants, too, boasted their private garden paradises. In 1605, the tradesman Andreas Scheler married Jacobina Cristell, daughter of an Augsburg family of merchants, thus receiving Augsburg citizenship. The couple had a house and garden in Jesuitengasse 14, containing rare flowers and plants (Stetten 1779: 122).[3] Many of these flora are portrayed in water colour drawings contained in an in-folio volume, a *florilegium* dating from 1615 and conserved in the Staats- und Stadtbibliothek Augsburg (Figs. 4 and 5).[4] The frontispiece shows a flower garden and the back front of a town house. The building reveals typical Augsburg features; the drawing probably does not represent a generalized architecture, but refers to Scheler's house itself. On the ground floor, a loggia, with an ornamental wall decoration in the lower part and landscape paintings in the upper part, opens to the garden with four arcades. In the middle of the flower beds, a fountain can be seen: it consists of a red marble basin on a marble pedestal and a bronze statuette spitting numerous thin water jets. The statuette represents a rider on a rearing horse. It is reminiscent of the famous bronze sculptures of the new fountain on the Rindermarkt in Munich, created for Duke Ferdinand by the ducal court sculptor Hubert Gerhard in 1587 (Diemer 2004: 1 and 194–206). In 1609, Duke Maximilian I had the fountain translocated to the Munich residence, omitting the central basin and the rider on horseback; the location of both after 1609 are unknown. It is unknown, as well, whether the artist who drew Scheler's *florilegium* introduced this sort of modern fountain creation—the bronze statuette of a rider on a jumping horse—as an element of fantasy. If Scheler really owned such a fountain—which we don't know—it must have echoed Gerhard's Munich sculpture. The arrangement of the flower beds in Scheler's garden perhaps refers to early seventeenth century prints and engravings, such as Crispijn de Passe's "Hortus floridus" (Arnheim 1614). The choice of species, which can be distinguished on the frontispiece and which are portrayed in the volume—tulips, iris, *fritillaria*, peonies, lillies, auricula and many others—perhaps refer to the "Hortus Eystettensis", published two years before, in 1613, by the apothecary and botanist Basilius Besler from Nuremberg (Besler 1613, 1613/1999). Besler, who is shown holding a basil plant as his "armorial bearings" (Fig. 6), had created the Eichstätt archbishop's residential garden together with Joachim Camerarius and Carolus Clusius, and for his book, consulted the physician and botanist Ludwig Jungermann, later a professor at the universities of Gießen and Altdorf.

Still, in the eighteenth century, the private gardens of Augsburg were considered and shown virtually as public attractions of the city. "Augspurgische Gartenlust", a

---

[3] "Andreas Scheler hatte gleichfalls einen vortrefflichen Garten. Er zog darinn die schönsten Blumen und Kräuter, und machte den Aufwand, die seltensten abmalen zu lassen, welche Sammlung noch vor einigen Jahren unter dem Vorrath von Seltenheiten des gelehrten Rathskonsulenten von Hertenstein gewesen ist."

[4] Staats- und Stadtbibliothek Augsburg, 2°Cod. H. 70: Blumenbuch des Andreas Scheler, Jesuitengasse 14.

Fig. 4 Augsburg, Andreas Scheler's Garden, frontispiece of his florilegium, 1615, water colour drawing, Staats- und Stadtbibliothek Augsburg. © Staats- und Stadtbibliothek Augsburg, 2° Cod. H. 70

Fig. 5 Augsburg, Andreas Scheler's Garden, fol. 157 of his florilegium, 1615, water colour drawings, Staats- und Stadtbibliothek Augsburg: Lathyrus vernus, Crocus flavus, Matthiola incana. © Staats- und Stadtbibliothek Augsburg, 2° Cod. H. 70

Fig. 6 Basilius Besler's armours, portrait and dedication, from: Hortus Eystettensis 1613, coloured copperplate engraving. © After: Besler, Basilius. 1613/1999. Hortus Eystettensis. Ed. Klaus Walter Littger, Werner Dressendörfer. Köln et al.: Taschen

series of twelve engravings by Carl Remshart after Johann Thomas Kraus, published in the 1720s by the editor Martin Engelbrecht, shows wealthy citizens's gardens, most of them set against the city silhouette (Fig. 7). In 1779, the chronicler Paul von Stetten, in his "Kunst-Gewerb- und Handwerks-Geschichte der Reichs-Stadt Augsburg", regrets the decline of horticulture in his city, at the same time emphatically praising the qualities of garden culture by a wealthy citizen: he can have great pleasure in cultivating a garden, if he avoids excess and if his expenses are reasonable; his industry and efforts will advance him, and the observation and study of the varieties of so many plants will elevate his soul to the Creator's glory (Stetten 1779: 124–125). Stetten's ideal of the citizen as a gardener is characterized by typical civic qualities guided by Christian faith: assiduity, activity, and moderation.

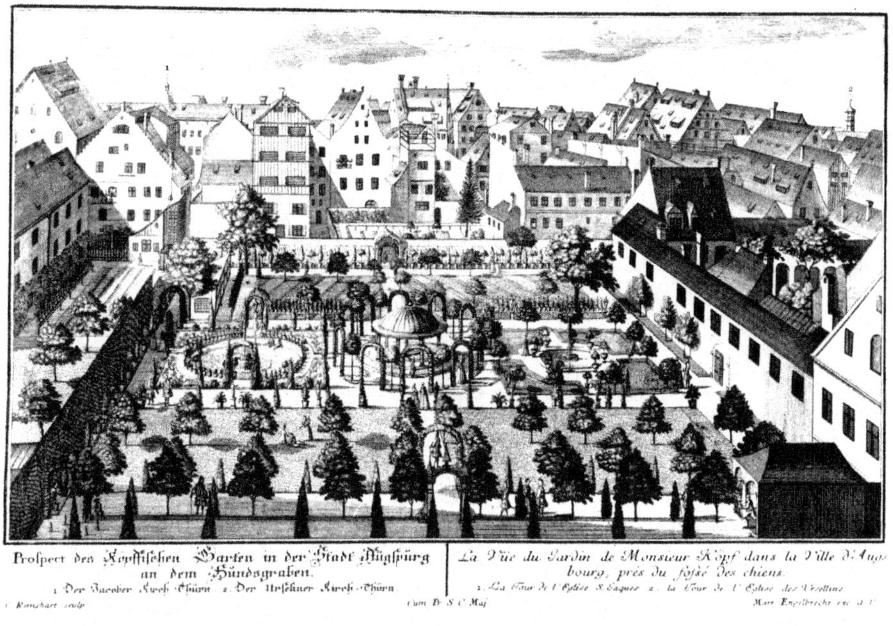

Fig. 7 Augsburg, Köpff's garden, engraving by Carl Remshart after Johann Thomas Kraus, Augspurgische Gartenlust, 1720s, Augsburg: Martin Engelbrecht. © Author

Breslau

Camerarius was one of the correspondents of Laurentius Scholz, another passionate gardener, a physician and botanist as well. Scholz's garden in Breslau is an extraordinary early example of the highly developed civic garden culture in late sixteenth Century in Germany, its international origins and learned inspirations (Oszczanowski 2004; Lauterbach 2008). A remarkable engraving dating from 1588 shows the garden, which followed Italian models, but does not hide its character as a typical Central European civic garden with utilitarian aspects such as quarters for vegetables (Fig. 8). The title of the print "Horti Scholziani Chalcographica Delineatio" emphasizes the innovative aspects: the new printing technique of the copper engraving (instead of woodcut) and also the fact itself of showing a private garden as a creation and a naturalist's "musaeum" of its own rights. From 1576 to 1579, Scholz had studied medicine at the universities of Padova and Bologna, and from there on had immediately travelled to Florence, Rome, and Naples. He undertook an intensive correspondence with Joachim Camerarius the Younger in Nuremberg, whom he sent seeds of rare plants and cuttings from Italy, by mail, in a box. Among the Camerarius documents, Scholz's correspondence is partly conserved in the famous Trew collection in the manuscript department of the Erlangen university library (Schmidt-Herrling 1940; Schnalke 1995). Back in his native city of Breslau, Scholz created his garden, which, from 1588, he opened to his

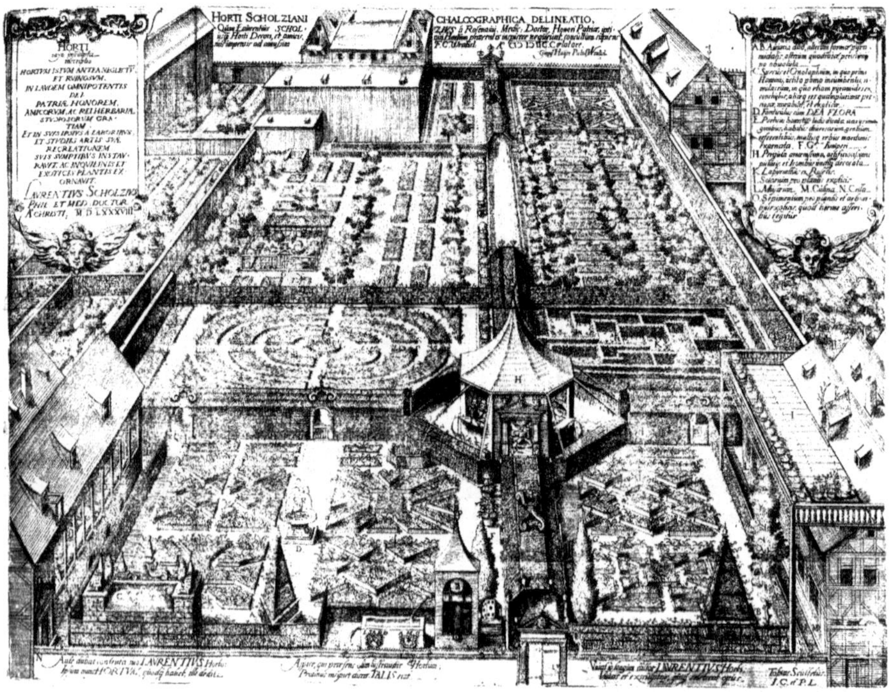

Fig. 8 Breslau, Laurentius Scholz's garden, „Horti Scholziani Chalcographica Delineatio, 1588, copperplate engraving by Tobias Scultetus. © After: Lauterbach, Christiane. 2008. Gartengesetze des Späthumanismus. Gärten und Parks als Lebens- und Erlebnisraum. Ed. Stefan Schweizer. Worms: Werner'sche

friends and guests. Several panegyrical texts by other authors describe the garden and the colloquial parties that took place in it. The engraving repeats a Latin text written on a stone plaque at the garden entrance which expressed the owner's gratitude to his native city and his friends, and self-confidently explained his concerns.

In a small publication, Scholz boasts of having remodelled a formerly neglected garden "ruin", creating a beautiful garden adorned with native and exotic plants, which are listed in his "Catalogus Arborum, Fruticum, ac Plantarum, tam indigenarum quam exoticarum, hortimedici D. Laurentii Scholzii Medici Vratisl. [aviensis]" (Breslau 1594). Following Scholz's preface, in which he underlines the importance of a botanical catalogue together with images, the illustrations help to commemorate the vegetation when the summer has gone and the garden is no longer blooming, and they virtually reassemble the garden owner's botanical treasures (Scholz 1594, s.p.).[5] The flower drawings, probably lost today, were

[5] "Hac intentione omnes etiam illas plantas, quae in hoc indice continentur, per Georgium Freibergerum, pictorem Vratislaviensem, non solum ut filius meus hyberno tempore, quo firmius memoriae ipsius inhaerent, has perlustrare posset, verum etiam, ut laborum meorum memoriam aliquam mihi ponerem, ad vivum depingicuravi."

executed by the Silesian painter Georg Freiburg. Scholz's Breslau garden features the characteristics of an important civic garden of the early period of horticulture: its owner and creator was a self-assured physician, a citizen proud of his international relations and social status and of his research and study. As a botanist and as a collector of botanical species, as well as of art objects, Scholz wanted his collection to be commemorated in a catalogue and represented in drawings.

Frankfurt, Hamburg, Nuremberg

In many cities, in the seventeenth to eighteenth centuries, the urban elite had a particular interest in horticulture. In 1641, in Frankfurt on the Main, the editor and engraver Matthäus Merian republished an enlarged version of his father-in-law, Johann Theodor de Bry's, *florilegium*: "Florilegium Renovatum et Auctum". In his dedication to Johann Schwind, mayor of Frankfurt, Merian underlines that Schwind, a passionate gardener, had "received a bright spark of inspiration during a journey in Italy undertaken a few years ago". Schwind dedicated himself to his own garden, where he cultivated native and foreign plants. Due to the mayor's "intelligence, love, application, and generosity" in the care of his garden, "even foreign visitors who were curious about rarities, and garden lovers, spread the name and celebrity and fame of the city"[6] (Merian 1641: 8).

In many printed views, the private gardens are set against the background of the city silhouette, houses, churches and towers. Though private, the gardens seem, as in Augsburg, a part of the public urban space. A fine example is Hans Simon Holtzbecker's view of the Anckelmann garden in Hamburg, from around 1669 (Fig. 9) (Heilmeyer 2003), when Caspar Anckelmann was an alderman of the city government. The same Holtzbecker, in the same period, made the beautiful watercolours for a comprehensive *florilegium* for the Hamburg mayor Barthold Moller, a garden lover, too (Horbas 2006; Roth 2007).

Nuremberg was the most important Free City of the Holy Roman Empire due to the fact that, until 1806, the Imperial Regalia were kept in the Franconian capital. The city of Nuremberg created a sort of botanical self-identity, on many levels: commercial, social, intellectual, artistic, and literal (Tschoeke 2008).

[6] Schwind habe "einen starcken Fewerfunken zur Entzündung dieses Geistes uff derer vor wenig Jahren gethanen Italianischen Raisen empfangen, in welchem Land, das etlich darein verliebte dem ersten erschaffenen Paradiß vergleichen wollen, es von köstlichen schönen Gärten eine geraume Anzahl gibet, und derer ohne Zweiffel mehr als bey andern Nationen zusehen seynd. Dem sey aber wie ihm wolle, so befindet sich auß dem Garten, welchen E.e. und Herrl. [Schwind] allhier von etlich Jahren hero uffgerichtet, von vielen underschiedlichen zumahl Landsfrembden Gewächsen und Früchten gepflantzet, mit statuis, wohldisponirten Durchgängen, Außtheilungen, Wasserwercken, und einem Lusthauß gezieret, und mit stätigem grossen Unkosten unterhalten, was bey dem Angeber, Uffrichter, Pflantzer und Erhalter derer Dingen für ein Verstand, Lieb, Fleiß, und Liberalität vorhanden, davon so gar die Statt selbsten, bey frembden, die der rariteten-vernünfftige inquirenten, und Liebhaber, auch Außbreiter seyn, einen sondern Namen, celebritet, und Ruhm erlanget, und bißdahero erhalten hat."

Fig. 9 Hamburg, Caspar Anckelmann's garden, gouache by Hans Simon Holtzbecker, about 1669, Staatliche Museen Berlin, Kupferstichkabinett. © After: Ein Blumengarten des Barock. 2003. Hortus Anckelmannianus, mit einem Essay von Marina Heilmeyer. München et al.: Prestel

Physicians and botanists, such as Joachim Camerarius the Younger and his circle of scientists, had a strong impact on the foundation of an academy, later university, in the city of Altdorf near Nuremberg, where, in 1626, a botanical garden was created (Fig. 1). From 1644, the members of the so-called Pegnitz Flower society (Pegnesischer Blumenorden), an academic corporation, dedicated themselves to the creation and purification of the German language as an idiom for scientific discussions and publications. Their academic signs and names were flowers. In the second half of the seventeenth century, a local school of flower painting had enormous success (Ludwig 1998). The famous naturalist and artist Maria Sibylla Merian (1647–1717), a native of Nuremberg, set standards for the beauty and exactness of scientific drawings representing fruits, flowers and insects. The sensational book about her research into the insects and flora of Suriname in South America—"Metamorphosis Insectorum Surinamensium"—was published in 1705.

Nuremberg: The Volkamer Family

The patricians and other rich citizens of Nuremberg, along with learned naturalists and scientists, did not dispose of large properties for their gardens. Several of them specialized in botany, which can be studied on a small scale. A well-known family of erudite merchants, as well as botanists and gardeners, in the city of Nuremberg was the Volkamer family—not to be confused with the Volckamers, a Nuremberg patrician family. In the seventeenth and eighteenth centuries, members of at least four generations of the Volkamers were passionate gardeners and had a high interest in botany (Lieb und Lust 2008; Lauterbach 2011; Volkamer 1708–1714/forthcoming). Johann Volkamer (1576–1661) achieved prosperity with a silk factory he established at Rovereto in northern Italy and by applying himself to trade, thence acquiring, in 1614, a property in Gostenhof, a suburb of Nuremberg, with a view to laying out a garden. An engraved portrait shows him holding a bunch of flowers (Fig. 10). His son Johann Georg Volkamer (1616–1693) was a physician and a scientific scholar of international renown, and also a member of the Pegnitz flower society, with a sunflower as his emblem. Two of his sons have an important place in the history of gardens and of botany: Johann Christoph Volkamer (1644–1720), who took over the family business and the Gostenhof garden (Fig. 11), and Johann Georg the Younger (1662–1744), who made a name for himself as a "famous medicus who was regarded as Germany's greatest botanist" (Will 1758). Johann Christoph's son Johann Magnus inherited the Gostenhof garden where, in 1726, an aloa americana bore its extremely rare flower (Fig. 12).

Johann Christoph Volkamer published two volumes of his famous book on citrus fruits—"The Nuremberg Hesperides"—in 1708 and 1714. A third volume, which the author was unable to complete before his death in 1720, is partly conserved (Volkamer 1708–1714/forthcoming; Lauterbach 2011; Martz 2011). Based on the state of knowledge at the time, Volkamer's work on the Hesperides provides an extraordinarily comprehensive classification and description of citrus trees in a lavishly illustrated text. It is the first systematic description in German of the genus *Citrus*. The copperplates show each fruit life-size at the top, sort of floating above *vedute* showing gardens, country houses, villas and landscapes of the Nuremberg region and Northern Italy.

The Volkamer family included successful merchants and famous physicians dealing and networking with many learned naturalists all over Europe. The author's family is the key to understanding and appreciating the scientific ambition and international scope of the work on the Hesperides.

Denominating his own garden in Latin, on the ancient Roman model—as *Viridarium Suburbaneum*[7]—Volkamer was laying claim to his international reputation. But with a strong impetus of local patriotism and accumulating several superlatives, Volkamer laid his own work on citrus at the feet of his native Nuremberg, "this Queen of German cities":

[7] Volkamer (1714: pl. VIIII).

Fig. 10 Portrait Johann Volkamer, etching by Bartholomäus Kilian after G. Strauch, 1660. © Stadtarchiv Nürnberg

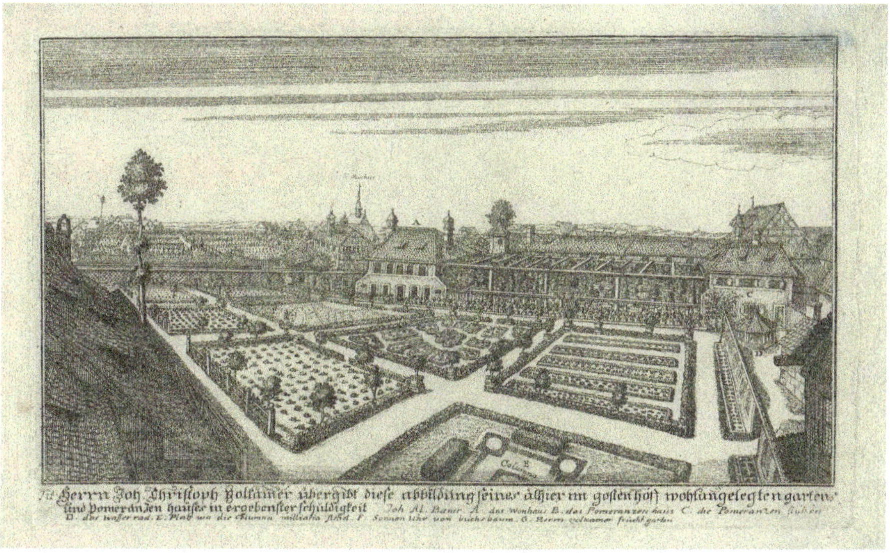

Fig. 11 Nuremberg, Johann Christoph Volkamer's garden in Gostenhof, about 1700, etching by Johann Alexander Boener. © Lust und Lieb hat mich beweget. 2008. Nürnberger Gartenkultur, ed. Jutta Tschoeke. Nürnberg: Museen der Stadt Nürnberg - Grafische Sammlung, cat. 23, 22

> Natives and foreigners alike will readily concede/to my Fatherland/noble Nuremberg/the reputation/of being the best-situated/most beautifully built city in the German Empire/and not just provided with all necessities/but also with all delights. Among these last are accounted/most justifiably/gardens for art and pleasure/which reveal themselves to be as exquisitely beautiful and lovely/as they are large and numerous/within and without the city walls[8] (Volkamer: vol. 1, preface).

Volkamer intended his work on citrus as a tribute to his native city. The topographical settings shown in the background of the title-page copperplates of all three volumes are, in fact, the city of Nuremberg and, in the second volume, Volkamer's own garden at Gostenhof. The plate featuring a lofty "*Cedro grosso Bondolotto*" above a *veduta* of the Free City of Nuremberg is a strong statement of local patriotism (Fig. 13). This is equally true of other botanical catalogues of the time, e.g., the Breslau citrus catalogue compiled by Caspar Wilhelm Scultetus (s. l. 1731) (Wimmer 2011), and publications dealing with Caspar Bose's garden in Leipzig (Wehmann 1723; Krausch and Wimmer 2000). Volkamer corresponded

[8] "Meinem Vatterland/ dem edlen Nürnberg/ geben In- und Ausländer/ungezwungen den Ruhm/ daß es die wolgelegenste/schönst-gebauteste/und nicht allein mit aller Bedürffniß/sondern auch mit allen Ergötzlichkeiten/best-versehenste Stadt im teutschn Reich seye. Zu diesen letztern rechnet man/mit dem besten Fug/die Kunst- und Lust-Gärten/ welche so auserlesen schön und anmuthig sich darstellen/als groß und zahlreich die Menge derselbigen/in- und ausserhalb der Ring-Mauren ist."

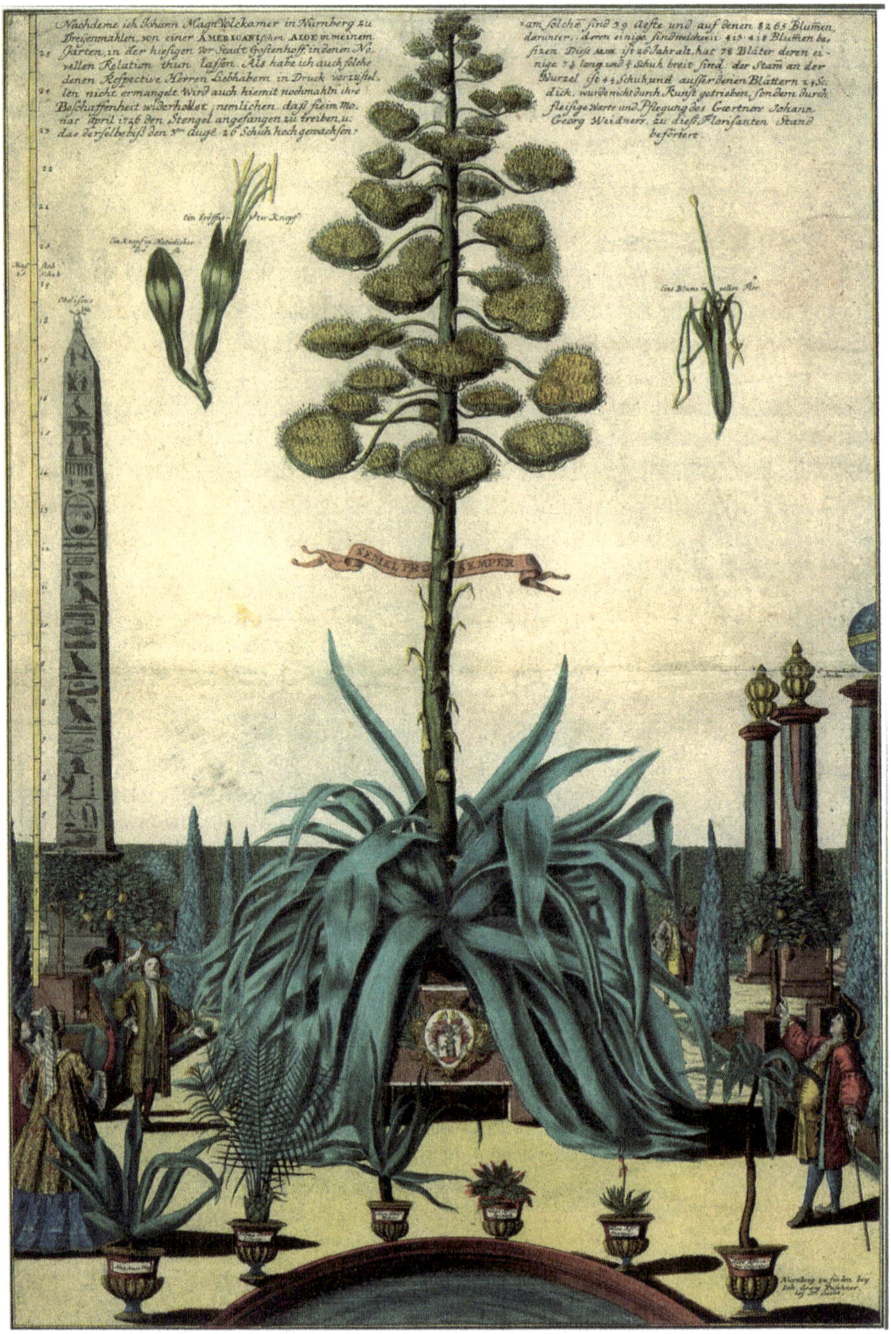

Fig. 12 Johann Georg Puschner, „Aloe americana florens in horto Volcameriano", 1726, coloured copperplate engraving. © Lust und Lieb hat mich beweget. 2008. Nürnberger Gartenkultur, ed. Jutta Tschoeke. Nürnberg: Museen der Stadt Nürnberg - Grafische Sammlung, cat. 23, 22

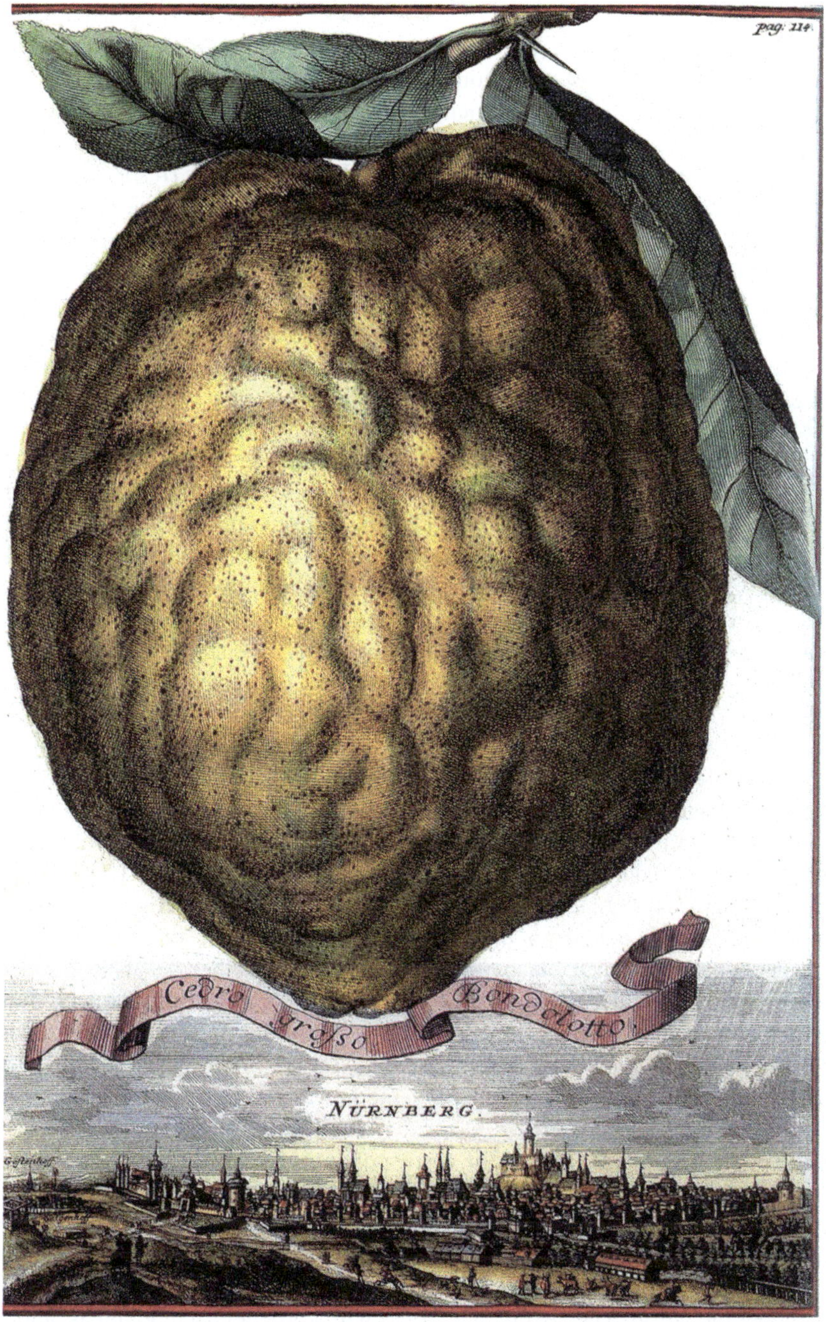

Fig. 13 Johann Christoph Volkamer, Nürnbergische Hesperides, Nürnberg 1708, vol. 1, pl. pag. 114, coloured etching and copperplate engraving. © After: Nürnbergische Hesperiden und Orangeriekultur in Franken (Orangeriekultur: Schriftenreihe des Arbeitskreises Orangerien in Deutschland e.V., 7). Petersberg: Michael Imhof Verlag, p. 94

with Caspar Bose (1645–1700), his equal as a successful merchant and a botanist at the same time, who was renowned for the rare and exotic plants he grew, including a valuable cherished citrus collection.

The named Volkamer family members were both prosperous merchants and internationally renowned scholars. Although they were not patricians, they were firmly rooted in the social life of the Free City. The *Nuremberg Hesperides* mirrors, in many respects, the interests, personal acquaintanceships and knowledge in various fields possessed not only by the author himself but also by his family. The Volkamer family shows how botanical specialization could be a sort of excellent solution for a lack of space, be it fruits or flowers, aloa, citrus or ananas: Johann Christoph Volkamer specialized in citrus fruits, his younger brother in flowers (auriculae) (Volkamer 1708); in 1700, Johann Georg had published the botanical catalogue "*Flora Noribergensis*" (^{2}1718); a near friend of the family, Michael Friedrich Lochner (1662–1720), a Nuremberg physician and naturalist, specialized in pineapples (Lochner 1716).

The *Nuremberg Hesperides* is the scholarly life's work of a wealthy merchant who was also a dedicated amateur with a strong desire to make a contribution of his own to a noble, scholarly competition, a case of familial and local, Nuremberg *aemulatio*. Johann Christoph Volkamer wrote in German for people like himself—amateur gardeners. The work is distinguished by its author's efforts to apply the scientific method consistently, and, on the other hand, his insistence on writing it in German in order to be understood by everyone. Volkamer wrote his work for garden owners who had some income and sufficient spare time to dedicate themselves intensively to their gardens and the study of botany.

Like his brother and father, Volkamer exchanged ideas constantly with numerous garden-owners and gardeners in Germany and abroad. He received valuable information from Schwöbber, Hamburg, Leipzig, Dresden, Breslau, Bohemia, Holland, Italy, Spain, South America (Suriname, Curaçao), the Cape of Good Hope and many other places near and far. Fruits, seedlings, seeds and even entire plants, with either naked roots or root balls, were sent to him. His Italian correspondents worked in places that his father and brother had also frequented while studying in Italy (including Padua, Bologna, Rome and Naples). Johann Christoph Volkamer's letters have not survived, but he often refers to the people with whom he corresponded and exchanged ideas, simply calling them "good friends". Botanical research in the early modern age operated through extensive international networks.

One of the largest collections of correspondence on medicine, botany and the natural sciences was assembled by Christoph Jakob Trew (1695–1769), a Nuremberg physician and botanist. He, too, was a member of a scholarly family and knew the Volkamers. The scientists Johann Georg Volkamer senior and junior are represented in Trew's collection by a copious correspondence with scholars throughout the world. They exchanged ideas with gardeners and botanists, and their gardens and places of work are also mentioned in the *Hesperides*. Checking such references systematically against the correspondence of his father and brother makes it possible to identify some of the "good friends" of Johann Christoph

Volkamer's. The gardener in Schwöbber, for instance, is Johann Friedrich Berner, whose correspondence from the years 1712 to 1717 has survived. Plant inventories and citrus catalogues are also mentioned as enclosures in letters written to the Volkamers. The Trew collection makes abundantly clear that Johann Christoph Volkamer used, for his research into citrus plants, the contacts that existed between his family, the Nuremberg scientific community and that of the world.

Bibliography

Sources

Besler, Basilius. 1613. Hortus Eystettensis. s. l.
Besler, Basilius. 1613/1999. Hortus Eystettensis. Ed. Klaus Walter Littger, Werner Dressendörfer. Köln et al.: Taschen.
Camerarius-Florilegium. 2004. Vollständige Faksimile-Ausgabe der Handschrift Ms. 2764 der Universitätsbibliothek Erlangen-Nürnberg, Einleitung: Dieter Vogellehner. Erlangen: Harald Fischer Verlag.
Gessner, Conrad. Conradi Gesneri Historia Plantarum. 1972–1980. Aus dem botanischen Nachlaß von Conrad Gesner (1516–1565). Universitätsbibliothek Erlangen. Eds. Zoller et al. 8 vols. Dietikon/Zürich: Graf.
Gessner, Conrad. 1561. De hortis Germaniae lieber. Straßburg.
Kraus, Johann Thomas. s.a. Augspurgische Gartenlust. Augsburg: Martin Engelbrecht.
Lochner, Michael Friedrich. 1716. Commentatio de Ananasa sive nuce pinea Indica. Nürnberg.
Lösel, Johann. 1654. Plantae in borussia sponte nascentes e manuscriptis parentis. Königsberg.
Merian, Maria Sibylla. 1705. Metamorphosis Insectorum Surinamensium. Amsterdam.
Merian, Matthäus. 1641. Florilegium Renovatum et Auctum. Frankfurt am Main: Merian.
Passe, Crispijn de. 1614. Hortus floridus. Arnheim.
Peschel, Johann. 1597/2000. Gartenordnung. Leipzig 1597. Ed. C. A. Wimmer. Nördlingen: Uhl.
Scholz, Laurentius. 1594. Catalogus Arborum, Fruticum, ac Plantarum, tam indigenarumquam exoticarum, horti medici D. Laurentii Scholzii Medici Vratisl. Breslau.
Stetten, Paul von. 1779. Kunst-Gewerb- und Handwerks-Geschichte der Reichs-Stadt Augsburg. Augsburg.
Stromer von Reichenbach, Wolf Albrecht. 1671-1673. Die edle Garten-Wissenschafft Aus Petri Laurembergii Rostochiensis Horticultura. 2 vols. Nürnberg: Endter.
Volkamer, Johann Christoph. 1708/1714. Nürnbergische Hesperides. Ed. Iris Lauterbach. Köln: Taschen (forthcoming).
Volkamer, Johann Christoph. 1708/1714. Nürnbergische Hesperides. 2 vols. Nürnberg: Endter.
Volkamer, Johann Georg. 1700. Flora Noribergensis. Nürnberg.
Wehmann, Achatz Friedrich. 1723. Hortus Caspar Bosianus. Leipzig.
Will, Georg Andreas. 1758. Nürnbergisches Gelehrten-Lexicon, 4. Nürnberg and Altdorf.

Secondary Literature

Diemer, Dorothea. 2004. Hubert Gerhard und Carlo di Cesare del Palagio. Bronzeplastiker der Spätrenaissance. (Jahresgabe des Deutschen Vereins für Kunstwissenschaft Berlin 2002/2003), 2 vols. Berlin: Deutscher Verlag für Kunstwissenschaft.
Goes, André van der (ed.). 2004. Tulpomanie. Die Tulpe in der Kunst des 16. und 17. Jahrhunderts. Zwolle: UitgeverijWaanders.

Heerwagen, Leo. 1915. Gartenbau und Gartenkunst in der Stadt Augsburg von den Anfängen bis zum heutigen Tage. Augsburg: Reichel.
Heilmeyer, Marina. 2003. Ein Blumengarten des Barock. Hortus Anckelmannianus, München et al.: Prestel.
Horbas, Claudia. 2006. Gartenlust und Blumenliebe. Hamburgs Gartenkultur vom Barock bis ins 20. Jahrhundert, Ostfildern-Ruit: Hatje Cantz.
Krausch, Heinz Dieter, and Clemens Alexander Wimmer. 2000. Zur Bedeutung des Boseschen Gartens in Leipzig für die Pflanzeneinführung. Zandera 15: 1–14.
Lack, H. Walter. 2001. Ein Garten Eden. Meisterwerke der botanischen Illustration. Köln et al.: Taschen.
Lauterbach, Christiane. 2008. Gartengesetze des Späthumanismus. Gärten und Parks als Lebens- und Erlebnisraum. Ed. Stefan Schweizer. Worms: Werner'sche.
Lauterbach, Iris. Mundus in litteris: Der Kaufmann als Gelehrter. Johann Christoph Volkamers Hesperidenwerk (Orangeriekultur im Bodenseeraum, Orangeriekultur: Schriftenreihe des Arbeitskreises Orangerien in Deutschland e.V., vol. 9), 130-159. Berlin: Lukas.
Lauterbach. Iris. 2011. Johann Christoph Volkamers Hesperidenwerk. In Die Frucht der Verheißung – Zitrusfrüchte in Kunst und Kultur, eds.Yasmin Doosry, Christiane Lauterbach and Johannes Pommeranz, 237–263. Nürnberg: Germanisches Nationalmuseum.
Lauts, Jan. 1979. Arnauld-Eloi Gautier Dagoty,"graveur de la Cour de Bade". Zum botanischen Sammelwerk der Markgräfin Karoline Luise von Baden. Jahrbuch der Staatlichen Kunstsammlungen in Baden-Württemberg 16: 95–106.
Ludwig, Heidrun. 1998. Nürnberger naturgeschichtliche Malerei im 17. und 18. Jahrhundert. Marburg: Basilisken-Presse.
Martz, Jochen. 2011.'Agrumen (...) zu einem dritten Theile von seinen Hesperidibus' – Zur unbekannten Kupferstichserie für einen dritten Band der ‚Nürnbergischen Hesperides'. Die Gartenkunst 23: 151–194.
Nissen, Claus. 1951. Die botanische Buchillustration. Stuttgart: Hiersemann.
Oszczanowski, Piotr. 2004. Wrocławski ogród Laurentiusa Scholtza St. (1552–1599) – Sceneria spotkán elityintelektualnej końca XVI wieku. Śląska Republika Uczonych. In Schlesische Gelehrtenrepublik, vol. 1, eds. Marek Hałub and Anna Mańko-Matysiak, 98–145. Warszawa.
Pavord, Anna. 1999. The tulip. London: Bloomsbury.
Roth, Dietrich (ed.). 2007. Das Moller-Florilegium. Hans Simon Holtzbeckers Blumenalbum für den Bürgermeister Barthold Moller, mit Beiträgen von Dietrich Roth und Thea Vignau-Wilberg.München: Hirmer.
Schmidt-Herrling, Eleonore. 1940. Die Briefsammlung des Nürnberger Arztes Christoph Jacob Trew (1695–1769) in der Universitätsbibliothek Erlangen. Erlangen: Universitätsbibliothek Erlangen.
Schnalke, Thomas. 1995. Natur im Bild. Anatomie und Botanik in der Sammlung des Nürnberger Arztes Christoph Jacob Trew. Eine Ausstellung aus Anlaß seines 300. Geburtstages. Erlangen: Universitätsbibliothek Erlangen.
Teget-Welz, Manuel. 2008. Martin Schaffner. Leben und Werk eines Ulmer Malers zwischen Spätmittelalter und Renaissance. Ed. Haus der Stadtgeschichte – Stadtarchiv Ulm (Forschungen zur Geschichte der Stadt Ulm, 32). Ulm, Stuttgart: Kommissionsverlag W. Kohlhammer.
Trauchburg, Gabriele von. 2001. Häuser und Gärten Augsburger Patrizier. München/Berlin: Deutscher Kunstverlag.
Tschoeke, Jutta (ed.). 2008. Lust und Lieb hat mich beweget. Nürnberger Gartenkultur. Nürnberg: Museen der Stadt Nürnberg – Grafische Sammlung.
Wettengl, Kurt (ed.). 1997. Maria Sibylla Merian 1647–1717. Künstlerin und Naturforscherin. Ostfildern-Ruit: Hatje-Cantz.
Wickert, Konrad. 1993. Universitätsbibliothek Erlangen-Nürnberg. Das Camerarius-Florilegium (Patrimonia 61). Berlin: KulturStiftung der Länder.

Wimmer, Clemens Alexander. 2011. Volkamers Rezeption in Schlesien. Ein Breslauer Zitruskatalog von 1731. Nürnbergische Hesperiden und Orangeriekultur in Franken (Orangeriekultur: Schriftenreihe des Arbeitskreises Orangerien in Deutschland e.V., vol. 7), 86–93. Petersberg: Michael Imhof Verlag.

Wimmer, Clemens Alexander, and Iris Lauterbach (eds.). 2003. Bibliographie der vor 1750 erschienenen deutschen Gartenbücher. Zusammengetragen von Clemens Alexander Wimmer unter Mitarbeit von Iris Lauterbach. Bücherei des Deutschen Gartenbaues e.V. Berlin. Nördlingen: Uhl.

Part V

Perspective

Landscape Design and the Natural Sciences in Germany and the United States in the Early Twentieth Century: "Reactionary Modernism"?

Joachim Wolschke-Bulmahn

Abstract

The increasingly scientific approach to garden culture and to professional ideas on designing gardens has repeatedly received special impulses over the centuries since the Early Modern Era. If previously, it was above all disciplines such as mathematics that found their expression in gardens and in forms of garden art perceived as modern, in the eighteenth and early nineteenth centuries, during the phase of the English landscape garden, particular links were established between scientific disciplines such as philosophy and aesthetics and garden art [see for example, regarding interdependences between philosophy and garden art (Lee 2007)]. Then, in the second half of the nineteenth century and the early twentieth century, it was scientific disciplines such as botany and biology, plant geography, plant sociology and ecology that exercised a particular influence on ideas about designing gardens and, finally, whole landscapes. As regards gardens, corresponding concepts for designing gardens were developed and published in Germany from 1900 on, firstly by the garden architect Willy Lange (1864–1941) under the term of the "nature garden". However, these not only received stimulus from the natural sciences, they were also based on nationalistic and racial notions about a supposed connection between man and nature and the landscape. At the time of National Socialism, such ideas were to become especially influential.

J. Wolschke-Bulmahn (✉)
Centre of Garden Art and Landscape Architecture, Leibniz University Hannover, Herrenhäuser Straße 8, 30419 Hannover, Germany
e-mail: wolschke-bulmahn@ila.uni-hannover.de

H. Fischer et al. (eds.), *Gardens, Knowledge and the Sciences in the Early Modern Period*, Trends in the History of Science, DOI 10.1007/978-3-319-26342-7_17

The Ideology of the Naturgarten (Nature Garden) in Germany

Theoretical concepts of garden design are an integral part of the intellectual history of a particular period and of social and political conditions, too. In Germany, time and again, garden architects have developed specific concepts of garden design in order to convey an explicit political message.[1] Around 1900, an ideology of the nature garden was developed for the design of gardens by garden architect Willy Lange (1864–1941). Although it claimed to be based on biology and other natural sciences, it was hostile to internationalism and was founded on nationalistic and racist ideas.[2] It paved the way, at an early stage, for National Socialist ideology in German garden design.

In 1870, the Irish garden author William Robinson (1838–1935) had already brought out the first edition of his idea of the "wild garden". Internationally, the "wild garden" was to be received very well by the specialized community of garden designers and landscape architects.[3] The two concepts of the "natural garden" and the "wild garden" would also be monopolised in hindsight, in the second half of the twentieth century, as an early "eco garden"—a one-sided and a historical perspective. 'Natural' and 'wild' trends in garden design were represented by their proponents and apologists as progressive and modern. This contribution, however, will include a discussion of whether or not adopting insights from the rapidly developing natural sciences for garden design, as it was first undertaken by Willy Lange in Germany, can also be called "progressive" and "modern" in social terms—with their positive connotations—or whether this trend should sooner be called part of "reactionary modernism" (Herff 1984).

In their remarkable book "An Introduction to the Study of Landscape Design", in the chapter "The Modern German Formal Style", Henry Vincent Hubbard and Theodora Kimball wrote:

> Previous to the very modern national consciousness of the German empire, there has been in Germany nothing which could fairly be called an independent historic style of landscape design [. . .] The modern German conscious seeking for national expression in every field has had its influence on German landscape architecture notably in the production of a formal style of landscape design, intentionally different from any style which has gone before. In many another style the artist has consciously adapted his means to his ends to express the ideal which seemed to him of most worth, but here for the first time landscape designers have gone deliberately to work to determine what their national ideal ought to be and then logically deduced what means should be employed for its attainment. (Hubbard and Kimball 1931: 51 f)

[1] Regarding the political and social dimensions of garden design, see Gröning and Wolschke-Bulmahn (1989).

[2] On the ideological dimension of Lange's Naturgarten, see, e.g., Wolschke-Bulmahn and Gröning (1992).

[3] On the comparative look at "wild" gardens and natural gardens, see Wolschke-Bulmahn (1992).

In all aspects "of landscape design" concerned with aesthetics, according to the authors, "the modern German feeling that a German must be different from other men in his nature and in his needs has found interesting expression" (ibid.: 53). Hubbard and Kimball link "modern" with "seeking for national expression" in garden design, and point out the differences to other national styles. Hubbard and Kimball then assign the term 'Modernism' in Germany to the formalist style that, from 1900, influenced by England, was propagated by architects such as Hermann Muthesius, Peter Behrens and others. It is remarkable, however, that they do not mention Willy Lange in their work "An Introduction to the Study of Landscape Design",[4] for if anyone was trying to develop a national style in Germany at that time, it was Lange.

But, in fact, the first three decades of the twentieth century in Germany as a whole saw developments in garden design that are worthy of a discussion that uses terms such as modernism and avant-garde—not only the formal style.[5] Thus, there were attempts to develop Jugendstil/art nouveau forms. For a few years, there were also efforts to express Expressionist ideas in garden forms. The garden architect Leberecht Migge was a pioneer in introducing technology to the garden, including from an aesthetic point of view. Particularly remarkable for Germany are the avant-garde tendencies that—influenced by the Dutch 'De Stijl' group of artists—made the design of gardens abstract to an extreme degree and sought to exclude both nature, in the form of naturalistic design and plantings, and romanticism from the garden. These developments, in particular, linked with such garden architects as Hans-Friedrich Pohlenz, Georg Pniower and Heinz Wichmann, were vilified under National Socialism, and most of the professional careers of their practitioners were terminated by blacklisting and other Nazi persecution measures. Most likely, all of them wanted to contribute to pointing the way to modernism in landscape architecture in the period under discussion here; some of them with a missionary zeal and in a doctrinaire way.

These significant developments on the theme of modernism, however, are not intended to be the subject of my observations on natural sciences; they have already been thoroughly discussed and publicized elsewhere—above all in a 2004 issue of Dora Wiebenson's journal *Centropa*, guest-edited by Steven Mansbach and myself, on the topic of the avant-garde.[6] In the present contribution, I would like to discuss the phenomenon that, concurrent with the fashion of the formal garden in Germany, ideas of the nature garden as an expression of modernity and national identity were being devised, based on the natural sciences, by the landscape architect and garden writer Willy Lange—ideas that were taken up by the American landscape architect Frank Albert Waugh and thenceforth also popularized in the USA.

[4] Lange's book "Gartenbilder" (1922) was, however, presented in the chapter "Selected list of references on landscape architecture", cf. Hubbard and Kimball (1931: 380).

[5] E.g., see the discussion by Gröning and Wolschke-Bulmahn (1989).

[6] Regarding "avant-garde and garden architecture in Germany", see the discussion in "Centropa" by Wolschke-Bulmahn (2004).

Garden Architect Willy Lange, the Naturgarten (Nature Garden) and the Natural Sciences

Ideas about the natural style in garden design were stimulated by developments in the modern natural sciences during the latter half of the nineteenth century, most strongly by such (emerging) scientific disciplines as plant geography, plant sociology and ecology. In numerous publications, Lange referred time and again to these disciplines, although not always explicitly and sometimes using terminology interchangeably. Thus, he referred, for example, to botany, ecology, plant geography and even ecological plant geography (cf. Lange 1909: 174) as guiding disciplines. With "ecological plant geography", Lange was referring to the Danish botanist Eugen Warming (1841–1924), who pursued ecologically-oriented vegetation geography in the 1890s, publishing his work "*Plantesamfund*" in 1895. It was translated into German in 1896 as "Lehrbuch der Ökologischen Pflanzengeographie. Eine Einführung in die Kenntnis der Pflanzenvereine" and into English in 1909 as "Oecology of Plants. An introduction to the study of plant communities".

The concept of 'ecology' was initially introduced by the German biologist Ernst Haeckel in his 1866 book "Generelle Morphologie der Organismen" ('General Morphology of Organisms'), in which he defined it as "the comprehensive science of the relationship of the organism to the environment" (General Morphology of Organisms; General Outlines of the Science of Organic Forms based on Mechanical Principals through the Theory of Descent as reformed by Charles Darwin). Lange was probably the first German garden architect to take up the concept of 'ecology'. He pursued the concept, however, not in the sense of preserving nature, but as "the science of communal living among entire groups" of similar species or, in other words, as a "physiognomic understanding of the plant world within nature" in the Humboldtian sense (Hermand 1997: 49 f).

As an introduction to Lange's ideas about the Naturgarten, natural garden design, a longer quotation from his most important work, "Gartengestaltung der Neuzeit", may serve to illustrate his ideas on the nature garden in regard to its significant relationship to the modern natural sciences:

> The garden is always a fenced-in piece of the landscape in which it lies. From the landscape we take, then, the design motifs, and will achieve a diversity of gardens concomitant with the diversity of the natural landscape characteristics. And everything in the garden will, through Art, be heightened compared to Nature. The plants of a German garden in particular will be unified according to German examples and German associations of character, but the single species are to be heightened in appearance inasmuch as we replace certain German types with similar but, in our view, more beautiful types (or cultivars) from non-German countries, under the condition that they belong in an identical or similar location in their home countries. In a word: the physiognomy of the German landscape is to be enhanced with similar non-German characteristics.
>
> With this call we find ourselves in harmony with the highest developments of the botanical sciences. Botany is, in the last analysis, the human attempt to relate, ordering and recognizing, to the plant world [...] Linné created his artificial system. De Candolle and others found familial relationships and united families, genii and species. It has taken the new botany to recognize plants in connection with their nurturing soils. On the

> teachings of ecology, i.e., knowledge of plant habitats, it founds the teachings of social coexistence of whole groups of species alien to each other. Out of this emerges a physiognomic comprehension of the plant world in Nature, that is related to artistic perception.
>
> The postulation of a re-creation of social coexistence among the plants of a garden, then, parallels the botanic sciences of our age [...]. If we can, in the garden design of our time, tap into the driving forces of our intellectual life, then the truly modern garden, the garden for our times, will be created (Lange 1909: 28 f).

That 'modernism' was not to be found in the *Neue Sachlichkeit* (New Objectivity) movement is made plain by Lange's polemic against the "apparitions that, as modern 'neue Sachlichkeit', disfigure the face of our homeland" (Lange 1929a: 214). In retrospect, he also distances himself from the romantic landscape garden with its "gentle undulations", calling it the "so-called modern garden art" (Lange 1900a: 73). Lange, by contrast, was seeking modernism in garden design through "beauty in the truth of nature. A scientific naturalism is our foundation" (Lange 1900a: 74). Lange defines as the "most modern aim" (1904a: 98) of his garden art: "Re-creation of plant societies on the example of the homeland nature in which the garden lies, but artistically heightened, taking account of the physiognomic similarities between German plants and foreign plants from appropriate locations" (Lange 1904a: 98).

The basis for this, wrote Lange, was the latest developments in botany:

> Seeking the natural relationships of the plants to their indigenous locations, and determining their social coexistences i.e., plant physiognomy and plant formation are becoming the newest branch of knowledge in botany—and in the garden? (Lange 1904a: 98).

> Garden design responds to the common striving for a new German art, when it touches on nature study of the native landscape in heightened artistic guise. Then it is German, then it is rooted in the native soil as no other art form can be. Then it is, with natural means, artistically modern (Lange 1904a: 98).

Alexander von Humboldt seems to have had a particularly strong influence on Lange's nature garden ideas. Humboldt was one of the first to take on the study of the relationship between organisms and their environment. Owing to his work "Ideen zu einer Geographie der Pflanzen nebst einem Naturgemälde der Tropenländer" (1807) ("Ideas for a Geography of Plants, together with a Natural Picture of the Tropics"), he is regarded as the founder of vegetation geography. Particularly influential on Lange were Humboldt's ideas about plant physiognomy. Humboldt described vegetation as the determining attribute of the 'Gestaltcharakter' of the diverse regions of the earth and introduced the physiognomic observation of plants. In 1806, he published "Ideen zu einer Physiognomik der Gewächse" ("Ideas for a Physiognomy of Plants") in which he also described what he considered the most important plant forms:

> Sixteen plant forms are the principle determinants of the physiognomy of nature [...] How interesting and instructive would a work be for a landscape painter, that offered the listed sixteen main forms, first individually and then showing their contrasts! (Humboldt, quoted after Kügelgen 1999: 155).

In the early nineteenth century, the concept of plant physiognomy was modern and could offer new perspectives for the understanding of plant associations. It even seemed to offer new insights into the relationship between the vegetation in particular regions of the world and those humans living there. Humboldt, in 1806, addressed the decisive impact made by the vegetation of a country on the development of individual cultures. According to Humboldt, "the knowledge of the natural character of different regions of the earth [is] most intimately connected with the history of the human race, and with its culture" (Humboldt 1806: 13). "The character of a people, gloomy or happy mood of humankind," Humboldt argued, was "mainly dependent on climate conditions [. . .] The poetic work of the Greeks and the harsher singing of the nordic primitive people owe their peculiar character in large part to the forms of plants and animals, to the mountain valleys which surrounded the poet, and the air, which blew around him." Thus, in Humboldt's philosophy, the outer appearance of plants, their physiognomy, gained considerable importance. He assumed that it could have an impact on even the character of a people.

Willy Lange was impressed by the writings of Humboldt, Warming, Haeckel and other scientists of the nineteenth century. Time and again, Lange emphasizes in his writings the importance of the natural sciences:

> Today we have a natural science that understands the history of development, that demystifies, that, where it treats of the interrelationships of living things to their homelands and their fellow beings, teaches understanding of the laws of life. Biology suffuses all previous external knowledge, biology—transferred to art—creates a new, a biological aesthetic: the teachings of appearances, which belong together biologically and which, in this sense—deeper than the word was previously meant—constitute a harmony; [. . .] today we can create plantings, design gardens according to the motifs of communities of plants in nature (Lange 1919: 29).[7]

"Biological garden design", according to scientifically worked out "examples from true nature", became an important characteristic in Lange's work in connection with references to "ecological-biological plant(ing) systems" (Lange 1922: 22). For Lange, it was certain that scientific progress would influence garden design. He saw the necessity for the "specialist who is thoroughly educated in plant geography and ecology" (Lange 1928: XIX). "German natural science, conveyed to the people," should show "the way to the heart, the way to a loving devotion to nature" (Lange 1928: 7). Science, for Lange, became *Weltanschauung*. He called for a nature garden "founded on a scientific weltanschauung in unison with the highest demands of the arts and the morality of our age, in unison with the

[7] Similarly, Lange had stated in 1913: "Today we have a natural science that is based on the history of development. It teaches us, as far as the interrelations between creatures with their homeland and their fellow creatures are concerned, to understand the laws of life. Biology penetrates all previous knowledge, which was only superficial. Biology, applied to art, establishes a new, a biological aesthetic" (Lange 1913: 29).

gospel of love" (Lange 1928: 8). Lange's ideas of an "ensoulment of the plants and of nature" contain resonances with Haeckelian monism:

> And the artistic concept of the ensoulment of plants and of nature requires that, in a distinctive garden situation, the plants should be arranged together that according to their physiognomy (appearance, garb) could appear to us in nature, united in distinctive societies (Lange 1928: 8).

The founder of the doctrine of Monism,[8] the German biologist Ernst Haeckel, who introduced the concept of ecology into science, was of special importance in popularizing social-Darwinistic thinking: "[T]hrough evolution he studied the world and everything in it including man and society as part of an organized and consistent whole. He therefore called his new evolutionary theory 'Monism'" (Gasman 1971).

Landscape Types and Natural Garden Design

A fundamental constituent of Lange's conception of culture, and of the National Socialist Blood-and-Soil-Ideology too, was the idea that societies on a high cultural level are 'rooted in the soil' (im Boden verwurzelt), that they live in a close union with the soil and, therefore, the landscape. This idea clearly had a racist and anti-Semitic background. For example, the Jewish people were often described by Volk groups in Germany as homeless and nomads, such terms indicating that Jews, Sinti, Roma and other people could not be rooted in the soil and, therefore, could not have a cultural standard.

A consequence of the idea that the German people were rooted in the soil was that every German was thought to require an appropriately designed spatial environment. If the Germans wanted to preserve their power and their cultural standard, then, Lange asserted, all artistic activities had to take as an example the local landscape—and this included the garden. The subordination of the garden to the landscape thus became an essential criterion of nature garden design. One important aspect of Lange's nature garden concept was therefore the idea of subordinating the garden to the surrounding landscape. The diversity of gardens, according to Lange, should correspond with the diversity of natural landscape characteristics. The German garden contained plant societies representative of the respective landscape in which it was situated. Scientific evidence for this garden ideology was sought in such emerging disciplines as phytogeography and plant sociology.

Applying the natural sciences to the 'national garden art' meant, to Lange, "not blindly imitating the German landscape but, according to the situation of the garden area, 're-creating'" (Lange 1900a: 74). Given the variety of what could become regarded as landscape in Germany, one might expect Lange to have suggested a matching number of natural garden designs. But for this, Lange believed three types

[8] Regarding reactionary ideas inherent in Haeckel's monism, see Gasman (1971).

of landscapes to be sufficient for natural garden design in Germany—he distinguished between "lowland, middle and highland landscapes" (Lange 1900a: 74).

This great significance of landscapes to which the gardens should conform in their design and selection of plants is plain from a discussion of Paul Graebner's "Botanischer Führer durch Norddeutschland". For Lange, it is the plant geography itself, "the artistic way of looking at the botanical sciences; for whoever is incapable of seeing things as a whole, for him the living interrelationships of the plant world will remain indecipherable" (Lange 1903: 182). The importance of plant geography for his nature garden is particularly plain when he demands that "this volume of indigenous plant geography of the lowlands should be followed by one on the German Mittelgebirge."

In his opinion, however, this type of knowledge about landscapes and their qualities and its significance for designing nature gardens, or rather, the necessary "feel for the landscape", as it was understood by Lange, was reserved for the Nordic type:

> The Nordic spirit knows how to connect the small with the infinite; the stream's bank with the seashore, the rockery plant with the highest mountains, [. . .] birch, juniper, gorse and heather with the endless heath; a society of spruce, foxglove and harebells evokes Thuringia. (Lange 1929b: 124)

But plant geography was not meant by Lange as a law for the garden designer. In his nature garden, Lange emphatically approves of the use of foreign plants that go with the indigenous plants in their physiognomy: "I am well aware that I am thus bringing together what ecological plant geography has carefully separated. But in a garden it is not ecological and scientific, but physiognomic and aesthetic principles that apply" (Lange 1909: 174).[9] In relation to the use of so-called 'exotic' or 'foreign' plants, Lange was thus much less doctrinaire than later representatives of the nature or "indigenous" garden, such as Alwin Seifert in Germany or Jens Jensen and Wilhelm Miller in the USA, who developed the prairie garden style.

Austrian Examples for Lange's Ideas About Natural Garden Design?

Lange's ideas on natural garden design are similarly comparable with Austrian and English publications about garden design which appeared one or two decades earlier. Plant geography or phytogeography apparently had already earlier become an objective for garden design.

Particularly remarkable in this context is the *Wiener Illustrierte Garten-Zeitung*. In this Austrian magazine, a series of articles published in the 1880s anticipates, in sometimes startling detail, the concept of the nature garden developed by Lange in

[9] Physiognomy was defined by Lange as the "expression of the living (natural) conditions in the physical shape" (Lange 1922: XII).

the first period of his literary activity, from about 1900 until 1906, in the magazines *Die Gartenwelt* and *Die Gartenkunst*.

The articles by Robert Gemböck, beginning in 1887, under the theme "The Reproduction of Nature in the Garden in Theory and Practice" are particularly noteworthy. Thus, Gemböck's remarks on the most effective way to realize the leitmotiv of a "part of mountainous wilderness" (Gemböck 1884: 357) could well have inspired Lange's "Pictures from the Mountainous Landscape" (1900/01). In this article, Lange develops corresponding ideas in a more comprehensive form. And Gemböck's concept of how to imitate a "meadow-wilderness" [Auwildnis] or—if there is clay soil—a swampy wood [Moorwald] in the garden (1884: 359), if the conditions of soil and water are suitable, could have given Lange the impulse for his article 'Water in the Landscape' (1901).

The understanding of art in Gemböck's and Lange's writings is also similar. Gemböck comments on the "flora of the marshy soil" as a leitmotiv "of a garden after the model of nature" in the following way:

> Just as the tones of a melody are effective not as by themselves but in relationship to the whole, so every plant species has its meaning in the great overall view of the original flora. If we focus therefore not on the single species but on their homogeneousness, we can gain from it an idea and in following an idea and in the reproduction of an idea we reach art. (Gemböck 1892: 15)

Lange's definition of art is nearly identical; he, too, defines as art the derivation of ideas from nature and their translation in an artistic manner into garden design. For him it is not the single plant, but the association of various plants that is the characteristic and determining factor. For Germany, Lange claims, the "artistic nature garden", by which he means the development of the beautiful out of the verity of nature, would be, as it were, a Goethian program (Lange 1904b: 168). Thus, art would be "the raising of nature to an idea" (1904b: 169).

In 1889, the Wiener Illustrierte Garten-Zeitung published an article 'Phytogeographical Groups in Gardens' ("Pflanzengeographische Gruppen in Gärten"), written by Beck von Mannagetta. It may have influenced Lange, too. The author complains of the menace to the local flora posed by expansion and traffic and by "plundering botanists." He therefore pleads that every garden must act in the interest of nature preservation "if he recognizes the associations of plants from nature which could then be imitated in an artificial way in his garden in accordance with local conditions" (Mannagetta 1889: 476).

The Naturgarten as an Expression of "Reactionary Modernism"?

Lange's "biological aesthetic" was modern in the sense that it applied science to design. He wanted to offer a scientific basis for garden design. But from a social perspective, it was reactionary, promoting dubious ideas about the assumed relationship between the German people and nature, and stoking up hostility towards the 'international' by excluding foreign plants with particular groups of the society. Or, to use the title of Jeffrey Herff's 1987 book, Lange's nature garden was a contribution to *Reactionary Modernism*, based on racist ideas about the close relationship of the German people to its landscapes.

For a 'true' German in those days, culture could only be national culture.[10] Consequently, Lange saw garden art as a constituent of national culture. He strongly rejected the idea that "art could be international", and proclaimed: "Let us find the national style for our gardens, then we will have art, German garden art. As long as different nations exist, there must exist different national styles" (1900c: 364). Lange's opposition to the architectural garden style as an expression of the anthropocentric and unnatural attitude of other cultures and of lower stages of cultural evolution culminated in his concept of a "nature garden". Thus, he added to the idea of nationality the idea of ranking. For him the superiority of the German people was part of their national identity. For Lange, the German people were rooted in the soil, and every German required and deserved an appropriate natural-spatial environment:

> Our feelings for our homeland should be rooted in the character of domestic landscapes; therefore it is German nature that must provide all ideas for the design of gardens. They can be heightened by artistic means, but we must not give up the German physiognomy. Thus, our gardens become German if the ideas for the design are German, especially if they are borrowed from the landscape in which the garden is situated. (1907: 358)

Such a view was not only popular during the Imperial Reich in Germany but continued through the Weimar Republic and became part of the state doctrine during National Socialism.

Lange understood his ideas about the nature garden as a contribution to the *Kulturkampf* in the field of garden art. He claimed that the ability to implement the highest stage of garden art was a racial characteristic of the German or Nordic people. He explained, for example, the contrast between the formal and the informal English garden style by different Weltanschauungen, and these again by the "differences between the souls of the two races". In the formal garden, according to Lange, the Nordic race "spiritually perished in the race-morast of the South" (Lange 1927: 5). In 1933, on the coming to power of National Socialism, he defamed the formal garden in the National Socialist magazine "Deutsche Kulturwacht" as characteristic of the "South-Alpine" race and as an attempt of this "un-Nordic race" to weaken the "Nordic race" and to strengthen international, anti-German forces. Lange furthered the idea of the superiority of the German people which was accepted by many Germans during the Imperial period and the Weimar Republic and which, finally, became a state doctrine under National Socialism. The following quotation gives an idea of what he stood for and at the same time of his understanding of cultural development as an evolutionary process:

[10] For various attempts to connect garden architecture and national identity in the nineteenth and twentieth centuries, see Gröning and Schneider (2001).

> History will call this new stage of garden style [the *Naturgarten*], which is firmly based on its precursors, the stage of the German garden style. Germany has been chosen to lend its name to this style in the history of gardens and to become once again an 'improver of the world'. (Lange 1922: 27)

Tendencies of Natural Garden Design in the United States: The Example of Frank A. Waugh

Ideas for nature gardens were developed in the United States, partially influenced by Willy Lange among others. Frank A. Waugh was a prolific writer on garden design and one among American garden architects to promote ideas about the natural style in landscape gardening. Waugh was deeply influenced by Lange's ideas. Waugh had studied garden design under Lange in 1910 at the Royal Horticultural School in Dahlem near Berlin (Waugh 1910: 604). In 1910, he published a series of articles on his German experiences in "The Country Gentleman".[11] In one of them, on "German Landscape Gardening", he enthusiastically endorsed Lange's work and its significance for landscape architecture. He particularly emphasized the fact that Lange's "biological-physiognomic method" would be better classified as the "ecological method" in the American vernacular.

> The best recent book on landscape gardening written in any language (and I cannot conscientiously except my own) is by a German. This man is Willy Lange, a landscape gardener in the suburbs of Berlin and a teacher in the Horticultural School in Dahlem [...] Herr Lange believes in what we in America call the natural style of gardening. In actual practice his work comes nearest to that of Mr. Warren H. Manning of Boston of any in our country. He has a method, fully worked out in scientific lines, in thoroughgoing German fashion. He calls it the biological-physiognomic method; but it would fit better to our use of language to call it the ecological method. Very roughly stated, this theory asserts that plants should be assembled in a garden in their natural relationships—placing together those plants which associate with one another in nature, placing such plant society in its proper soil and on its proper geologic formation (Waugh 1910: 790).[12]

Lange repeatedly expressed the fact that his ideas on plant physiognomy were an important component of his teachings at the Royal Gardeners' Teaching Institute (Königliche Gärtnerlehranstalt) in Berlin-Dahlem. "Natural science sets the rules—love leads the rules to art [...] and so at the school today we pursue rural nature studies, plant physiognomy" (1905: 115).

Numerous publications indicate that Waugh, early in his career, was deeply attracted to the natural sciences in their possible applications for landscape design. During his time in Germany, he also published on the Royal Botanical Garden in

[11] I am grateful to Christopher Vernon, who provided me with copies of these articles in the 1990s.

[12] Similarly, in his 1917 book Waugh referred to Willy Lange: "This ecological principal is the one most clearly elucidated by Willy Lange in his important work, *Die Gartengestaltung der Neuzeit*." (Waugh 1917: 52).

Dahlem near Berlin and how it was shaped by famous botanists such as Dr. Engler. Waugh's statement reveals a deep affinity for Humboldtian thinking.

> This great scientist is best known in America through his modern classification of plants. Ever since the scientific world has adopted the theory of evolution, it has been felt that plants ought to be classified according to their genetic relationships [. . .] The 'systematic' part of the garden is properly arranged according to Dr. Engler's theory of the evolution of plants, and thus becames [sic] all the more interesting and valuable to botanists.
>
> The truly leading feature of the garden, however—the principal talking point, as a traveling man of world would say—is of much greater interest to the practical plantsman and landscape gardener. This is what is called the 'geographical section' of the garden, though in the final result geography takes second place. This occupies the largest, choicest and best-developed part of the garden. Here is the effort made to show all the flora of all the geographical regions of the north temperate and sub-arctic zone [. . .] Besides assembling the plants of particular regions, the more heroic attempt is made to present them in their natural physical, geographical and botanical surroundings. Thus when we come to the flora of the Austrian Alps, we find the plants placed on great artificial mountains made out of rocks like those in the Austrian Alps [. . .] These natural associations of species with species are of the greatest importance in the knowledge of landscape gardening, at least in the modern natural style; for no expert artist in the field would think of planting together species which do not associate in nature; neither would he plant any species on a geologic formation unsuited to it. So the modern student of landscape art is bound to study plants from this particular point of view and in this special way—the way in which they are shown in the botanic garden at Dahlem. For that reason the place is unusually interesting to me (Waugh 1910: 470 f).

Time and again, one finds references in Waugh'spublications showing that he was well familiar with the discussions on design of gardens and the cultural landscape in Germany. Specialists who had influenced the discussion in Germany, such as Willy Lange, Fürst von Pückler,[13] Hermann Jäger,[14] Leberecht Migge and also, in Waugh's book "Rural Improvement", Robert Mielke (Waugh 1914: 57 and 180), were quoted as exemplary, and Waugh´s readers were referred to their books. In his later publications, too, Waugh referred to Willy Lange, in 1917, for instance, in "The Natural Style in Landscape Gardening".[15] Nevertheless, despite his references to the 'ecological method', Waugh seems markedly more reserved with regard to the exact application of the natural sciences.

Waugh was a prolific writer. One of his most fascinating, although also quite ambivalent, publications is "The Natural Style in Landscape Gardening", published in 1917. In it, one can find a plea for tolerance and democracy, fascinating ideas

[13] See the quotation taken by Waugh from Pückler-Muskau, "One principle should, above all, underlie the art of park design; namely, the creation, from the material in hand, out of the place as it stands, of a concentrated picture having Nature as its poetical ideal; the same principle which, embodied in all other spheres of art, makes of the true work of art a microcosm, a perfect, self-contained world in little" (Waugh 1928, 143).

[14] See the reference to Jaeger's "Lehrbuch der Gartenkunst" (1877) (Waugh 1928: 230).

[15] "This ecological principle is the one most clearly elucidated by Lange in his important work, 'Die Gartengestaltung der Neuzeit'" (Waugh 1917: 52).

about landscape design in its probable relations to other fields such as music, as well as a highly doctrinaire undertone and—as with Willy Lange—dubious ideas about "races" and their connections to landscapes.

For a long time, Waugh seems to have been seriously underestimated by garden historians as a landscape architect of the early twentieth century—despite being one of the most interesting.[16] His ideas about, for example, composing music and designing gardens deserve further investigation. Waugh was so convinced of the significance of music for landscape design that he stated in his 1917 book on the "Natural Style in Landscape Gardening": "After a good many years of study and teaching, however, I am inclined to believe that music offers the readiest approach to a spiritual interpretation of the landscape" (1917: 56).

It is difficult to describe Waugh's own ideas about plant use exactly. On the one hand, he called for a "spiritual interpretation of the landscape" (1917: 17); on the other, he referred to the new "branch of botanical science known as ecology" (1917: 51) and claimed: "If we as landscape gardeners desire to preserve the whole aspect of nature, with all its forms intact, we will keep all plants in their proper social grouping" (1917: 51).

Unlike Lange, Waugh apparently did not refer so explicitly to the natural sciences, to botany, plant geography, etc., but favored a general relation to ecology. Waugh raises the question: "What then is the informing spirit of the natural style? Is it not the spirit of the natural landscape?" (1917: 21). And by answering his own question, he makes the 'landscape' the determining factor for natural garden design.

> It is not difficult to see that this spirit of the landscape is different from the spirit of architecture. Thus anyone who is capable of a spiritual conception of any sort can readily accept the principle that, while the formal garden should be animated by the architectural spirit, the informal garden should live by the spirit of the landscape (1917: 22).

And a few pages later:

> Whether our foregoing definition of the natural style is adequate or defective, it must be plain that any naturalistic style of landscape gardening is largely dependent on the native landscape. The ideas, motives, and methods must come mainly from nature (1917: 25).

Considering the broad varieties of different landscapes in the United States, it would have been quite difficult to subordinate the informal garden to the spirit of the landscape. Waugh's remarks on this problem for garden design resemble closely Willy Lange's ideas about the nature garden. One important aspect of

[16] In 2003 Katherine Crewe introduces her article "The rural landscape of Frank Waugh" in the following way: "While Frank Waugh's contributions to landscape architecture are acknowledged today, they are vaguely understood. Though he is broadly associated with his contemporaries on the East Coast such as F. L. Olmsted, Jr. and Warren Manning, or with the Prairie Landscape Architects in the mid-West, he differs from both groups. Waugh may best be identified by his strong preoccupation with small rural towns, villages and farms, the subject of a lifetime of teaching and publications." (Crewe 2003: 126).

Lange's nature garden concept was the idea of subordinating the garden to the surrounding landscape. Lange believed three types of landscapes to be sufficient for natural garden design in Germany: "a mountainous landscape" (Gebirgslandschaft), "an intermediate landscape" (Mittellandschaft) and "a level or plain landscape" (Ebenlandschaft) (Lange 1900b: 436). Not only did the plants need to submit themselves to these types of landscape ["as the landscape's most excellent contributors to its character" (Lange 1900b: 435)], the architectural objects should do this, too, in view of the natural influences and "on the basis of the character of the lineage of the landscape's inhabitants" (Lange 1900b: 436).

Waugh agreed with Lange that the landscape should be the determining factor for the design of the natural garden. The natural style of landscape gardening for him "was a method of landscape gardening in which the natural forms of landscape are used and imbued with the spirit of the native landscape" (1917: 43). He referred to the discussion in Germany and to the use of the term *Kultur-Landschaft* (cultural landscape), and stated: "Doubtless, we in America will presently come to a similar understanding of the beauty of well farmed country, and will learn to love the farm landscape and to realize its deeper spiritual significance" (1917: 39 f). "For this reason it is desirable to consider what are the principle landscape types" (1917: 34).

Waugh, similar to Lange's landscape types, distinguished just four "main types of landscapes" for the United States (1917: 37):

> It is reasonably accurate to say, though there is nothing scientific in classification, that the four great types of landscape are the sea, the mountains, the plains, and the forests. These great types everyone should know. Certainly every man who professes to be a landscape architect should assimilate into his own life these fundamental landscape forms. (1917: 34)

But in Waugh's writings about natural garden design, there is also, similar to those of Lange, a particular belief in the 'race' as an important influence on gardens, when he claims:

> In landscape gardening, on the other hand, styles are national—perhaps, more strictly speaking, racial. The Japanese style embodies the garden characteristics of a whole race. The Italian style does the same. Every style which ever had a name was called by the name of the race or nation which practiced it; and one of the questions now before the house is whether we shall ever have an American style.
>
> We may therefore define style, as used in this particular art, as being the expression of the national, racial or ethnic quality in landscape gardening. (1917: 19)

Ideas about the supposed connection between 'race', cultural landscapes and gardens was later to become expressed much more aggressively in the USA, e.g., by the landscape architect Jens Jensen.

William Miller and Jens Jensen: Advocates of the Prairie Style of Landscape Gardening

In 1915, landscape architect Wilhelm Miller (1869–1938) published "The Prairie Spirit in Landscape Gardening. What the people of Illinois have done and can do toward designing and planting public and private grounds for efficiency and beauty", a publication that was to receive quite a lot of attention among garden designers in the next two decades (Fig. 1).[17] The "'prairie spirit' was to be expressed fundamentally through the use of plants indigenous to the prairie landscape [...] Miller urged his readers to compose native plant groupings in ecologically appropriate 'plant societies' or associations" (Vernon: 274). He emphasized a "strong national or regional character" (1915: 1) as an objective for landscape design in the Middle West. "Systematic botany" and ecology, "a new and fascinating branch of botany that deals with plant societies" (1915: 18) were, for Miller, important tools for the garden designer. Miller himself, presumably on account of anti-German sentiment during the First World War, "changed his first name 'Wilhelm' to 'William' in 1919. Shortly thereafter, he retired to southern California and later died there in obscurity on 16 March 1938" (cf. Vernon 1995: 275).

The Danish-American landscape architect Jens Jensen (1860–1951) was much more successful and more 'sustainable' than Miller in propagating his ideas about the 'prairie garden' and regional modes of garden design. Jensen opposed the use of non-native plants in American parks and gardens and promoted the regional prairie style. To understand Jensen's work, his plea for the use of native plants and his concept of garden design, one has to consider his ideas about the relationship between people, 'races', and nature. Jensen believed that ideas about nation, race, and the natural environment are closely interwoven. In an article about the art of landscaping in the German journal "Die Gartenschönheit" (Garden Beauty), Jensen drew, in 1923, a parallel between races and plant species:

> Perhaps it may be too restricted to design a landscape picture only by the means of simple indigenous plants. But please consider that it was them amongst whom we grew up, that they taught us a particular language, without interruption since the earliest days of our tribe, that they are interwoven with the soul of our race and, indeed, no art of landscape gardening will be called true art and will be able to reflect the soul of a tribal people, if it does not take its means of expression from the environment of these people. (Jensen 1923: 68)

These ideas were further developed by Jensen in his 1939 book *Siftings*. In drawing a connection between race and landscape, Jensen stated: "Thus speaks the soul of our native landscape. Nothing can take its place. It is given to us when we are born, and with it we live" (1939: 9). With respect to the art of garden design, he ruled: "Art must come from within, and the only source from which the art of

---

[17] For more detail about Miller and the "Prairie Spirit", see Vernon (1995: 271 ff); see a critical discussion of Miller's and Jensen's ideas about the "Prairie Spirit" by Gröning (1997: 235 f).

Fig. 1 Cover, Wilhelm Miller, "The Prairie Spirit in Landscape Gardening" (Urbana 1915)

landscaping can come is our native landscape. It cannot be imported from foreign shores and be our own" (Jensen 1939: 63). For Jensen, "no plant is more refined than that which belongs. There is no comparison between native plants and those imported from foreign shores which are, and shall always remain so, novelties" (1939: 40 f). The fact that Jensen himself was born in Denmark and had lived there for more than

20 years, yet could presume to develop the American "Prairie Style", might cast some doubt on the idea of any congenital relationship of humans to particular landscapes.

It is remarkable that Jensen, as Willy Lange did earlier, stressed the alleged existence of differences between Scandinavian races and Latin races: "The influence of the Mediterranean, where the Latin races live, has created a distinct character quite different from that of the Germanic races" (1939: 25). In 1939, at the highpoint of National Socialist power in Germany, Jensen wrote a letter to his German colleague Camillo Schneider in Berlin. His letter discussed Alwin Seifert (1892–1972), another German landscape architect.[18] Seifert, a fanatical anti-Semite, fully supported National Socialism and argued, e.g., in his 1937 article "Natur und Technik im deutschen Straßenbau" ("Nature and technology in German road construction") in the magazine "Naturschutz" ("Nature Conservation"), that landscape architecture in Nazi-Germany must stand by the "fate-determined poverty of plants" in its landscape design (cf. Gröning and Wolschke-Bulmahn 1987: 149). But where Seifert still allowed for a few less native plants in a garden, Jensen took a more uncompromising position:

> Seifert seems to distinguish between the garden inside an enclosure and the landscape—here he submits to compromise. [...] I cannot see how you can compromise on a difference between the garden and its surroundings, then the garden remains a stranger in its own land. To be true to yourself, I mean true to your native landscape is a very fundamental issue—it is to be, or not to be. In the garden you give assent to one idea and outside its boundary to another. Strange things, grotesque things, usually attractive to the novice will creep in and the purity of thoughts in garden making suffers. Freaks are freaks and often bastards—who wants a bastard in the garden, the out of door shrine of your home?[19]

Finally, to take up the title of Jeffrey Herff's book "Reactionary Modernism", Jensen makes it very clear in his publications that his own ideas about regional garden design and the "prairie spirit" need to be classified as belonging to "reactionary modernism", at least in their tendency. That racism was an important motive in his plea for native plants is also shown by his article "Die 'Lichtung'" (The 'Clearing') published in the magazine "Die Gartenkunst" in 1937:

> The gardens that I created myself shall [...] be in harmony with their landscape environment and the racial characteristics of its inhabitants. They shall express the spirit of America and therefore shall be free of foreign character as far as possible [...] The Latin and the Oriental crept and creeps more and more over our land, coming from the South, which is settled by Latin people, and also from other centers of mixed masses of immigrants. The Germanic character of our race, of our cities and settlements was overgrown by foreign character. Latin spirit has spoiled a lot and still spoils things every day (Jensen 1937: 177).

[18] For biographical and bibliographical information about Camillo Schneider, Alwin Seifert, Willy Lange and other German landscape architects mentioned in this essay, see Gröning and Wolschke-Bulmahn (1997).

[19] Jens Jensen, letter of April 15, 1939, to Camillo Schneider, Jens Jensen Collection, Sterling Morton Library, Lisle, Illinois.

Let me wind up by remarking that corresponding ideological links between ideas about humans and ideas about gardens have been ignored by garden historians in the U.S. for a long time.[20]

Other American landscape architects and garden writers of the early twentieth century could be included in this discussion as to how, towards the end of the nineteenth and beginning of the twentieth centuries, the natural sciences influenced the design of gardens and landscapes in Germany, the USA and other countries.

Among them, the landscape architect Elsa Rehmann would certainly be particularly important. Together with Edith A. Roberts, a plant ecologist, she propagated, in their book "American Plants for American Gardens" (1929), the use of indigenous plants and natural plant communities. Roberts, "who was considered a pioneer in the new science of plant ecology" (Wurman 2012: 115), was Chair of the Botany Department at Vassar College (ibid.: 115). She received her training at the University of Chicago, where she "undoubtedly" must have been influenced "by the 'Prairie School' design philosophy of Jens Jensen, and his friend, Dr. Henry Cowles, professor of botany" (ibid.: 115 f). In the book "American plants for American gardens" and in other publications, Rehmann, using the science of plant ecology as her departure point, called for indigenous vegetation to be used in American garden design. On the importance of plant ecology as a new scientific discipline, she makes the following introductory remarks:

> Plant ecology, a comparatively new study of plants in relation to their environment, contributes toward a keener understanding of this [plants native to America] natural vegetation and its use in garden making. It draws attention to the native plants as they appear in the landscape and suggests their inherent appropriateness to grounds and gardens (Roberts and Rehmann 1929: 1).

Rehmann reveals quite a narrow perspective on "native" plants, when she calls for studies into the plant ecology of the various regions of the USA, "remembering that a plant is no longer native if it is transferred to a locality in which it is not really indigenous or, in a real ecological sense, to a location not environmentally correct" (Rehmann 1933: 130).

Roberts and Rehmann do not take up any new strands, but rather pursue trends in the development of a national garden culture, of the American garden, in a narrow form. With their book, they seek to show "the use of ecology in selecting American plant material for American grounds and gardens" (Roberts and Rehmann 1929: 3).

A closer and more detailed investigation needs to be undertaken to determine whether and to what extent these ideas can be understood on a purely scientific basis or whether they reflect ideological notions relating no longer just to 'plants' and 'nature' but also to humans: ideas such as Rehmann's notions and those of other American landscape architects, ideas as understood by representatives of the ecological movement in Germany in the 1980s and 1990s, and ideas currently being expressed about natural gardens and ecological gardens, as well as the call

---

[20] See, for example, Grese (1992).

for removing so-called non-indigenous, exotic plants from gardens and the corresponding landscapes.[21] The historical examples of Willy Lange and Jens Jensen, to name just one example each from Germany and the USA, are certainly indicative of other possibilities.

Bibliography

Sources

Gemböck, R. 1884. Der Garten der Zukunft. Wiener Illustrierte Garten-Zeitung 9: 356–359.

Gemböck, R. 1892. Die Flora des Moorbodens als billiges Material zu einem "Garten nach dem Vorbilde der Natur". Wiener Illustrierte Garten-Zeitung 17: 15.

Hubbard, Henry Vincent, and Theodora Kimball. 1931. An Introduction to the Study of Landscape Design. New York: The Macmillan Company.

Humboldt, Alexander von. 1806. Ideen zu einer Physiognomik der Gewächse. Tübingen.

Humboldt, Alexander von. 1807. Ideen zu einer Geographie der Pflanzen nebst einem Naturgemälde der Tropenländer.

Jensen, Jens. 1923. Die Landschaftsgärtnerei. Eine Kunst. Die Gartenschönheit 4, 4: 68–69.

Jensen, Jens. 1937. Die "Lichtung". Die Gartenkunst 50, 9: 177–180.

Jensen. Jens. 1939. Siftings. Reprint by Johns Hopkins University Press. With a foreword by Charles E. Little and an afterword by Darrel G. Morrison. Baltimore/London, 1990.

Lange, Willy. 1900a. Bilder aus der Gebirgslandschaft. 1. Hochgebirgsvegetation. Die Gartenwelt 5, 7: 73–75.

Lange, Willy. 1900b. Gartengestaltung und Aesthetik. Die Gartenwelt 4, 37: 433–436.

Lange, Willy. 1900c. Garten und Weltanschauung. Die Gartenwelt 4: 364.

Lange, Willy. 1900/01. Bilder aus der Gebirgslandschaft. Die Gartenwelt 5, 7: 73–75, 326–328.

Lange, Willy. 1901. Das Wasser in der Landschaft. Die Gartenwelt 5: 438–440, 459–461, 557–559.

Lange, Willy. 1903. Discussion of: Paul Graebner, Botanischer Führer durch Norddeutschland. Die Gartenkunst 5, 10: 182.

Lange, Willy. 1904a. In welcher Weise kann der modernen Kunstrichtung in der Gartenkunst praktisch Rechnung getragen werden?. Die Gartenkunst 6, 5: 98.

Lange, Willy. 1904b. Die Pflanzung im Garten nach physiognomischen Grundsätzen. Die Gartenkunst 6: 168.

Lange, Willy. 1905. Meine Anschauungen über die Gartengestaltung unserer Zeit. Die Gartenkunst 7, 7: 113–116.

Lange, Willy. 1907. Gartengestaltung der Neuzeit. Leipzig: Verlagsbuchhandlung von J. J. Weber.

Lange, Willy. 1909. Gartengestaltung der Neuzeit. Second, amended and expanded edition. Leipzig: Verlagsbuchhandlung von J. J. Weber.

Lange, Willy. 1913. Der Garten und seine Bepflanzung. Kosmos. Gesellschaft der Naturfreunde: Stuttgart.

Lange, Willy. 1919. Der Garten und seine Bepflanzung.

Lange, Willy. 1922. Gartenbilder. Leipzig: Verlagsbuchhandlung J. J. Weber.

Lange, Willy. 1927. Gartenpläne. Leipzig: Verlagsbuchhandlung J. J. Weber.

Lange, Willy. 1928. Gartengestaltung der Neuzeit. Leipzig: Verlagsbuchhandlung J. J. Weber.

[21] See on this discussion i.a. the article by Gröning and Wolschke-Bulmahn "Some notes on the mania for native plants in Germany" (1992), and the reactions to it, e.g., Sorvig (1994); see also the article Gröning and Joachim Wolschke-Bulmahn (1994).

Lange, Willy. 1929a. Discussion on E. Ewald, Das Gesicht der deutschen Heimat, Landschaft und Baukunst. Die Sonne 6, 5: 214.
Lange, Willy. 1929b. Nordische Gartenkunst. Die Sonne 6, 3: 124.
Mannagetta, G. Ritter Beck von. 1889. Pflanzengeographische Gruppen in Gärten. Wiener Illustrierte Garten-Zeitung 14: 476.
Miller, Wilhelm. 1915. The Prairie Spirit of Landscape Gardening. What the people of Illinois have done and can do toward designing and planting public and private grounds for efficiency and beauty. University of Illinois, College of Agriculture, Urbana.
Rehmann, Elsa. 1933. An ecological approach. Landscape Architecture, July 1933. 239–245. Reprinted in Women in landscape architecture. Essays on history and practice, eds. Louise A. Mozingo and Linda Jewell, 129–133. Jefferson/North Carolina, and London: McFarland & Company, Inc., Publishers.
Roberts, Edith A., and Elsa Rehmann. 1929. American Plants for American Gardens. Reprint 1996, with a foreword by Darrel G. Morrison. Athens/London: The University of Georgia Press.
Waugh, Frank A. 1910. A Horticultural School. The Country Gentleman. 23 June 1910: 604.
Waugh, Frank A. 1910. German Landscape Gardening. The Country Gentleman. 25 August 1910: 790.
Waugh, Frank A. 1910. The Royal Botanic Garden. The Country Gentleman. May 12 1910: 470 f.
Waugh, Frank A. 1914. Rural Improvement. The Principles of Civic Art Applied to Rural conditions, including Village Improvement and the Betterment of the Open Country. New York: Orange Judd Company.
Waugh, Frank A. 1917. The Natural Style in Landscape Gardening. Richard G. Badger: Boston.
Waugh, Frank A. 1928. Book of Landscape Gardening. Treatise on the General Principles Governing Outdoor Art; with Sundry Suggestions for their Application in the Commoner Problems of Gardening. New York: Orange Judd Publishing Company.

Secondary Literature

Crewe, Katherine. 2003. The rural landscapes by Frank Waugh. Landscape Journal 22, 2: 126–139.
Gasman, Daniel. 1971. The Scientific Origin of National Socialism. Social Darwinism in Ernst Haeckel and the German Monist League. London/New York: MacDonald and American Elsevier Inc.
Grese, Robert E. 1992. Jens Jensen. Maker of Natural Parks and Gardens. Baltimore, Maryland: The Johns Hopkins University Press.
Gröning, Gert. 1997. Ideological aspects of nature garden concepts in late twentieth-century Germany. In Nature and ideology. Natural garden design in the twentieth century, Dumbarton Oaks Colloquium on the History of Landscape Architecture, vol. XVIII, ed. Joachim Wolschke-Bulmahn, 221–248. Washington D.C.
Gröning, Gert, and Joachim Wolschke-Bulmahn. 1987. Der Drang nach Osten. Zur Entwicklung der Landespflege im Nationalsozialismus und während des Zweiten Weltkrieges in den 'eingegliederten Ostgtebieten'. München: Minerva Publikation.
Gröning, Gert, and Joachim Wolschke-Bulmahn. 1989. Changes in the philosophy of garden architecture in the 20th century and their impact upon the social and political environment. Journal of Garden History 9, 2: 53–70.
Gröning, Gert, and Joachim Wolschke-Bulmahn. 1992. Some notes on the mania for native plants in Germany. Landscape Journal 11, 2: 116–126.
Gröning, Gert, and Joachim Wolschke-Bulmahn. 1994. Response: If the shoe fits, wear it!. Landscape Journal 13, 1: 62 f.

Gröning, Gert, and Joachim Wolschke-Bulmahn. 1997. Grüne Biographien. Biographisches Handbuch zur Landschaftsarchitektur des 20. Jahrhunderts in Deutschland. Berlin and Hannover: Patzer Verlag.

Gröning, Gert, and Uwe Schneider. 2001. Gartenkultur und nationale Identität. Grüne Reihe - Quellen und Forschungen zur Gartenkunst. Volume 19. Wernersche Verlagsgesellschaft. Worms, Germany.

Herff, Jeffrey. 1984. Reactionary Modernism: Technology, Culture and Politics in Weimar and the Third Reich. New York and Cambridge: Cambridge University Press.

Hermand, Jost. 1997. Rousseau, Goethe, Humboldt: Their Influence on Later Advocates of the Nature Garden. In Nature and ideology. Natural garden design in the twentieth century, Dumbarton Oaks Colloquium on the History of Landscape Architecture, vol. XVIII, ed. Joachim Wolschke-Bulmahn, 35–58. Washington D.C. Dumbarton Oaks Research Library and Collection.

Huber, Otto. 1999. Die "Geographie der Pflanzen". In: Alexander von Humboldt. Netzwerke des Wissens, exhibition catalog, published by the Kunst- und Ausstellungshalle der Bundesrepublik Deutschland GmbH. Bonn.

Kügelgen, Helga von. 1999. Pflanzenformen aus der Physiognomie der Naturals Programm für den Landschaftsmaler. Ein Brief Humboldt an Schinkel für Rugendas. In Alexander von Humboldt. Netzwerke des Wissens, exhibition catalog, published by the Kunst- und Ausstellungshalle der Bundesrepublik Deutschland GmbH. Bonn.

Lee, Michael. 2007. The German "Mittelweg": Garden Theory and Philosophy in the Time of Kant. London: Routledge.

Morrison, Darrel G. 1996. Foreword, to: Edith A. Roberts and Elsa Rehmann, American Plants for American Gardens. 1929. Athens/London: The University of Georgia Press.

Sorvig, Kim. 1994. Natives and Nazis: An imaginary conspiracy in ecological design. Commentary on G. Groening and J. Wolschke-Bulmahn's "Some notes on the mania for native plants in Germany". Landscape Journal 13, 1: 58–61.

Vernon, Christopher. 1995. Wilhelm Miller and The Prairie Spirit in Landscape Gardening. In Regional Garden Design in the United States, eds. Therese O'Malley and Marc Treib, 271–276. Dumbarton Oaks Colloquium on the History of Landscpae Architecture, vol. XV. - Washington D.C.

Wolschke-Bulmahn, Joachim. 1992. The "Wild Garden" and the "Nature Garden" – Aspects of the garden ideology of William Robinson and Willy Lange. Journal of Garden History 12, 3: 183–206.

Wolschke-Bulmahn, Joachim. 2004. The Avant-Garde and garden architecture in Germany. On a forgotten phenomenon of the Weimar period. Centropa 4, 2: 101–109.

Wolschke-Bulmahn, Joachim, and Gert Gröning. 1992. The ideology of the nature garden. Nationalistic trends in garden design in Germany during the early twentieth century. Journal of Garden History 12, 1: 73–80.

Wurmann, Dorothy. 2012. Elsa Rehmann, ecological pioneer: "A patch of ground". In Women in landscape architecture. Essays on history and practice, eds. Louise A. Mozingo and Linda Jewell, 113–128. Jefferson/North Carolina and London: McFarland & Company, Inc., Publishers.

Index

A

Abu Zacaria (Abu Zacaria Yahya ibn Muhammad ibn Ahmad Ibn al-Awwam Al-Ishbili), 309
Adonis, 184, 189, 204
Aeolus, 165
Agricola, Georg Andreas, 314
Agustí, Miquel, 305, 309, 312
Alamanni, Luigi, 270, 279, 281, 309
Alberti, Leandro, 275, 281, 283, 286
Alberti, Leon Battista, 177, 178, 312
Aldini, Tobia, 257
Aldrovandi, Ulisse, 195, 196, 198, 200
Alessandro d'Este, Cardinal, 78, 84
Alexander VII, Pope, 125, 127, 129. *See also* Chigi, Fabio
Alfonso II d'Este, Duke, 78
al-Jazarî, Badî' az-Zamân Abû l-'Izz ibn Ismâ'îl ibn ar-Razzâz, 81
Alonso de Herrera, Gabriel, 305, 306, 309–316
Alpinus, Prosper, 257
Alvares, Rodrigo, 310
Anaximenes of Miletus, 115
Anckelmann, Caspar, 256, 331
Anckelmann, Eberhard. I., 256
Andreae, Johann Valentin, 154, 160, 170, 171, 175
Anguillara, Luigi, 197
Antonini, Antonio, 113
Apelles, 184
Apiarius, Mathias, 186
Apollo, 56–58, 71, 72, 120, 184, 186, 204
Arndt, Johann, 175
Arrieta, Juan de, 316
Aubriet, Claude, 229
Audebert, Nicolas, 84, 85, 91, 93–96
Augustus, Roman Emperor, 105, 127
Azevedo Pita Lima, Paulo de, 306, 312

B

Bacon, Francis, 16, 20, 30, 31, 35–40, 46, 51, 154, 170, 176
Baglione, Giovanni, 106
Baldinucci, Francesco Saverio, 213, 214, 216, 221–223
Balthassar Beck, 189
Bandini, Ottaviano, Cardinal, 114
Banū Mūsa, 81
Barberini family, 110, 114, 120
Barberini, Francesco, Cardinal, 112
Barberini, Maffeo, 120, 121. *See also* Urban VIII, Pope
Basseus, Nicolas, 204
Bauhin, Caspar, 204, 206, 256–258
Bauhin, Jean, 206, 215, 225, 256–258, 300
Beale, John, 46, 49, 50
Beck of Strasbourg, Reinhard, 186
Behrens, Peter, 347
Bernardo, Giovanni Maria, 309
Berner, Johann Friedrich, 339
Bernini, Gianlorenzo, 125, 127
Besler, Basilius, 257, 258, 328
Bettini, Mario, 10, 12, 15, 21
Beverningh, Hieronymus van, 299, 300
Bimbi, Bartolomeo, 212–217, 221–227, 229
Biringucci, Oreste Vannocci, 85–88, 92, 97–100
Blasi, Luca, 114
Blondel, Jacques-François, 14, 15, 308
Boccaccio, Giovanni, 281
Bock, Hieronymus, 184, 191, 256
Boëll, Charles, 138, 140
Bol, Hans, 200, 201, 204
Bonacolsi, Rinaldo, called il Passerino, 262
Bonnefond, Nicolas de, 309, 315
Borghini, Raffaello, 279
Borromini, Francesco, 110, 119

H. Fischer et al. (eds.), *Gardens, Knowledge and the Sciences in the Early Modern Period*, Trends in the History of Science, DOI 10.1007/978-3-319-26342-7

Bose, Caspar, 335
Bosse, Abraham, 17
Boyceau, Jacques, 12, 142, 315
Boyle, Robert, 20, 33, 51
Bracci, Ignazio, 122
Bradley, Richard, 314
Bramante, Donato, 142
Breyne, Jacob, 290, 292, 294–296, 298–300
Breyne, Johann Philipp, 296
Briseux, Charles-Étienne, 308
Brosse, Guy de la, 241
Bruegel, Pieter, the Elder, 198, 199
Brunfels, Otto, 184, 186, 189, 191, 198, 204, 256
Bruyn, Nicolas de, 201
Bry, Johann Theodor de, 331
Bumaldi, Giovanni Antonio, 255, 257. *See also* Montalbani, Ovidio
Buonarroti, Michelangelo, 82
Buontalenti, Bernardo, 85, 88, 91, 284–286
Bussato, Marco, 270, 278, 279, 283
Bussin (architect), 140

C

Caetano de Sousa, Manuel, 306
Calzolari, Francesco, 198
Cambyses, 160
Camerarius, Joachim, the Younger, 202, 256, 257, 319–321, 325, 329, 332
Campanella, Tommaso, 120, 121, 154, 175
Cappello, Bianca, Grand Duchess of Tuscany, 285
Carapinha, Aurora, 306, 307, 311
Carballo y Sampayo, Diego de, 309
Carita, Helder, 306, 307
Carlo III Borromeo, 168
Carmenta, 167
Cartari, Vincenzo, 312
Casabona, Giuseppe, 280
Casaubon, Isaac, 247
Castille, Nicolas Jeannin de, 140
Castro, Petro de, 257, 262
Caus, Salomon de, 21, 308
Celsus, Aulus Cornelius, 275
Cerceau, Jacques Androuet du, 138, 140, 141, 145, 146, 150
Cesalpinus, Andrea, 256
Cesi, Federico, 114, 122
Charles II of England, 119
Charles V, Emperor, 81
Charleton, Walter, 29, 35, 45, 46
Chastillon, Claude de, 149–150
Chemnitz, Johann, 257–259
Chigi, Fabio, 125, 127. *See also* Alexander VII, Pope
Cibo, Gherardo, 204
Clarici, Paolo Bartolomeo, 226
Clavena, Nicolao, 257
Clavius, Christoph, 15, 110, 115, 117, 122, 125, 129
Clerico, Luca (Leclerc, Luc), 79
Clouck, Andries, 236
Clusius, Carolus, 198, 200, 240, 241, 243, 246, 247, 249, 257, 320, 325
Cluyt, Dirk Outgaertsz, 240, 243
Cocchi, Antonio, 221
Cock, Hieronymus, 198
Cock, Symon, 193
Coignet, Renaud, 81
Colbert, Jean-Baptiste, 56, 57, 61
Coler, Johann, 316
Collaert, Adriaen, 200, 201, 204
Colleoni, Bartolommeo, 167
Columna, Fabius, 300
Commandino, Federico, 85
Commelijn, Caspar, 219
Commelijn, Jan, 219, 226
Compton, Henry, 298
Constantine VII, Emperor, 308–309
Copernicus, Nicolaus, 121, 123, 126
Cordus, Euricius, 191, 198, 257
Cornut, Jacques-Philippe, 257
Cort, Cornelis, 198, 199
Cortuso, Giacomo Antonio, 198
Cosimo I de' Medici, Grand Duke of Tuscany, 192, 195, 213, 215, 216, 218
Cosimo III de' Medici, Grand Duke of Tuscany, 213–219, 223, 226, 227, 229, 255
Court, Benoît de, 257
Cowles, Henry, 362
Crescenzi, Pietro, 270, 271, 274–276, 278, 280, 309
Crewe, Katherine, 357
Cristell, Jacobina, 325
Crivelli, Giovanni Angelo, 168
Cupani, Francesco, 215, 225

D

Dahuron, René, 263
Daléchamps, Jacques, 257
Dalle Greche, Domenico, 196

Danti, Egnazio, 106
d'Aquino, Carlo, 309, 315
Darwin, Charles, 348
Davanzati, Bernardo, 270, 278
Da Vinci, Leonardo, 184
De Candolle, 349
De Filiis, Angelo, 113
Dell'Antonio, Andrew, 130
Della Porta, Giambattista, 88, 257
Del Papa, Giuseppe, 215, 217–218, 223
Del Riccio, Agostino, 270, 275, 278–281, 283, 284
Descartes, René, 31
Desmarets, Daniel, 292, 293
Dézallier d'Argenville, Antoine-Joseph, 13, 16, 21, 22, 308
Diderot, Denis, 25
Digby, Kenelm, 42, 257
Diodorus Siculus, 168
Dioscorides, 184, 186, 193, 198, 204, 248
Dodoens, Rembert, 193, 200, 204. *See also* Dodonaeus, Rembertus
Dodonaeus, Rembertus, 241, 246–249, 257, 258
d'Olbreuse, Eleonore, 263
Doni, Anton Francesco, 154, 170, 175
Doosry, Yasmin, 217, 219, 226
Dorsten, Theodor, 257
Dos Santos, Eugénio, 306
Doublet, Philips, 17
Duchesne, Antoine-Nicolas, 308
Du Mont, Andreas, 17
Dunphy Wind, Geraldine, 125, 126
Dupérac, Étienne, 145–147
Du Pinet, Antoine, 200
Dürer, Albrecht, 184, 186

E
Earl of Sandwich, ambassador to Spain, 316
Eberlin von Günzburg, Johann, 177
Eckhardt, Johann Georg von, 264
Egenolph, Christian, 191, 193, 199
Egenolph, Paul, 195
Engelbrecht, Martin, 328
Engler, Adolf, 356
Este family, 78
Estienne, Charles, 309, 312–314
Euler, Leonard, 17, 19
Evelyn, John, 15, 20–22, 29–51, 151, 218, 314, 316

F
Fabricii, Principio, 106
Fagel, Gaspar, 299
Fagiolo, Marcello, 120
Falcone, Giuseppe, 309
Falda, Giovanni Battista, 108, 114, 120–122
Falloppia, Gabriele, 195
Federico II Gonzaga, Duke of Mantua, 262
Fénelon, François, 178
Ferdinand, Duke of Bavaria, 325
Ferdinand II of Tyrol/Austria, Archduke, 198
Ferdinand I of Habsburg, Holy Roman Emperor, 193
Ferrari, Giovanni Battista, 225, 226, 309
Ferreira, Simoes, 306
Ferro, Giovanni, 112
Filarete, Antonio Averlino, 153, 154, 157, 159, 162–164, 166, 168, 170, 172, 175, 177, 178
Filarete, Ercole, 202
Fillassier, Jean Jacques, 310
Finé, Oronce, 189
Floris, Frans, 198, 199
Fogel, Martin, 253–264
Fontana, Carlo, 168
Fouquet, Nicolas, 62
Francesco I de'Medici, Grand Duke of Tuscany, 198, 200
Francini, Alessandro, 147
Francini, Tommaso, 142, 145
François I of France, 141
Frederick the Great, 17
Freiburg, Georg, 331
Friedrich I., Duke of Württemberg, 170, 206
Fries, Lorenz, 189
Frizimelica, Hieronymus, 255
Fuchs, Leonhart, 191–193, 198, 200, 204, 206
Fugger, Jakob, 322
Fugger, Raymund, 323
Füllmaurer, Heinrich, 191
Furttenbach, Joseph, 163, 165, 262

G
Gady, Alexandre, 137, 139
Gaismair, Michael, 177
Galen of Pergamon, 186, 285
Galilei, Galileo, 31, 113, 114, 119–122, 126, 129, 130
Galle, Philips, 200, 201
Gallo, Agostino, 270, 276, 278

Galluzzi, Riguccio, 217–219, 226, 227
Galvani, Giovan Alberto, 78
Ganay, Ernest de, 139, 140
Garrido, Joao António, 305, 309
Gemböck, Robert, 353
Gerard, John, 204
Gerhard, Hubert, 325
Gessner, Conrad, 197, 257–259, 321–323
Gheyn II, Jacques de, 244
Ghini, Luca, 192, 195
Ghinucci, Tommaso, 78–79
Gilabert, Francisco, Don, 309, 313
Giorgio Martini, Francesco di, 164, 177
Gobelin, Abbe', 217
Goethe, Johann Wolfgang von, 353
Goltzius, Julius, 200, 201
Gott, Samuel, 176
Gozzoli, Benozzo, 167
Graebner, Paul, 352
Grassi, Orazio, 125, 126
Gregory XIII, Pope, 84, 106–108, 112, 113, 117
Greuter, Bartholomew, 115, 118, 123
Grew, Nehemiah, 29, 46–49, 51, 257
Grienberger, Christoph, 15–17, 125
Grien, Hans Baldung, 186
Grifoni, Alessandro, 196
Groen, Jan van der, 15
Grosser, Martin, 315
Guazzo, Stefano, 129
Guillaume, Jean, 143
Gutbier, Aegidius, 255

H
Hachtmann, Margaret, 255
Haeckel, Ernst, 348, 350, 351
Hájek, Tadéas, 197
Hall, Francis, 119. *See also* Line, Francis, 119
Handsch, Georg, 197
Hartlib, Samuel, 32–33, 35, 36, 49, 50
Henri IV of France and Navarre, 65, 137, 141, 145, 147
Hercules, 167, 184, 189, 216–220, 222
Heresbach, Conrad, 315
Herff, Jeffrey, 354, 361
Hergot, Hans, 177
Hermann, Paulus, 290, 298
Hero of Alexandria, 79, 81–82, 85, 88, 90, 92
Hertenstein, Ludwig Bartholmä von, 325
Hesse, Heinrich, 13, 15
Heurnius, Johannes, 247
Heurnius, Otto, 247
Heyden, Pieter van der, 198
Hill, Thomas, 199, 315
Hirschfeld, Christian Cay Lorenz, 308
Hobbes, Thomas, 33, 42, 51
Hoffmann, Moritz, 254, 257
Hoghelande, Johan van, 241
Holtzbecker, Hans Simon, 256, 331
Horstius, Gisbertus, 257
Hörwarth, Johann Heinrich, 325
Hubbard, Henry Vincent, 346, 347
Hubert, Hans, 213, 221, 224
Humboldt, Alexander von, 349, 350, 356
Huygens, Christiaan, 17

I
Ibn Al-Awwam Al-Ishbili, Abu Zakariya Yahya ibn Muhammad ibn Ahmad, 313. *See also* Abu Zacaria, 311
Ippolito II d'Este, Cardinal, 78, 81, 82
Isaiah, Prophet, 115
Isingrin, Michael, 191, 192

J
Jäger, Hermann, 356
Jeannin, Pierre, 137, 138, 140
Jensen, Jens, 352, 358–363
Jesu Maria, Theobaldo de, 309
Joachim II, Margrave of Brandenburg, 192, 206
Joanna of Austria, Princess, 197
Johann Friedrich, Duke of Brunswick and Lüneburg, 254
Joncquet, David, 257
Jonston, Jan (John), 300
Julien, Pascal, 313
Jungermann, Ludwig, 257, 325
Jungius, Joachim, 254, 255, 258

K
Kandel, David, 184
Karoline Luise, landgravess of Baden, 320
Kimball, Theodora, 346, 347
Kircher, Athanasius, 21, 31, 45, 110, 114, 117, 127, 128
Kliemann, Julian, 216
Knoop, Johann Hermann", 13
Konig, Joannis, 204
Kraus, Johann Thomas, 328

L

La Court Van der Voort, Pieter, 314
Lamb, Carl, 82–83
Lambeck, Heinrich, 255
Lamy, Bernard, 21
Lange, Willy, 345–358, 361, 363
Langley, Batty, 21, 22
La Quintinie, Jean de, 56, 58–72, 314
La Salle de l'Etang, Simon-Philibert de, 309, 315
Las Casas, Gonzalez de, 316
Lasne, Michel, 147
Lauremberg, Peter, 257
Lawson, William, 315
Lazzaro, Claudia, 287
Leclerc, Sébastien, 22
Leibniz, Gottfried Wilhelm, 17, 19, 254, 258, 264
Lemercier, Jacques, 137–147, 149–151
Le Nôtre, André, 12, 13, 17, 21, 137, 144, 151, 311
Liberale da Udine, Giovanni, 193
Liberale, Giorgio, 197, 198
Liébault, Jean, 309, 313
Liger, Louis, 309, 314, 316
Ligorio, Pirro, 78, 80, 82, 92
Ligozzi, Jacopo, 200
Ligustri, Tarquinio, 117
Line, Francis, 119. *See also* Hall, Francis
Linnaeus, Carl, 296, 349
Lipsius, Justus, 206
L'Obel, Matthias de, 200, 204, 257
Lochner, Michael Friedrich, 338
London, George, 298, 299
Longhi, Martino, 106
Lonitzer, Adam (Lonicer, Adam), 193, 195, 257, 258
Lord Burghley (Sir William Cecil), 204
Lorenzi, Stoldo, 286
Lorrain, Claude, 112
Lösel, Johann, 322
Los Rios, Gregorio de, 305, 310, 311, 316
Louis XIII of France, 12
Louis XIV of France, 22, 56, 57, 59–61, 64, 71, 72, 229
Lucretius, 29, 30, 42–44, 48–51
Luigi d'Este, Cardinal, 84

M

Maccarone, Curzio, 78–79
Madruzzo, Cristofano, Cardinal, Prince of Trento, 193
Maelcote, Odo van, 125
Magalotti, Lorenzo, 255
Major, Johann Daniel, 257
Malvasia, Innocenzo, Monsignor, 114
Managetta und Lerchenau, Günter Ritter Beck von, 353
Manesson-Mallet, Alain, 22
Manning, Warren H., 355, 357
Mansart, François, 137, 151
Mansbach, Steven, 347
Maranta, Bartolomeo, 195
Marcgrave, Georg, 300
Marot, Daniel, 13
Marquis of Castelo Melhor, 312
Marquis of Pombal (Carvalho e Mello, Sebastiao José de), 306
Martellange, Étienne, 138, 139
Mattioli, Pietro Andrea, 193, 195–198, 204, 206, 257, 258
Max III Joseph, Elector of Bavaria, 320
Maximilian I, Duke of Bavaria, 325
Maximilian II of Habsburg, Holy Roman Emperor, 197
Medici, Catherine de, 145, 147
Medici family, 85, 200, 213, 216, 217, 222, 223, 229, 271, 272, 281, 283, 285
Medici, Francesco Maria de', Prince Cardinal, 223
Medici, Gian Gastone de', 227
Medici, Maria de', 137
Melo, Francisco Manuel de, Dom, 312, 313
Mendez de Torres, Luiz, 316
Merian, Maria Sibylla, 331
Merian, Matthäus, 331
Mersenne, Marin, 21, 31
Meursius, John, 257
Meyer, Albrecht, 191
Meyer, Cornelis, 103–105, 127
Meyerpeck, Wolfgang, 197
Micheli, Pier Antonio, 214, 217, 219, 221, 225–228
Michiel, Pietro Antonio, 196, 197
Mielke, Robert, 356
Migge, Leberecht, 347, 356
Miller, Mara, 291
Miller, Philip, 310, 314
Miller, Wilhelm, 352, 359–363
Milton, John, 33, 49, 50
Mindequia, Francisco Martinez, 146
Moller, Barthold, 331
Moller, Johann, 255
Mollet, André, 151
Mollet, Claude, 142, 315

Momper, Joos de, 200, 201
Montaigne, Michel de, 85, 88
Montalbani, Ovidio, 255, 257
Morandini, Francesco, called il Poppi, 198
Morel, Jean-Marie, 308
More, Thomas, 154, 158, 159, 170, 171, 175
Morison, Robert, 256, 257, 264, 300
Munting, Abraham, 257
Muthesius, Hermann, 347
Mylius, Georg, 257

N
Newton, Isaac, 33
Nicholas V, Pope, 168
Nini, Giacomo Filippo, Cardinal, 127
Nunez, Luca Pablo, 312

O
Oelhafen, Nikolaus, 257
Offerd, Karl, 262, 263
Oldenburg, Henry, 255
Olmsted, Frederick Law, Jr., 357
Oughtred, William, 21

P
Paleotti, Gabriele, Cardinal, 196
Palladius, 271, 283, 308
Parkinson, John, 256–258
Pasqualigo, Filippo, 195
Passe, Crispijn de, 200, 201, 206, 277, 325
Paulli, Simon, 257
Pauw, Pieter, 241–247
Pepys, Samuel, 33
Perronet, Henry, 263
Petiver, James, 296
Pfautzius, Joannes, 257
Philip III of Spain, 313
Pilleterius, Caspar, 257
Pinelli, Giovanni Vincenzo, 200
Piso, Willem, 300
Plantin, Cristophe, 200
Plukenet, Leonard, 299, 300
Pniower, Georg, 347
Poggio Braccioloni, Gianfrancesco, 168
Pohlenz, Hans-Friedrich, 347
Pollaiuolo, Antonio del, 168
Pollaiuolo, Piero del, 168
Pona, Pierre, 200, 257
Porro, Girolamo, 201, 204, 245
Pozzeserrato, Lodovico (Toeput, Lodewijck), 200
Prothero, Rowland Edmund, 313
Pückler-Muskau, Hermann Ludwig Heinrich Fürst von, 356

Q
Quiccheberg, Samuel, 224

R
Radi, Agostino, 110
Ranke, Leopold von, 160
Ray, John, 33, 256, 257
Raymond, John, 218
Réaux, Gédéon Tallement de, 137
Rebelo, Joao, 315
Rebelo, Manuel Joaquim, 315
Redi, Francesco, 217–219, 223, 255
Rehmann, Elsa, 362
Remshart, Carl, 328
Rendella, Prospero, 257
Riccio, Agostino del, 270, 275, 278–281, 283, 284, 315
Richelieu, Cardinal, 137
Richeome, Louis, 115, 117, 119, 120, 122, 123
Robert II, Count of Artois, 81
Roberts, Edith A., 362
Robinson, William, 346
Rosselli, Stefano, 196
Rossi, Teodosio, 109, 110
Roux, Agustin, 309, 315
Rovere, Francesco Maria II della, Duke of Urbino, 274
Rowley, William, 37, 40, 46
Royer, Johann, 256, 257
Rozier, François, 309, 315
Rucellai, Giovanni, 270, 279
Rudolf II of Habsburg, Holy Roman Emperor, 197
Ruel, Jean, 186, 189

S
Sacchi, Andrea, 120
Sadeler I, Johan, 200, 201
Sagredo, Diego, 312
Saint-Simon, Louis de Rouvroy de, 19
Salinas, Diego de, 316
Salviati, Antonio, 285
Salviati, Filippo, 114

Sapidus, Johannes (Witz, Hans), 184
Schaffner, Martin, 324
Scheiner, Christoph, 114
Scheler, Andreas, 325, 326
Schenck, Johann Heinrich Rudolf, 257
Schickhardt, Heinrich, 88–90, 170, 175
Schlegel, Paul Marquard, 255
Schnabel, Johann Gottfried, 177
Schneider, Camillo, 361
Schoeffer, Peter, 186
Scholz, Laurentius, 329, 330
Schomberg, Henri de, 137
Schott, Gaspar, 10, 12, 15, 21
Schott, Johann, 184, 186, 192
Schwenckfeld, Caspar, 257
Schwind, Johann, 331
Scultetus, Caspar Wilhelm, 335
Segewaart, Cornelis, 293
Seifert, Alwin, 352, 361
Serres, Olivier de, 59, 62, 65, 66, 68, 306, 312, 313
Sforza, Francesco, 156, 160
Sixtus V, Pope, 126
Sloane, Hans, 292, 293
Soavi, Sebastiano, 198
Soderini, Giovan Vettorio, 257, 270, 271, 274–281, 284, 285
Speckle, Veit Rudolph, 191
Spigelius, Adrianus, 257
Stetten, Paul von, 328
Stiblin, Caspar, 154
Strocchi, Maria Letizia, 213, 226
Stromer von Reichenbach, Wolf Albrecht, 12, 13, 19, 321–322
Swanenburg, Willem Isaacsz van, 236, 237
Switzer, Stephen, 21, 22

T
Tabernaemontanus (Dietrich, Jacob, of Bergzabern), 204, 257
Taegio, Bartolomeo, 270, 278, 315
Targioni Tozzetti, Antonio, 214, 221, 222
Targioni Tozzetti, Giovanni, 214–216, 219, 221, 222, 227
Tassius, Johann Adolph, 255
Tasso, Torquato, 112
Taylor, Jeremy, 50
Tedaldi, Giovambatista, 270, 278
Temple, William, 43
Thevenot, Melchissedech, 255
Thurneisser, Leonhart, 200
Tilli, Michelangelo, 218–220, 222, 227, 229
Toar da Silveira, Afonso, 309
Torre y Ocón, Francisco de la, 314
Tournefort, Joseph Pitton de, 214, 222, 225, 229
Treveris, Peter, 186
Trew, Christoph Jakob, 338
Tribolo, Niccolò, 216
Trichet du Fresne, Raphaël, 257
Turner, William, 199

U
Ucello, Paolo, 167
Ulisse Aldrovandi, 200, 224, 225
Urban VIII, Pope, 106–113, 120, 121, 129. *See also* Barberini, Maffeo
Ursinus, Leonhard, 256, 257
Utens, Justus, 271, 273, 281, 282, 286

V
Valckenborg, Lucas van, 200
Valgrisi, Vincenzo, 193, 197
van Beverningh, Hieronymus, 290, 294–296, 298
Vasari, Giorgio, 216, 285, 286
Veiras, Denis, 177
Venard, Claude, 3, 6, 79, 84
Venuto, Antonino, 270, 276–279
Vernon, Christopher, 355
Verrocchio, Andrea del, 167
Vesalius, Andreas, 192
Vesling, Johann, 257
Visconti, Bianca Maria, 160, 162
Vitoria, Baltazar de, 312
Vitruvius, 81, 125, 126, 271
Viviani, Vincenzo, 114
Vogel, Martin senior, 255
Volkamer family, 333–339
Volkamer, Johann Christoph, 225, 226, 333–339
Volkamer, Johann Georg senior, 333, 335
Volkamer, Johann Georg the Younger, 333, 335
Volkamer, Johann Magnus, 333
Vorstius, Adolphus, 257
Vos, Maarten de, 200, 201
Vries, Hans Vredeman de, 200

W
Waller, Edmund, 30, 43, 44, 48, 50
Waller, Johann, 314

Warming, Eugen, 348, 350
Waugh, Frank Albert, 347, 355–359
Weiditz, Hans, 184, 189
Welschius, Gottfried, 255
Wevetzer, Martin, 261
Wevetzer, Valentin, 255
Whately, Thomas, 308
Wichmann, Heinz, 347
Wiebenson, Dora, 347
Wierix, Hieronymus, 200
Wild, John, 258–259
Wilhelm IV, landgrave of Hessen-Kassel, 320
William III of Orange, Prince, 292, 299
Woodbridge, Kenneth, 137
Wotton, William, 15
Woudt, Jan Cornelisz van 't, 236, 237

Lightning Source UK Ltd.
Milton Keynes UK
UKOW06n1417081116
287153UK00002B/32/P

9 783319 263403